Collins

OCR Gateway GCSE 9-1 Combined Science Higher

Revision Guide

Fran Walsh, Eliot Attridge and David Brodie

About this Revision Guide & Workbook

Revise

These pages provide a recap of everything you need to know for each topic.

You should read through all the information before taking the Quick Test at the end. This will test whether you can recall the key facts.

Practise

These topic-based questions appear shortly after the revision pages for each topic and will test whether you have understood the topic. If you get any of the questions wrong, make sure you read the correct answer carefully.

Review

These topic-based questions appear later in the book, allowing you to revisit the topic and test how well you have remembered the information. If you get any of the questions wrong, make sure you read the correct answer carefully.

Mix it Up

These pages feature a mix of questions for all the different topics, just like you would get in an exam. They will make sure you can recall the relevant information to answer a question without being told which topic it relates to.

Test Yourself on the Go

Visit our website at **collins.co.uk/collinsGCSErevision** and print off a set of flashcards. These pocket-sized cards feature questions and answers so that you can test yourself on all the key facts anytime and anywhere. You will also find lots more information about the advantages of spaced practice and how to plan for it.

Workbook

This section features even more topic-based questions as well as Higher tier practice exam papers, providing two further practice opportunities for each topic to guarantee the best results.

ebook

To access the ebook revision guide visit

collins.co.uk/ebooks

and follow the step-by-step instructions.

Contents

HT **Higher Tier Content**

HT Higher Tier Content

Chemistry

Paper 3 Particles

Paper 3 Elements, Compounds and Mixtures

HT Higher Tier Content

Contents   5

HT Higher Tier Content

Paper 4   Global Challenges

Physics

Paper 5   Matter

Paper 5   Forces

HT Higher Tier Content

HT Higher Tier Content

HT Higher Tier Content

Review Questions

Recap of KS3 Biology Key Concepts

1 Name **three** structures found in both plant and animal cells. [3]

2 Why do plant cells have a cell wall? [1]

3 Why do plant cells contain chloroplasts? [1]

4 What is an enzyme? [1]

5 What is the name of the enzyme that breaks down starch? [1]

6 The enzymes we use to break down food are proteins.

How do high temperatures affect these proteins? [1]

7 What are the sections of DNA, found on chromosomes, which are responsible for inherited characteristics called? [1]

8 Which of the following characteristics in plants and animals are likely to be inherited?

   A flower colour in plants
   B fur colour in rabbits
   C weight in humans
   D blood group in humans
   E number of branches on a tree [3]

9 Where in the cell is DNA found? [1]

10 DNA exists as a coiled structure. What is the name we give to this structure? [1]

11 Respiration occurs in all living things. What is the purpose of respiration? [1]

12 Write down the equation for respiration. [2]

13 Plants carry out photosynthesis. Write the equation for photosynthesis. [2]

14 Plants can store the glucose produced in photosynthesis in their leaves. What is the glucose converted to for storage? [1]

15 Plants can also use the glucose produced by photosynthesis to build other molecules.

Name **two** molecules made by the plant. [2]

16 What is diffusion? [1]

17 In the lungs, where does diffusion of oxygen happen? [2]

18 How are the structures in the lungs adapted for efficient diffusion?   [3]

19 Put the following in order of size, starting with the smallest:

organ   cell   organ system   tissue   [2]

20 Construct a food chain using the organisms below:

caterpillar   leaf   fox   bird   [2]

21 What do the arrows in a food chain show?   [1]

22 Which gas is taken in by plants during respiration?   [1]

23 Which gas is taken in by plants during photosynthesis?   [1]

24 Which gas is the main contributor to global warming?   [1]

25 What does 'extinct' mean?   [1]

26 What is a balanced diet?   [3]

27 What are the **five** food groups?   [5]

28 Give **three** examples of recreational drugs.   [3]

29 Which organ is primarily affected by excessive alcohol consumption?   [1]

30 What is the name of the substance found in red blood cells that carries oxygen?   [1]

31 The diagram below shows the human respiratory system:

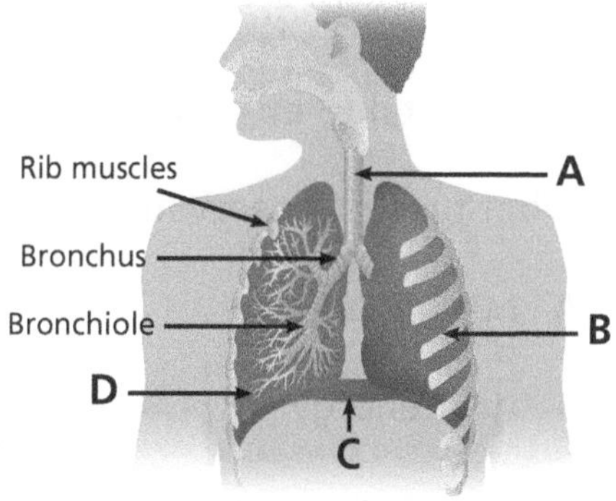

a) Which letter refers to the diaphragm?   [1]

Rib muscles

b) Which letter refers to the trachea?   [1]

Bronchus

c) Which letter refers to the ribs?   [1]

Bronchiole

D

d) Oxygen and carbon dioxide are exchanged at structure **D**.

A

B

C

What is the name of this structure?   [1]

e) By what process are oxygen and carbon dioxide exchanged?   [1]

f) The respiratory system contains cells which produce mucus. What is the function of
the mucus?   [1]

Total Marks _____ / 56

Review Questions

Recap of KS3 Chemistry Key Concepts

1 Write down whether each substance in the table is an **element, mixture** or **compound**.

Name of Substance	Type of Substance
Distilled water	
Gold	
Glucose	
Salt water	

[4]

2 **a)** Look at the following chemical reactions and tick (✓) the **two** reactions that are written correctly.

    A magnesium + sulfuric acid → magnesium nitrate + hydrogen

    B copper oxide + hydrochloric acid → copper chloride + water

    C iron + phosphoric acid → iron phosphate + hydrogen

    D lithium oxide + nitric acid → lithium nitrate + hydrogen   [2]

  b) When magnesium is burned in air, it produces a bright white light and a white, powdery residue.

    Write the **word equation** for the reaction.   [2]

3 Draw **three** diagrams to show the arrangement of particles in a **solid**, a **liquid** and a **gas**.   [3]

4 For a metal to increase in temperature, its atoms must gain energy.

What type of energy must the metal atoms gain?

  A nuclear energy   **C** chemical energy

  B heat energy   **D** kinetic energy   [1]

5 Linda is stung by a bee.
The bee's sting is acidic.

  a) What chemical is the opposite of an acid?   [1]

  b) Which of the following could neutralise an acid?

    A water   **C** vinegar

    B baking soda   **D** magnesium   [1]

6 Gold and platinum are among a few metals that can be found in their elemental form on Earth.

a) Explain why these metals are found in their elemental form. [2]

b) Suggest why aluminium was discovered much later in human history than gold and platinum. [4]

7 Ruth is using universal indicator to identify acids and bases.

Suggest what colour universal indicator would turn for each of the examples below.

a) Stomach acid [1]

b) Juice of a lime [1]

c) Tap water [1]

d) Hand soap [1]

8 The diagram below shows the arrangement of atoms in four different substances.

A 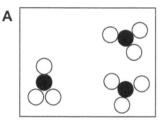   C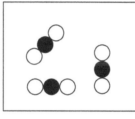

B 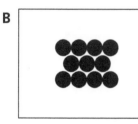   D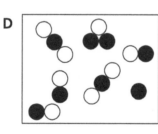

a) Which box has a substance that is a mixture of compounds? [1]

b) Which box has a substance that could be carbon dioxide? [1]

c) Which box has a substance that could be a pure element? [1]

9 In winter fish are able to survive in a fish pond, even when the fish pond freezes over.

Explain what property of water enables the fish to survive. [2]

Total Marks .... / 29

Review Questions

Recap of KS3 Physics Key Concepts

1 In what form do hydroelectric generating systems store energy?

   A gravitational potential energy   **C** thermal energy

   B elastic potential energy   **D** internal energy   [1]

2 Which of these is **not** a unit of energy?

   A joule   **B** kilojoule   **C** kilowatt   **D** kilowatt-hour   [1]

3 Which of these **cannot** be detected by the human senses?

   A infrared radiation   **C** sound waves

   B radio waves   **D** visible light   [1]

4 Which of these is **not** part of the electromagnetic spectrum?

   A infrared radiation   **C** sound waves

   B radio waves   **D** visible light   [1]

5 Which of the following is a contact force?

   A electric force   **C** gravitational force

   B force of friction   **D** magnetic force   [1]

6 Which of these is a renewable energy resource?

   A coal   **B** gas   **C** oil   **D** wind   [1]

7 Which of these **always** happens during refraction?

   A waves pass from one medium into another   **C** light is absorbed

   B energy transfers   **D** frequency changes   [1]

8 What does an ammeter measure?

   A electric charge   **C** electric potential difference

   B electric current   **D** electrical resistance   [1]

9 Which of the following do atoms contain?

A only electrons

C only nuclei

B electrons and nuclei

D either electrons or nuclei [1]

10 What is the name of the device that is designed to transfer energy from one circuit to another?

A ammeter   B battery   C diode   D transformer [1]

11 The density of water is 1000kg/m³ and the density of aluminium is 2400kg/m³.

a) What is the mass of:

i) 1m³ of water?

ii) 1m³ of aluminium? [2]

b) A bucket has a volume of 0.01m³. It is full to the brim with water.
What is the mass of the water? [3]

c) If the same water is in the same bucket on the Moon, and none of it evaporates, what will be the mass of the water? [1]

d) The bucket of water will be easier to carry on the Moon. Explain why this is. [2]

e) What does evaporation of water from the bucket do to i) the mass of water,
ii) the volume of water, and iii) the density of water? [3]

12 Match the words with their meanings by writing a list of pairs of letters and numbers.
For example: A2 is a correct answer.

A respiration   1 the space occupied by an object, measured in m³

B power   2 the process of getting energy from food and oxygen

C d.c.   3 small particle from which all material is made

D reflection   4 the rate of transfer of energy

E resistance   5 an electric current that stays in the same direction

F atom   6 sound with frequency too high for human hearing

G volume   7 opposition to electric current

H ultrasound   8 the return of waves when hitting a surface [8]

Total Marks _____ / 29

Cell Structures

You must be able to:

- Describe how light microscopes and staining can be used to view cells
- Explain how the main subcellular structures are related to their functions
- Explain how electron microscopy has increased understanding of subcellular structures.

The Light Microscope

- Light microscopes are useful for viewing whole cells or large subcellular structures.
- The specimen is placed on a glass slide, covered with a cover slip and placed on the stage of the microscope.
- The eyepiece and objective lenses magnify the object.
- A lamp provides illumination.
- **Magnification** is calculated by multiplying the magnification of the eyepiece lens by the magnification of the objective lens.
- Typical magnification is between 40 and 2000 times larger with a **resolution** of about 0.2 micrometres (μm).
- **Stains** can be used to colour whole cells and structures within cells, e.g. the nucleus, to make them easier to see.
- Sometimes a **mordant** is used, which fixes the stain to the structures.

Stained Bacteria Viewed using a Light Microscope

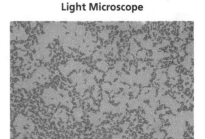

The Electron Microscope

- Electron microscopes are useful for viewing subcellular structures, such as ribosomes, mitochondrial membranes and nuclear membranes, in detail.
- They use a beam of electrons instead of a lamp.
- The specimen is placed inside a vacuum chamber.
- Electromagnets are used instead of lenses.
- The image is viewed on a TV screen.
- Typical magnification is 1 to 2 million times larger with a resolution of 2 nanometres (nm).

Bacteria Viewed using an Electron Microscope

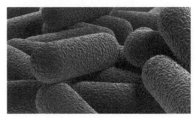

SI Units and Interconverting Units

1 metre (m) = 1 000 000 micrometres (μm)
$1\mu m = 10^{-6}m$
1 metre (m) = 1 000 000 000 nanometres (nm)
$1nm = 10^{-9}m$
To convert m to mm, multiply by 1000.
To convert mm to μm, multiply by 1000.
To convert μm to nm, multiply by 1000.

Typical Animal Cell

10–100μm

Bacteria

1–10μm

Chloroplast

0.5μm

Virus

80nm

Ribosome

25nm

Cell Membrane Thickness

20nm

Subcellular Structures

- The following structures are common to both animal and plant cells:
  - **nucleus** – controls the cell and contains genetic material in the form of chromosomes
  - **cytoplasm** – where most chemical reactions take place
  - **cell membrane** – a barrier that controls the passage of substances into and out of the cell and contains receptor molecules
  - **mitochondria** – contain the enzymes for cellular respiration and are the site of respiration.
- Additionally, plant cells contain:
  - **cell wall** – made from cellulose and provides structural support
  - **vacuole** – contains cell sap, which provides support
  - **chloroplasts** – contain chlorophyll and are the site of photosynthesis.

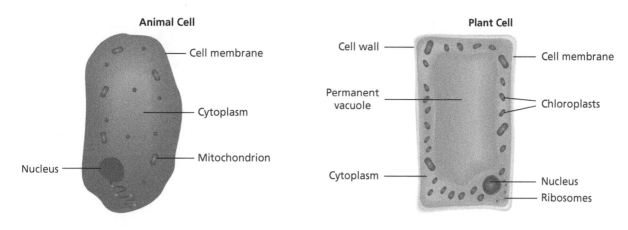

Types of Cells

- **Prokaryotes**, e.g. bacteria, have no nucleus – the nuclear material lies free within the cytoplasm.
- They may contain additional DNA in the form of a plasmid.
- **Eukaryotes**, e.g. human cheek cell, amoeba, plant cell, have a nucleus bound by a nuclear membrane.
- In eukaryotes, the cell wall is only present in plant cells.

A Prokaryote

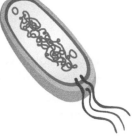

Quick Test
1. The eyepiece lens of a light microscope is ×10 and the objective lens is ×4. What is the total magnification?
2. Which part of a cell is:
a) the site of respiration?
b) the site of photosynthesis?
c) a barrier controlling what passes in and out of the cell?
d) where most chemical reactions take place?
3. Name **two** structures found in plant cells, but not in animal cells.

Key Words
magnification
resolution
stain
mordant
receptor molecules
chlorophyll
prokaryote
plasmid
eukaryote

What Happens in Cells

You must be able to:

- Describe the structure of DNA
- Explain how enzymes work.

DNA

- **DNA** (deoxyribonucleic acid) is a **polymer**.
- It is made of two strands of nucleic acid.
- The two strands are in the form of a **double helix**.
- DNA is found in the nucleus of the cell.
- It provides a code for making different proteins.

Double Helix

> ### Key Point
>
> There is genetic material, in the form of DNA, inside every cell.

Enzymes

- Enzymes are biological catalysts and increase the rate of chemical reactions inside organisms.
- Enzymes are made of proteins and the amino acid chain is folded to make a shape into which **substrates** (substances) can fit.
- The place where the substrate fits is called the **active site**.
- Enzymes are specific and only substrate molecules with the correct shape can fit into the active site. This is called the 'lock and key' hypothesis.
- Enzymes are **denatured** when they lose their shape. The substrate no longer fits and the enzyme does not work.

> ### Key Point
>
> When enzymes are denatured, they change shape, so the substrate can no longer fit in the active site.

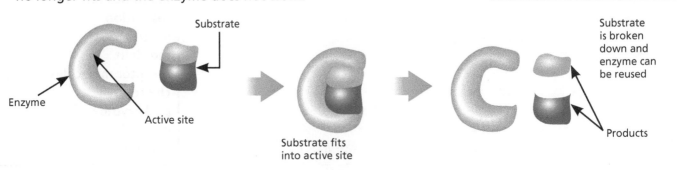

Substrate

Enzyme

Active site

Substrate fits into active site

Substrate is broken down and enzyme can be reused

Products

Factors Affecting Enzyme Action

- The rate of enzyme action is affected by temperature, pH, substrate concentration and enzyme concentration.
- High temperatures and deviation from the optimum (ideal) pH cause enzymes to lose their shape and become denatured.
- Low temperatures slow down the rate of reaction.
- Different enzymes have different optimum pH levels depending on where they act in the body.
- Many human enzymes have an optimum temperature of 37°C as this is normal human internal body temperature.
- As substrate concentration increases, the rate of enzyme activity increases to the point where all the enzymes present are being used.
- As enzyme concentration increases, the rate of enzyme activity increases to the point where all the substrate present is being used.

> **Key Point**
>
> When interpreting graphs about enzyme activity, remember – the steeper the line, the more rapid the increase or decrease in rate of enzyme activity.
>
> When explaining the shape of a graph, relate the explanation to scientific knowledge.

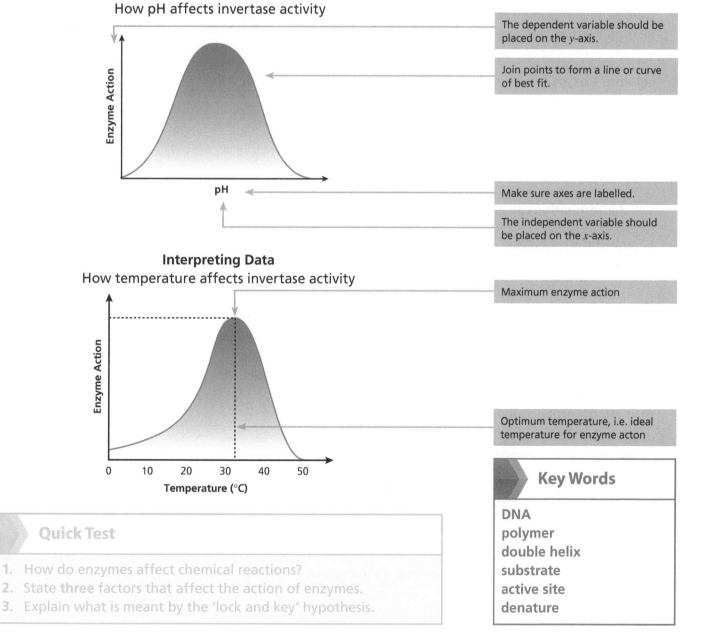

Presenting Data
How pH affects invertase activity

The dependent variable should be placed on the *y*-axis.

Join points to form a line or curve of best fit.

Make sure axes are labelled.

The independent variable should be placed on the *x*-axis.

Interpreting Data
How temperature affects invertase activity

Maximum enzyme action

Optimum temperature, i.e. ideal temperature for enzyme acton

Quick Test

1. How do enzymes affect chemical reactions?
2. State **three** factors that affect the action of enzymes.
3. Explain what is meant by the 'lock and key' hypothesis.

> **Key Words**
>
> **DNA**
> **polymer**
> **double helix**
> **substrate**
> **active site**
> **denature**

Respiration

You must be able to:

- Describe cellular respiration
- Compare aerobic and anaerobic respiration
- Describe the synthesis and breakdown of large molecules such as proteins, carbohydrates and lipids.

Cellular Respiration

- Cellular respiration happens inside the cells of all plants and animals. It occurs continuously and is controlled by enzymes.
- It is an **exothermic** reaction and releases energy in the form of a high energy molecule called ATP (adenosine triphosphate).

Aerobic Respiration

- **Aerobic respiration** (in the presence of oxygen) happens in almost all organisms all the time.

LEARN

$$glucose + oxygen \longrightarrow carbon\ dioxide + water\ (+ energy\ released)$$

$$C_6H_{12}O_6 + 6O_2 \longrightarrow 6CO_2 + 6H_2O\ (+ energy\ released)$$

Anaerobic Respiration

- Cellular respiration can also happen anaerobically (without oxygen).
- In animal cells, **anaerobic respiration** produces lactic acid:

$$glucose \longrightarrow lactic\ acid\ (+ energy\ released)$$

- Anaerobic respiration occurs when oxygen cannot be delivered to the cells fast enough, for example during vigorous activity.
- When exercise stops there is an **oxygen debt**, which must be paid back to remove the lactic acid which has accumulated in the cells.
- This is why the breathing rate is so fast.
- In yeast, anaerobic respiration produces ethanol:

$$glucose \longrightarrow carbon\ dioxide + ethanol\ (+ energy\ released)$$

> **Key Point**
>
> Anaerobic respiration in yeast is called fermentation.
>
> In everyday life, ethanol tends to be called alcohol.

	Where it Occurs	Oxygen	Breakdown of Glucose	ATP Yield per Mole of Glucose	Energy Released per Mole of Glucose
Aerobic	Mitochondria	Needed	Complete	38	2900kJ
Anaerobic	Cytoplasm	Not needed	Incomplete	2	120kJ

Breakdown of Biological Molecules

- We take in carbohydrates, proteins and lipids in our diet.
- These are all large molecules called **polymers**.
- During digestion, they are broken down by enzymes into smaller molecules, called **monomers**.
- Fats need to be broken down into small droplets to make them more digestible. This is called **emulsification**.
- In the small intestine, bile is responsible for emulsifying fats.

Name of Enzyme	Substrate it Acts Upon	Name of Monomers Produced
Carbohydrase	Carbohydrate	Glucose
Protease	Proteins	Amino acids
Lipase	Lipids (fats)	Fatty acids, glycerol

Carbohydrate, e.g. starch   Glucose   Lipid   1 glycerol and 3 fatty acids

Carbohydrase   Lipase

Protein   Amino acids

Protease

Synthesis of Biological Molecules

- Glucose, amino acids, glycerol and fatty acids are transported by the blood to the cells in the body.
- Inside the cells:
  - Glucose can be used in respiration.
  - Amino acids can be used to make useful proteins, e.g. enzymes and hormones.
  - Glycerol and fatty acids can be used to make useful lipids.
- Breakdown of proteins, carbohydrates and lipids produces the building blocks to synthesise the many different types of molecules that our bodies need to function.

> **Key Point**
>
> Polymers are long-chain molecules that can be broken down into smaller units called monomers.

> **Quick Test**
>
> 1. What is the word equation for aerobic respiration?
> 2. Where does aerobic respiration happen?
> 3. What are the products of fermentation?
> 4. What are the names of the enzymes that break down:
>   a) proteins?
>   b) fats?
>   c) carbohydrates?

> **Key Words**
>
> exothermic
> aerobic respiration
> anaerobic respiration
> oxygen debt
> polymer
> monomer
> emulsification

Photosynthesis

You must be able to:

- Name organisms that photosynthesise and state why they are important
- Describe photosynthesis
- Describe experiments to investigate photosynthesis
- **HT** Explain factors that affect the rate of photosynthesis.

Photosynthesis

- Green plants and algae make their own food using **photosynthesis**.
- They trap light from the Sun to fix carbon dioxide with hydrogen (from water) to form glucose.
- The glucose is used by the plants and algae to build larger molecules, such as complex carbohydrates and proteins.
- These are then passed onto animals in the food chain.

- The equation for photosynthesis is:

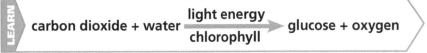

LEARN

$$\text{carbon dioxide} + \text{water} \xrightarrow[\text{chlorophyll}]{\text{light energy}} \text{glucose} + \text{oxygen}$$

- Photosynthesis is an **endothermic** reaction.
- Photosynthesis occurs inside the chloroplasts, which contain **chlorophyll**.

Investigating Photosynthesis

- Watching gas bubble up from *Cabomba*, a type of pond weed, is a good visual demonstration of photosynthesis.
- The experiment can be used to investigate how the amount of light or the temperature of the water affects the rate of photosynthesis.

A student carried out an investigation into the effect of light on photosynthesis. They placed the apparatus shown on the right at different distances from a desk lamp and counted the number of bubbles produced in 5 minutes. The results are shown in the table.

Distance of Lamp (cm)	0	5	10	15	20	25
Number of Bubbles	108	65	6	2	0	0

Key Point

Photosynthetic organisms are the main producers of food and, therefore, biomass for life on Earth.

Palisade Cell

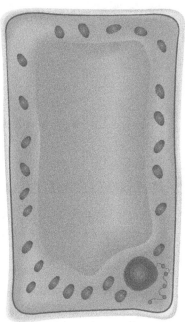

Packed with chloroplasts for photosynthesis.

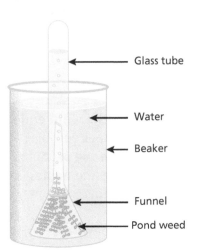

Glass tube

Water

Beaker

Funnel

Pond weed

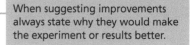

Evaluate the method and suggest possible improvements for further investigations.

Evaluation:

- It would have been better to use smaller distance intervals, e.g. 2.5cm, because there were not many bubbles after 5cm.
- It would have been difficult to count the bubbles at 0 because they were quite fast. It would have been better to collect them and measure the volume of gas produced.
- Another improvement would be to count for a longer time as there were so few bubbles at 10 and 15cm.

When evaluating the method, consider:
- Was the equipment suitable?
- Were there enough results?
- Was it a fair test?
- Were the results reliable?

When suggesting improvements always state why they would make the experiment or results better.

Factors Affecting Photosynthesis

- The optimum temperature for photosynthesis is around 30°C.
- At 45°C, the enzymes involved start to become denatured.
- Increased carbon dioxide levels increase the rate of photosynthesis.

HT In the second graph on the right, the rate stops increasing because either temperature or light becomes the limiting factor.

- Increasing the light intensity increases the rate of photosynthesis.

HT In the third graph on the right, the rate stops increasing because temperature or carbon dioxide becomes the limiting factor.

- Distance of the lamp from the plant can be used to change light intensity. To get a measure of the light intensity, use the inverse number of the distance squared. Light intensity obeys an inverse square law – if you double the distance, you quarter the light intensity.

HT Relative light intensity $= \frac{1}{d^2}$, where d is the distance. Taking three distances, 10, 20 and 30, the relative light intensity is $\frac{1}{100}$, $\frac{1}{400}$ and $\frac{1}{900}$ (or 0.01, 0.0025 and 0.00111) respectively. These can be multiplied by 100 for ease of reading, i.e. 1.0, 0.25 and 0.111. Doubling the distance leads to a quarter of the light intensity; trebling it a ninth; and so on.

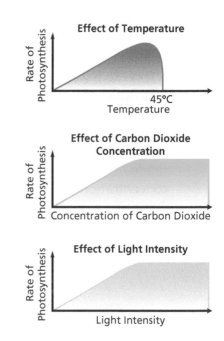

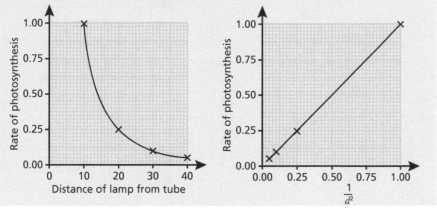

The graph on the left shows how distance affects the rate of photosynthesis. By using the relative light intensity ($\frac{1}{d^2}$), the curve is transformed into a straight line on the graph on the right.

Quick Test

1. What are a) the reactants and b) the products of photosynthesis?
2. Where in the cell does photosynthesis occur?
3. Why does photosynthesis stop at temperatures over 50°C?

Key Words

photosynthesis
endothermic
chlorophyll
HT limiting factor

Supplying the Cell

You must be able to:

- Explain how substances are transported by diffusion, osmosis and active transport
- Describe the stages of mitosis
- Give examples of specialised cells
- Explain the role of stem cells.

Transport of Substances In and Out of Cells

- Substances need to pass through cell membranes to get in and out of cells. This can happen in one of three ways:
  - **Diffusion** – the net movement of particles from an area of high concentration to an area of low concentration, i.e. along a concentration gradient.
  - **Active transport** – molecules move against the concentration gradient, from an area of low concentration to an area of high concentration. This requires energy.
  - **Osmosis** – the net movement of water from a dilute solution to a more concentrated solution through a **partially permeable** membrane.

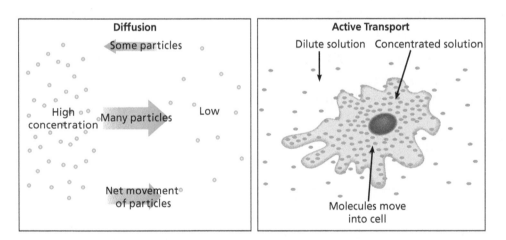

Diffusion

Some particles

High concentration — Many particles → Low

Net movement of particles

Active Transport

Dilute solution   Concentrated solution

Molecules move into cell

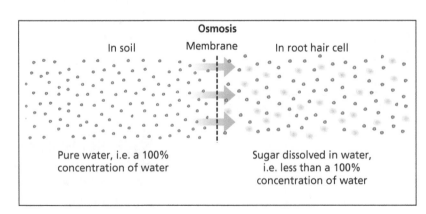

Osmosis

In soil   Membrane   In root hair cell

Pure water, i.e. a 100% concentration of water

Sugar dissolved in water, i.e. less than a 100% concentration of water

Key Point

Oxygen and glucose move into cells by diffusion. Carbon dioxide moves out of cells by diffusion.

Oxygen
Glucose

Carbon dioxide

Waste products

Key Point

A concentration gradient is a measurement of how the concentration of something changes from one place to another.

Key Point

Glucose is moved from the intestine to the blood stream by active transport. Minerals from the soil move into root hair cells by active transport.

Key Point

A dilute solution has a high water potential. A concentrated solution has a low water potential. In osmosis, water diffuses from a region of high water potential to an area of low water potential.

Mitosis

- Mitosis is the division of cells to produce new cells.
- New cells are needed to replace those that die or are damaged.
- Mitosis allows certain organisms to reproduce asexually.
- Mitosis plays a big part in increasing the size of organisms.
- The growth and division of cells is called the **cell cycle**.
- The stages of mitosis are:
    1. Parent cell contains chromosomes.
    2. Chromosomes are copied.
    3. Chromatids (the two sets of chromosomes) pull apart.
    4. Two new cells created, each identical to parent.

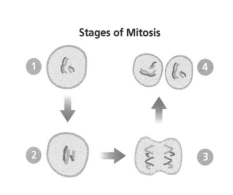

Stages of Mitosis

Cell Differentiation

- Cell differentiation is when one cell changes into another type of cell.
- In animals this usually happens at an early stage of development and occurs to create specialised cells. In mature animal cells, differentiation is mostly restricted to replacement and repair.
- Many plant cells, however, maintain the ability to differentiate.
- Here are some examples of specialised cells:

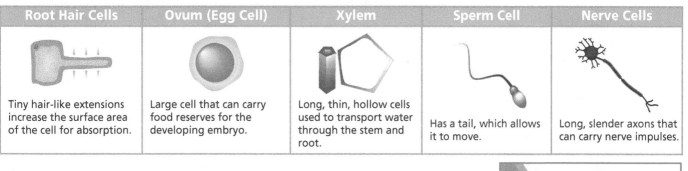

Root Hair Cells	Ovum (Egg Cell)	Xylem	Sperm Cell	Nerve Cells
Tiny hair-like extensions increase the surface area of the cell for absorption.	Large cell that can carry food reserves for the developing embryo.	Long, thin, hollow cells used to transport water through the stem and root.	Has a tail, which allows it to move.	Long, slender axons that can carry nerve impulses.

Stem Cells

- Stem cells are undifferentiated cells that can differentiate to form specialised cells, such as muscle cells or nerve cells.
- Human stem cells can come from human **embryos**, in umbilical cord blood from new born babies or from adult **bone marrow**.
- Embryonic stem cells can differentiate into any type of cell.
- Adult stem cells can only differentiate into the cells of the type of tissue from which they came.
- Plant stem cells are found in **meristematic** tissue, which is usually in the tips of shoots and roots.
- Stem cells have the potential to provide replacement cells and tissues to treat Parkinson's disease, burns, heart disease and arthritis. The tissues made will not be rejected by the body.
- Stem cells can also be used for testing new drugs.

Key Point

There are ethical issues surrounding the use of stem cells, for example: Is destroying embryos taking a life? Is it right to dispose of human embryos? And who decides which people should benefit from such a treatment?

Key Words

diffusion
active transport
osmosis
partially permeable
cell cycle
embryos
bone marrow
meristematic

Quick Test

1. What are the differences between diffusion and active transport?
2. Name **two** molecules that diffuse into cells.
3. Why is mitosis important?
4. Where do **a)** embryonic and **b)** adult stem cells come from?

The Challenges of Size

You must be able to:

- Explain why organisms need exchange surfaces and transport systems
- Describe substances that need to be exchanged and transported
- Describe the human circulatory system.

Exchange Surfaces and Transport Systems

- Exchange surfaces allow efficient transport of materials across them by mechanisms such as diffusion and active transport.
- In simple **unicellular** organisms, the cell membrane provides an efficient exchange surface.
- **Multicellular** organisms, which have a smaller surface area to volume ratio, have developed specialised exchange surfaces.
- The following substances all need to be exchanged and transported:
  - oxygen and carbon dioxide
  - dissolved food molecules, e.g. glucose and minerals
  - **urea** (waste product from breakdown of proteins)
  - water.
- An efficient exchange system should have:
  - a large surface area to volume ratio
  - membranes that are very thin so diffusion distance is short
  - a good supply of transport medium (e.g. blood, air, etc.).

Two groups of students investigated how quickly coloured water diffused into agar cubes with a surface area of 24cm².

They repeated each test three times. Here are the results of one of their tests:

	Time Taken to Diffuse in Seconds			Average (Mean)
	1	2	3	
Group 1	122	124	126	124
Group 2	136	128	150	138

The data from Group 1 is **repeatable**. This is because all three times recorded are near to each other.

The data is not **reproducible**. This is because the data from Group 1 is different to Group 2.

The data from Group 1 is more **precise** as all the times are near the mean.

A Single Alveolus and a Capillary

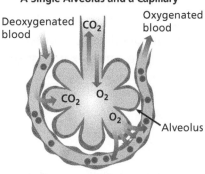

A single villus

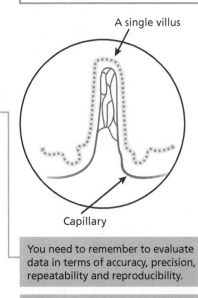

Capillary

You need to remember to evaluate data in terms of accuracy, precision, repeatability and reproducibility.

Measuring in second intervals is more accurate than measuring in minutes.

The Human Circulatory System

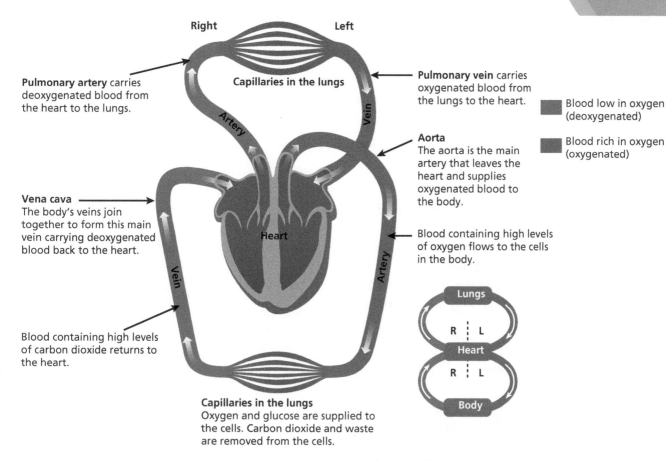

Right

Left

Pulmonary artery carries deoxygenated blood from the heart to the lungs.

Capillaries in the lungs

Artery

Vein

Pulmonary vein carries oxygenated blood from the lungs to the heart.

Blood low in oxygen (deoxygenated)

Blood rich in oxygen (oxygenated)

Aorta
The aorta is the main artery that leaves the heart and supplies oxygenated blood to the body.

Vena cava
The body's veins join together to form this main vein carrying deoxygenated blood back to the heart.

Heart

Vein

Artery

Blood containing high levels of oxygen flows to the cells in the body.

Blood containing high levels of carbon dioxide returns to the heart.

Lungs

R L

Heart

R L

Body

Capillaries in the lungs
Oxygen and glucose are supplied to the cells. Carbon dioxide and waste are removed from the cells.

- Humans have a double circulatory system with two loops:
  - one from the heart to the lungs
  - one from the heart to the body.
- The advantage of a double circulatory system is that it can achieve a higher blood pressure and, therefore, a greater flow of blood (and oxygen) to tissues.
- Substances transported by the circulatory system include oxygen, carbon dioxide, dissolved food molecules, hormones, antibodies, urea and other waste products.
- The right side of the circulatory system carries deoxygenated blood. The left side carries oxygenated blood.

Quick Test

1. What are **three** features of an efficient exchange system?
2. Why do multicellular organisms need specialised exchange systems?
3. What are the advantages of a double circulatory system?

Key Words

unicellular
multicellular
urea
repeatable
reproducible
precise
deoxygenated
oxygenated

The Heart and Blood Cells

You must be able to:

- Explain how the structures of the heart and blood vessels are adapted to their functions
- Explain how red blood cells and plasma are adapted to their functions.

The Heart

- The heart pumps blood to the lungs and around the body.
- It is made mostly of muscle.
- The left ventricle needs to pump blood round the whole body and, therefore, has thicker, more muscular walls.
- The valves between **atria** and **ventricles**, and ventricles and blood vessels, are to prevent blood flowing backwards.

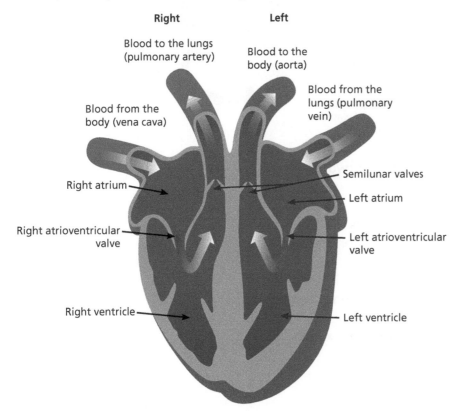

Right

Left

Blood to the lungs (pulmonary artery)

Blood to the body (aorta)

Blood from the body (vena cava)

Blood from the lungs (pulmonary vein)

Right atrium

Semilunar valves

Left atrium

Right atrioventricular valve

Left atrioventricular valve

Right ventricle

Left ventricle

Flow of Blood Through the Heart

From body to right atrium

↓

To right ventricle

↓

To lungs via pulmonary artery

↓

From lungs to left atrium via pulmonary vein

↓

To left ventricle

↓

To body via aorta

Red Blood Cells

- Red blood cells carry oxygen.
- They have a biconcave disc shape that maximises the surface area for absorbing oxygen.
- They contain **haemoglobin**, which binds to oxygen in the lungs and releases it at the tissues.
- Red blood cells do not have a nucleus, which means there is more space to carry oxygen.

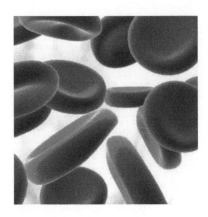

Blood Vessels

- There are three types of blood vessel: arteries, veins and capillaries.

Artery   Vein   Capillary

Valve

Lumen   Lumen

Capillaries are much smaller than veins or arteries.

- Arteries carry blood away from the heart.
- Veins carry blood to the heart.
- Capillaries deliver nutrients to cells and remove waste products from them.
- Arteries have thick outer walls with thick layers of elastic and muscle fibres because they have to carry blood under high pressure.
- Veins have a large lumen (cavity or opening) and thin walls, since blood is under low pressure.
- Veins also have valves to stop blood flowing backwards.
- Capillaries have very thin, permeable walls to allow substances to easily pass into and out of tissues.

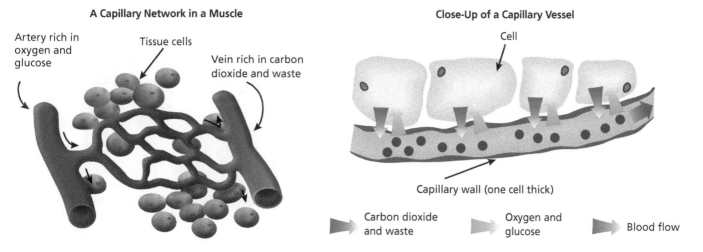

A Capillary Network in a Muscle

Artery rich in oxygen and glucose

Tissue cells

Vein rich in carbon dioxide and waste

Close-Up of a Capillary Vessel

Cell

Capillary wall (one cell thick)

Carbon dioxide and waste   Oxygen and glucose   Blood flow

Plasma

- Plasma is the pale coloured liquid part of blood. It transports:
  - hormones
  - antibodies
  - nutrients, such as glucose, amino acids, minerals and vitamins
  - waste substances, such as carbon dioxide and urea.

Key Words

atria
ventricles
haemoglobin
lumen

Plants, Water and Minerals

You must be able to:

- Explain how plants take in water and mineral ions
- Describe the processes of translocation and transpiration
- Explain how xylem and phloem are adapted to their functions
- Explain how water uptake in plants is measured and the factors that affect it.

Mineral and Water Uptake in Plants

- Plants need to take in water for photosynthesis and minerals for general health. These are taken in through the roots.
- Root hair cells have a large surface area to maximise absorption of water and minerals.
- Their cell membrane is thin, which also helps absorption.
- Water enters root hair cells by osmosis.
- Minerals enter root hair cells by active transport.
- There are three essential minerals that plants need to be healthy:
  - nitrates
  - phosphates
  - potassium.

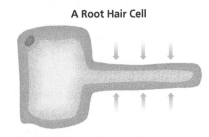

A Root Hair Cell

Transport Systems in Plants

- Plants have two types of transport tissue: xylem and phloem.
- The xylem and phloem are found in the stem of the plant.
- Movement of glucose from the leaf to other parts of the plant by phloem tissue is called **translocation**.

Cross-Section of a Stem

Phloem vessels carry food substances up and down the plant

Xylem

Phloem

Xylem vessels carry water up from the roots

Tissue Properties	Xylem Vessels	Phloem Vessels
What does it transport?	Xylem transports water and minerals from the soil to other parts of the plant.	Phloem transports the glucose made in the leaf by photosynthesis to other parts of the plant.
How is it adapted to its function?	Xylem vessels are made of dead cells. They have a thick cell wall and hollow lumen. There are no cell contents and no end cell walls, therefore, there is a continuous column for water to move up.	Phloem vessels are made of living cells. They have lots of mitochondria to release energy to move substances by active transport.

Transpiration

- Transpiration is the upward flow of water from roots to leaves, from where it evaporates into the atmosphere.
- The more water a plant loses via the leaves, the more water it will need to take in through the roots.

$$\text{rate of transpiration} = \frac{\text{volume of water lost}}{\text{time}}$$

- A **potometer** can be used to measure water uptake.
- As water is lost from the leaves, the air bubble moves to the left.
- Water is lost through **stomata** in the leaves.
- Plants can close stomata to reduce water loss. However, this closure also reduces the intake of carbon dioxide, which will limit photosynthesis.

Water Uptake and Transpiration

- The factors that affect water uptake and transpiration are:
  - wind velocity
  - temperature
  - humidity.
- You need to analyse distributions of data and predict the shape of graphs.

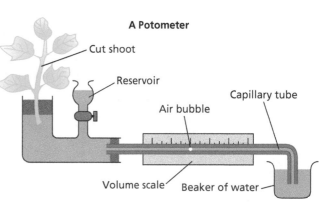

A Potometer

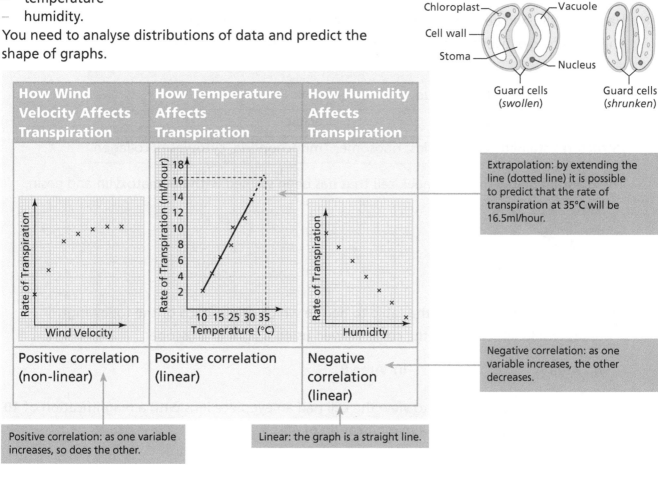

How Wind Velocity Affects Transpiration	How Temperature Affects Transpiration	How Humidity Affects Transpiration
Positive correlation (non-linear)	Positive correlation (linear)	Negative correlation (linear)

Extrapolation: by extending the line (dotted line) it is possible to predict that the rate of transpiration at 35°C will be 16.5ml/hour.

Negative correlation: as one variable increases, the other decreases.

Positive correlation: as one variable increases, so does the other.

Linear: the graph is a straight line.

Quick Test

1. Name the **three** essential minerals plants need for good health.
2. What is transported by:
   a) xylem?
   b) phloem?
3. Give **three** factors that affect the rate of transpiration.

Key Words

translocation
potometer
stomata

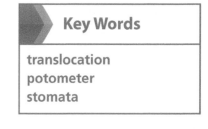

Cell Structures

1. Complete the table below using a tick or cross to indicate if the structures are present or absent in each of the cells.

Structure / Cell	Cell Wall	Cytoplasm	Nuclear Membrane	Cell Membrane	Chloroplasts
Plant					
Animal					
Bacterium					

[3]

2. Haematoxylin and eosin is a common stain made of two components.
The table below gives some information about each of the components.

Component	Haematoxylin	Eosin
Colour	black / blue	pink
Type of stain	basic	acidic
Structures stained	nucleic acids, ribosomes	mitochondria, cytoplasm, collagen

The diagram below shows a cheek cell that has been stained with haematoxylin and eosin.

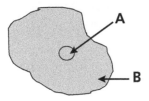

a) Use the information from the table to decide which colour each of the following areas would appear when the cell is viewed using a light microscope:

   i) Area A   ii) Area B   [2]

b) The light microscope used to view the cell had an eyepiece lens with a magnification of 10 and an objective lens with a magnification of 40.

   What is the total magnification?   [1]

3. The table below gives the diameter of two different cells.

Cell	Diameter
A	3000nm
B	4.5µm

Which cell is larger?   [1]

Total Marks _____ / 7

What Happens in Cells

1 The figure on the right shows a DNA molecule.

   a) The two chains of the DNA molecule form a coiled structure.

   What is the name of this structure? [1]

   b) Where in the cell is DNA found? [1]

2 The diagram below is about enzyme action.

   a) Label the enzyme, substrate and active site. [3]

   b) Describe what will happen to the enzyme at temperatures above 50°C. [2]

   c) Name a factor, other than temperature, that affects enzyme activity. [1]

Total Marks _____ / 8

Respiration

1 The table below compares aerobic and anaerobic respiration.

Complete the table using words from the list.

complete   incomplete   cytoplasm   low   high   mitochondria

	Aerobic	Anaerobic
Where it occurs		
Energy release		
Breakdown of glucose		

[6]

2 Complete the equation for aerobic respiration:

_____ + _____ → _____ + _____ + energy [4]

Total Marks _____ / 10

Photosynthesis

1 Complete the word equation for photosynthesis:

_____ + _____ $\xrightarrow{\text{light energy}}$ _____ + _____   [4]

2 The cell on the right is packed with subcellular structures that contain the enzymes for photosynthesis.

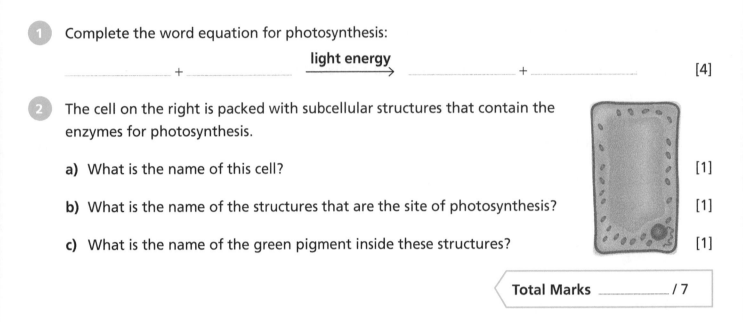

a) What is the name of this cell?   [1]

b) What is the name of the structures that are the site of photosynthesis?   [1]

c) What is the name of the green pigment inside these structures?   [1]

Total Marks _____ / 7

Supplying the Cell

1 The drawing below shows a single celled organism called an amoeba.

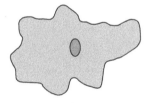

a) Name **two** substances that pass through the cell membrane into the amoeba.   [2]

b) By what process will the carbon dioxide produced by respiration move out of the amoeba?   [1]

c) Choose the correct word to complete the sentence below.

Amoeba is a **unicellular / multicellular** organism.   [1]

2. Complete the sentences by choosing the best words from the selection below.

You do **not** have to use all the words.

low   between   chloroplasts   energy   uneven   against   net   high   slow

Diffusion is the _____ movement of particles from an area of _____

concentration to an area of _____ concentration. Active transport moves particles

_____ a concentration gradient and requires _____ .   [5]

3. Which of the following statements best describes osmosis?

   A Osmosis is the movement of particles from an area of high water potential to an area of low water potential.

   B Osmosis is the movement of particles from an area of low water potential to an area of high water potential.

   C Osmosis is the movement of water from an area of high water potential to an area of low water potential.

   D Osmosis is the movement of water from an area of low water potential to an area of high water potential.   [1]

4. Mitosis is a form of cell division that produces new cells.

   Give **two** reasons why the body needs to produce new cells.   [2]

5. Human stem cells can be made to differentiate into different types of cells.

   From where are adult stem cells obtained?   [1]

6. Many people are against using embryonic stem cells to treat human diseases.

   Give **one** ethical reason why people may be against this.   [1]

   Total Marks _____ / 14

The Challenges of Size

1. The diagram below shows an exchange system in humans.

    a) In what organ of the body would you find this exchange system? [1]

    b) What structure is letter **A**? [1]

    c) What structure is letter **B**? [1]

    d) Movement of what substance is represented by arrow **C**? [1]

    e) Movement of what substance is represented by letter **D**? [1]

Deoxygenated blood

Oxygenated blood

A

C

D

B

Total Marks _____ / 5

The Heart and Blood Cells

1. The diagram below shows the human heart.

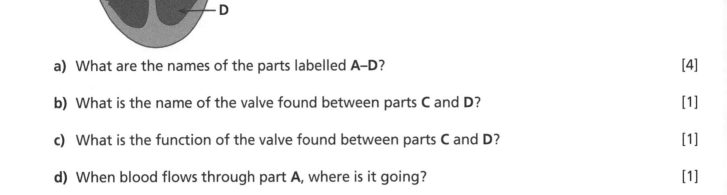

A

B

C

D

    a) What are the names of the parts labelled **A–D**? [4]

    b) What is the name of the valve found between parts **C** and **D**? [1]

    c) What is the function of the valve found between parts **C** and **D**? [1]

    d) When blood flows through part **A**, where is it going? [1]

2 Match each blood vessel on the left to **two** statements on the right with straight lines.

Blood Vessel   **Statements**

Capillary

Has valves
Smallest blood vessels
Thick muscular walls
Small lumen
Thin permeable walls
Large lumen

Artery

Vein
[6]

3 Name **three** substances carried by plasma. [3]

Total Marks _____ / 16

Plants, Water and Minerals

1 Complete each statement below by naming the correct process.

a) Minerals move into root hair cells by… [1]

b) Water moves into root hair cells by… [1]

c) Oxygen moves into the alveoli by… [1]

d) Glucose moves from the intestine to the blood by… [1]

2 The guard cells to the right are found on the underside of a leaf.

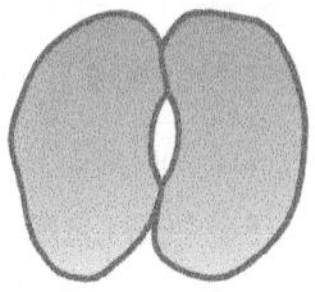

What is their function? [2]

3 Transpiration is movement of water in a plant from roots to leaves.

For each of the following changes, state whether it will **increase** or **decrease** the rate of transpiration.

a) Increase in wind velocity [1]

b) Increase in temperature [1]

c) Increase in humidity [1]

Total Marks _____ / 9

Coordination and Control

You must be able to:

- Describe the structure of the nervous system
- Explain how the body produces a coordinated response
- Explain a reflex arc and its function.

The Nervous System

- The nervous system is composed of two parts:
  - the **central nervous system** – the brain and spinal cord
  - the **peripheral nervous system** – all the other nerve cells that connect to the central nervous system.
- The nervous system receives **stimuli** from the environment, via **receptors** in sense organs, and coordinates a response.

Neurones

- Messages are carried to different parts of the body as electrical impulses via **neurones**.

Sensory Neurone   Impulse travels towards cell body

Relay Neurone   Impulse travels first towards and then away from cell body

Motor Neurone   Impulse travels away from cell body

- Sensory neurones carry impulses from receptors to the central nervous system.
- Relay neurones pass the impulse to a motor neurone.
- Motor neurones send the impulse to an **effector**, which is a muscle or gland that produces a response.
- The connection between two neurones is a gap called a **synapse**.
- When the impulse reaches the synapse, it stimulates the release of neurotransmitter chemicals from the neurone.
- The chemicals diffuse across the gap and bind to receptors on the next neurone.
- This triggers a new electrical impulse.
- At the synapse the message goes from electrical to chemical and back to electrical.

Coordinated Responses

- In a voluntary (coordinated) response, there may be a number of possible responses.
- The impulse is sent to the brain, which makes a decision as to the best responses.

Sense Organs

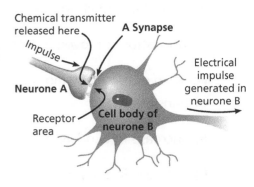

Sense Organ	Receptors Detect
Eye	Light
Ear	Sound
Nose	Smell
Tongue	Taste
Skin	Touch, pressure, pain, temperature change

Chemical transmitter released here

A Synapse

Impulse

Neurone A

Receptor area

Cell body of neurone B

Electrical impulse generated in neurone B

The Reflex Arc

- Reflexes are automatic responses to certain stimuli. You do not think about them.
- The function of a reflex is to protect the body.
- The response doesn't involve the brain – the pathway goes via the spinal cord, so it is quicker.
- A reflex response happens faster than a coordinated response. Example reflexes include touching a hot object, the pupil of the eye getting smaller in bright light and the 'knee jerk' reaction.

Touching a Hot Object

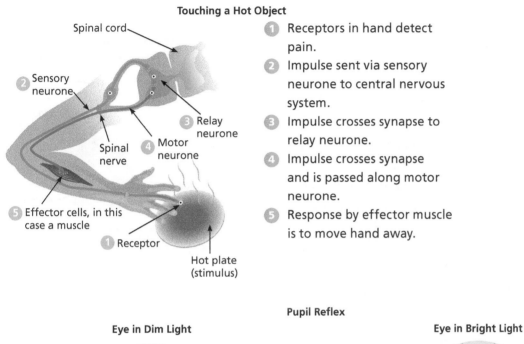

1. Receptors in hand detect pain.
2. Impulse sent via sensory neurone to central nervous system.
3. Impulse crosses synapse to relay neurone.
4. Impulse crosses synapse and is passed along motor neurone.
5. Response by effector muscle is to move hand away.

Pupil Reflex

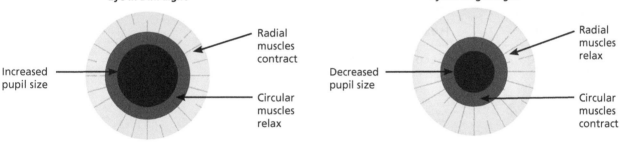

Eye in Dim Light

Radial muscles contract

Increased pupil size

Circular muscles relax

Eye in Bright Light

Radial muscles relax

Decreased pupil size

Circular muscles contract

The Endocrine System

You must be able to:

- Describe how hormones can coordinate a response
- **HT** Explain negative feedback with reference to thyroxine and adrenaline production
- Describe the role of hormones in the menstrual cycle.

The Endocrine System

- Hormones are chemical messengers produced by glands.
- They are released directly into the blood and travel to the target organ.
- The cells of the target organ contain receptors to which the hormone can bind.

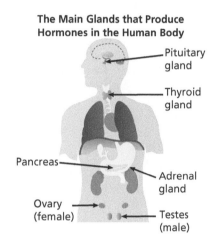

The Main Glands that Produce Hormones in the Human Body

Gland	Hormone(s) Produced	Function / Target Organ
Pituitary	LH, FSH	Involved in reproduction
Pituitary	ADH	Controls water content of blood
Pituitary	Growth hormone	Stimulates growth
Thyroid	Thyroxine	Controls metabolism
Pancreas	Insulin and glucagon	Controls blood glucose levels
Adrenal	Adrenaline	Fight or flight
Ovaries	Oestrogen and progesterone	Reproduction and secondary sexual characteristics
Testes	Testosterone	Secondary sexual characteristics

HT Negative Feedback Systems

- Hormone production is often controlled by centres in the brain by the mechanism of negative feedback.
- Negative feedback mechanisms act like a thermostat in a home:
  - The temperature drops and the thermostat turns on the heating.
  - The thermostat detects the rise in temperature and turns off the heating.

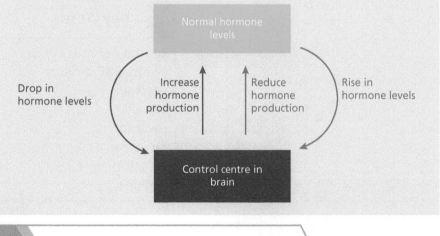

HT Thyroxine

- Thyroxine is produced by the thyroid gland and regulates metabolism.
- The thyroid gland is controlled by the pituitary gland.
- When the level of thyroxine drops, the pituitary produces thyroid-stimulating hormone (TSH).
- TSH stimulates the thyroid gland to produce thyroxine.
- Increasing levels of thyroxine cause the production of TSH to decrease.
- Thyroxine levels return to normal.
- This is an example of negative feedback.
- Production of thyroxine negates the decreasing levels of thyroxine in the blood.

HT Adrenaline

- Adrenaline is called the 'fight or flight' hormone and is produced by the adrenal glands in response to exercise, anxiety or fear.
- Adrenaline increases heart and breathing rates, increases the rate of blood supply to the muscles and raises blood glucose levels in preparation for fight or flight.

> **HT Key Point**
>
> The pituitary gland is often called the 'master gland' because it controls several other glands.
>
> It produces TSH, which acts on the thyroid gland.
>
> It also produces adrenocorticotrophic hormone (ACTH), which acts on the adrenal glands.

Hormones in the Menstrual Cycle

- Three hormones each play a different role in the menstrual cycle.
- **Oestrogen** – causes build-up of the uterus wall.
- **Progesterone** – maintains the lining of the womb.
- **Follicle-stimulating hormone (FSH)** – stimulates release of an egg because it stimulates the ovaries to produce oestrogen.

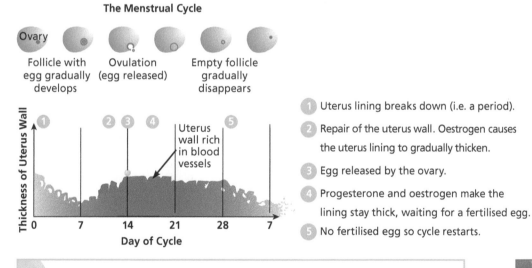

The Menstrual Cycle

Ovary

Follicle with egg gradually develops

Ovulation (egg released)

Empty follicle gradually disappears

1. Uterus lining breaks down (i.e. a period).
2. Repair of the uterus wall. Oestrogen causes the uterus lining to gradually thicken.
3. Egg released by the ovary.
4. Progesterone and oestrogen make the lining stay thick, waiting for a fertilised egg.
5. No fertilised egg so cycle restarts.

Uterus wall rich in blood vessels

Thickness of Uterus Wall

Day of Cycle

> **Key Words**
>
> gland
> HT negative feedback
> oestrogen
> progesterone
> follicle-stimulating hormone (FSH)

Hormones and Their Uses

You must be able to:

- Explain the use of hormones as contraceptives and their advantages and disadvantages
- HT Explain the interactions of hormones in the menstrual cycle
- HT Explain the use of hormones to treat infertility.

Hormones as Contraceptives

- Oestrogen can be used as a method of contraception.
- When taken daily, the high levels produced inhibit production of follicle-stimulating hormone (FSH) so egg development and production eventually stop.
- Progesterone also reduces fertility by causing a thick sticky mucus to be produced at the cervix, which prevents sperm from reaching the egg.
- There are over 15 methods of contraception.
- Hormonal contraceptives like the pill are very effective but can cause side effects and do not protect against sexually transmitted diseases (STDs)
- Barrier forms of contraception like condoms, may protect against STDs. They can be less reliable than hormonal methods if they are not used correctly.

> **Key Point**
>
> Most contraceptive pills combine progesterone and oestrogen.
>
> The progesterone-only pill is less effective.

	The Contraceptive Pill	Condoms
Advantage (High priority)	• More than 99% effective	• 98% effective • Protect against sexually transmitted diseases
Advantage (Low priority)	• Can reduce the risk of getting some types of cancer	
Disadvantage (High priority)	• Does not protect against sexually transmitted diseases	
Disadvantage (Low priority)	• Can have side effects	• Can only be used once

> **Key Point**
>
> You need to be able to evaluate risks and make decisions based on the evaluation of evidence. Decisions should always be justified.

> When making a decision, divide the information into advantages and disadvantages and prioritise them.

HT Hormones in the Menstrual Cycle

- The hormones involved in the menstrual cycle interact with each other.
- FSH stimulates oestrogen production by the ovaries.
- Oestrogen causes the pituitary to produce **luteinising hormone (LH)** and to stop producing FSH.
- LH causes an egg to be released.

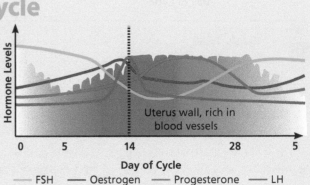

HT Treating Infertility

- Fertility drugs work by increasing levels of certain hormones in the body.
- If women have low levels of FSH, they can be given drugs containing FSH or LH to stimulate the release of eggs.
- Fertility drugs are also given to couples undergoing **in vitro fertilisation (IVF)** procedures to stimulate egg production.
- This allows several eggs to be collected at one time.
- The eggs are then fertilised outside the body.
- The cells divide to form an **embryo**, which is then inserted into the woman's uterus.

Scientific developments create issues

Social issues – the effect on people

Ethical issues – who has the right to decide which embryos 'live' or are destroyed

Economic issues – is the cost of developments worth the rewards?

Advances in IVF

New techniques offer the possibilities of improving a patient's odds of having a baby through in vitro fertilisation.

A single IVF cycle has about a 32% chance of resulting in a live birth and, to improve the odds, doctors often implant several embryos in the uterus during a single IVF cycle. This leads to high rates of twins and triplets, which can impact on the health of both mother and baby.

Chromosome abnormality is the main cause of miscarriage. Screening chromosomes involves taking cells from the embryo at day five to see if the normal number of chromosomes is present. Embryos with extra or missing chromosomes are not used. More research is needed on screening chromosomes, with a larger number of patients, to definitively determine the degree of benefit, especially since the cost of screening is around £2000 per patient.

Monitoring cell divisions is a technique that takes thousands of pictures of the embryos dividing at an early stage. If the division is atypical, the embryos are not used. Monitoring cell division has the advantage that it costs less and is less invasive.

Science has limititations

Science does not have an answer for everything – we still do not know enough about why the incidence of failure in IVF is so high.

There are also questions that science cannot answer, such as when is an embryo a person? Is destroying an embryo the same as taking a life? Different people will have different opinons.

Key Point

You need to evaluate the social, economic and ethical implications of scientific developments and understand the power and limitations of science.

Key Point

Scientific developments have the potential to improve the quality of life for many people but they also create issues.

Quick Test

1. What **two** hormones are commonly used in the contraceptive pill?
2. Suggest **two** advantages of using hormonal methods of contraception.
3. Suggest **two** disadvantages of using hormonal methods of contraception.
4. HT Explain how hormones can be used to treat infertility.
5. What do you understand by the terms **social**, **economic** and **ethical**?

Key Words

HT **luteinising hormone (LH)**
HT **in vitro fertilisation (IVF)**
HT **embryo**

Maintaining Internal Environments

You must be able to:

- Explain why it is important to maintain a constant internal environment
- Explain how blood sugar levels in the body are controlled
- Compare Type 1 and Type 2 diabetes.

Homeostasis

- **Homeostasis** means keeping the internal body environment constant.
- The body needs to control levels of water, glucose and salts to ensure chemicals can be transported effectively into and out of cells by osmosis and active transport.
- It also needs to maintain a constant temperature, since chemical reactions in the body are catalysed by enzymes that function best at their optimum temperature.

> **Key Point**
>
> Enzymes within the human body work best at 37°C.

Control of Blood Sugar Levels

- Sugar in the blood comes from the food we eat.
- The blood sugar (glucose) levels are controlled by the **pancreas**.

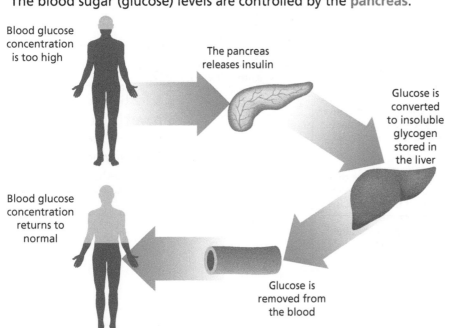

Blood glucose concentration is too high

The pancreas releases insulin

Glucose is converted to insoluble glycogen stored in the liver

Blood glucose concentration returns to normal

Glucose is removed from the blood

> **Key Point**
>
> Excess glucose is converted to glycogen in the liver.

- The pancreas produces **insulin**, which causes cells to become more permeable to glucose. Glucose moves from the blood into cells.
- Insulin also causes the liver to turn excess glucose into glycogen.

HT The pancreas also produces the hormone **glucagon** when blood glucose levels fall, e.g. during exercise.

HT Glucagon causes the liver to convert glycogen back into glucose and release it into the blood stream.

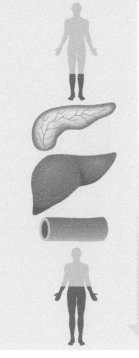

If the blood glucose concentration is too low...

the pancreas releases **glucagon**.

Insoluble glycogen from the liver is then converted to glucose...

which is released into the blood.

The blood glucose concentration returns to normal.

> **HT** **Key Point**
>
> Make sure you understand the difference between glucose, glucagon and glycogen.

Diabetes

- Diabetes is a condition that causes a person's blood sugar level to become too high. There are two types of diabetes.

	Type 1	Type 2
Incidence	Less common	Most common
Onset	Develops suddenly, often at a young age	Develops gradually, often in people over 40 who may be overweight
Cause	Pancreas does not produce insulin	Not enough insulin is produced or cells are not affected by insulin
Treatment	Insulin injections for life, plus eating sensibly	Healthy eating, exercise, possible medication

> **Key Point**
>
> Type 2 diabetes cannot be treated with insulin injections.

Key Words

homeostasis
pancreas
insulin
HT glucagon

Recycling

You must be able to:

- Give examples of recycled materials and explain why recycling of materials is important
- Recall the key points in the carbon and nitrogen cycles
- Explain the role of microorganisms in recycling of materials.

The Importance of Recycling

- Living things are made of substances they take from the world around them.
- Plants take in elements such as carbon, oxygen, nitrogen and hydrogen from the air or from the soil.
- The plants turn the elements into complex carbohydrates, which are eaten by animals.
- These elements must be returned to the environment so they can be used by other plants and animals.
- Elements are recycled through **biotic** (living) components of an ecosystem (animals, plants, decomposers) and also **abiotic** (non-living) components (oceans, rivers, atmosphere).

> **Key Point**
>
> Carbon, nitrogen and water are vital substances for all living things and must be recycled.

The Water Cycle

- All living organisms need water to grow and survive.
- Water cycles through the atmosphere, soil, rivers, lakes and oceans.
- The water cycle is important to living organisms because it influences climate and maintains habitats.
- It also ensures there is flow of fresh water into lakes, rivers and the sea, and carries important nutrients.

The Nitrogen Cycle

- Nitrogen in the air must be turned into a form that plants can use by **nitrogen fixation**.
- Lightning 'fixes' nitrogen in the air and turns it into nitrates in the soil.
- Nitrogen-fixing bacteria in plant roots turn nitrogen into nitrates.
- The Haber process uses nitrogen to make fertilisers.
- Decomposition is the process by which dead animals and plants are turned into ammonium compounds by putrefying bacteria.
- Ammonium compounds are turned into nitrates by nitrifying bacteria.
- Denitrifying bacteria turn nitrates in the soil into nitrogen gas.

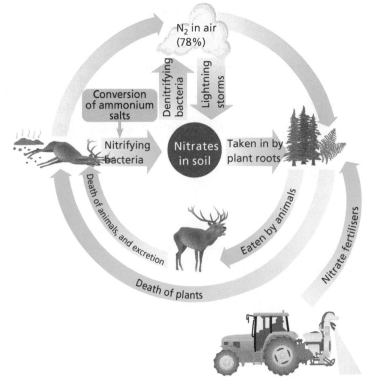

The Carbon Cycle

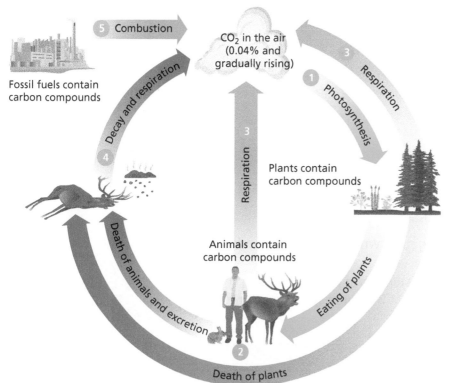

① Carbon dioxide is taken out of the atmosphere by plants when they photosynthesise.
② The carbon in plants passes to animals when they are eaten.
③ Animals and plants respire, releasing carbon dioxide back into the air.
④ Microorganisms **decompose** dead organisms and waste, and release carbon dioxide when they respire.
⑤ Combustion releases carbon dioxide into the atmosphere.

Decay

* Microorganisms such as bacteria and fungi are called **decomposers**.
* They cause decay by releasing enzymes that break down compounds.
* They need water and a warm temperature (usually 20-40°C) to survive.
* **Detritivores**, such as earthworms and woodlice, break down decaying material into small pieces, increasing its surface area and speeding up decay.

> ### Key Point
>
> Carbon dioxide is put back into the atmosphere by three types of respiration – animal, plant and microbial – along with combustion, and is removed by photosynthesis.

Quick Test

1. Why is the water cycle important to living organisms?
2. Name the **four** types of bacteria that are involved in the nitrogen cycle.
3. How is carbon removed from the atmosphere?

Key Words

biotic
abiotic
nitrogen fixation
decompose
decomposers
detritivores

Interdependence

You must be able to:

- Describe the different levels of organisation in an ecosystem
- Explain how factors can affect communities
- Describe the importance of interdependence and competition in the community.

Levels of Organisation within an Ecosystem

- An ecosystem is a place or habitat together with all the plants and animals that live there.
- The plants and animals within an ecosystem form the **community**.
- A **population** is all the animals or plants of the same species within a community.

Word	Definition
Producer	All plants – they use the Sun's energy to produce food
Herbivore	An animal that eats only plants
Consumer	All animals – they consume food
Carnivore	An animal that eats only animals
Omnivore	An animal that eats both animals and plants

> Remember to use scientific vocabulary, terminology and definitions.

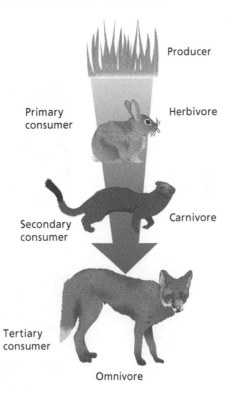

Producer

Primary consumer — Herbivore

Secondary consumer — Carnivore

Tertiary consumer

Omnivore

> **Key Point**
>
> Producers use energy from the Sun to make food.

Factors Affecting Communities

- Biotic factors are living organisms or things that influence or affect ecosystems, for example:
  - number of predators
  - food availability
  - number of insects to pollinate plants
  - disease
  - human activity, such as mining, which may destroy habitats.
- Abiotic factors are non-living conditions or things that affect ecosystems and the organisms in them, for example:
  - temperature
  - light intensity
  - moisture levels
  - pH of soil
  - salinity (salt levels) of water.

Interdependence and Competition

- The organisms within a community may depend on other organisms for food (**predation**) or shelter.
- Sometimes an organism lives on another organism, e.g. mistletoe lives on trees and absorbs nutrients from the tree.
- This type of dependence is known as **parasitism**.
- **Mutualism** is a state in which two organisms depend on each other and both benefit, e.g. a bee feeds from a flower and the flower is pollinated.
- Organisms within a community will compete with each other:
  - animals compete for food, space, water, mates
  - plants compete for space, water, light, minerals.

> **Key Point**
>
> Mutualism benefits both organisms. Parasitism only benefits one organism and harms the other organism.

Buffalo and Oxpecker Bird – Mutualistic Relationship

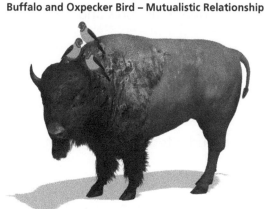

> **Quick Test**
>
> 1. Explain the difference between a community and a population.
> 2. Suggest **two** biotic factors and **two** abiotic factors that affect a community.
> 3. Name **three** things for which animals may compete.

> **Key Words**
>
> community
> population
> predation
> parasitism
> mutualism

Review Questions

Cell Structures

1. Draw a line from each cell part to the correct description.

Cell Part		Description
Cell membrane		Contains chlorophyll
Mitochondria		Gives support
Chloroplast		Controls movement of substances in and out of the cell
Cell wall		Contains the enzymes for respiration

[3]

Total Marks _____ / 3

What Happens in Cells

1. Pectinase is an enzyme that catalyses the breakdown of pectin, a component of the cell wall in fruits. By breaking down the cell wall, pectinase releases the juice from within the cells.

A group of students wanted to investigate the effect of temperature on pectinase activity.
The students chopped an apple into small pieces and divided it into seven beakers.
They added pectinase to each beaker.
Each beaker was then placed into a different temperature water bath for 30 minutes.
After 30 minutes, the students filtered the apple-pectinase mixtures and collected the juice produced.

a) What is the independent variable in this experiment? [1]

b) What is the dependent variable? [1]

c) To make the test fair, suggest **two** variables the students should have controlled. [2]

The students recorded their results in a table.

Temperature in °C	10	15	20	25	30	35	40
Amount of Juice Collected in cm³	5	7	9	12	16	17.5	15

d) Draw a graph of the students' results. [3]

e) What conclusions could the students draw from this experiment? [1]

f) Suggest **one** way in which the experiment could be improved. [1]

Total Marks _____ / 9

Respiration

1 Tom is running up a hill. Every few minutes he stops because he has to catch his breath by breathing deeply and his legs muscles feel painful.

a) What is the name of the substance that has built up in Tom's muscles and is causing pain? [1]

b) Why does Tom get out of breath? [4]

c) When Tom reaches the top of the hill he rests for 10 minutes and then begins to walk slowly back down.

i) What kind of respiration is occurring in Tom's cells as he starts his descent and where in the cell does it occur? [2]

ii) Write the equation for this type of respiration. [2]

2 Steve wants to make some homemade strawberry wine.
He picks some strawberries and adds water and yeast to them.

a) What kind of microorganism is yeast? [1]

b) What substance does the yeast ferment to produce alcohol? [1]

c) After a few days, Steve notices that his wine is bubbling. Suggest why it is bubbling. [1]

Total Marks _____ / 12

Photosynthesis

1 Some students wanted to carry out an experiment to investigate the effect of temperature on photosynthesis in pond weed.
The diagram shows the apparatus they used.

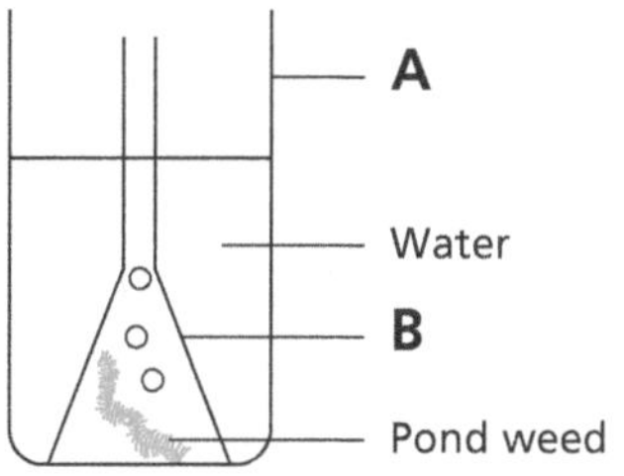

A
Water
B
Pond weed

a) What are the names of the pieces of apparatus labelled **A** and **B**? [2]

b) Explain how the students could use the apparatus above to carry out their experiment. [2]

c) Name **two** pieces of equipment not shown in the diagram that the students would also need. [2]

d) HT The students notice that increasing the temperature from 30°C to 35°C does **not** result in more bubbles being produced. Suggest why this is. [1]

Total Marks _____ / 7

Review Questions

Supplying the Cell

1. Root hair cells absorb water from the soil. How are they adapted for absorbing water? [1]

2. The diagram shows a working muscle cell.

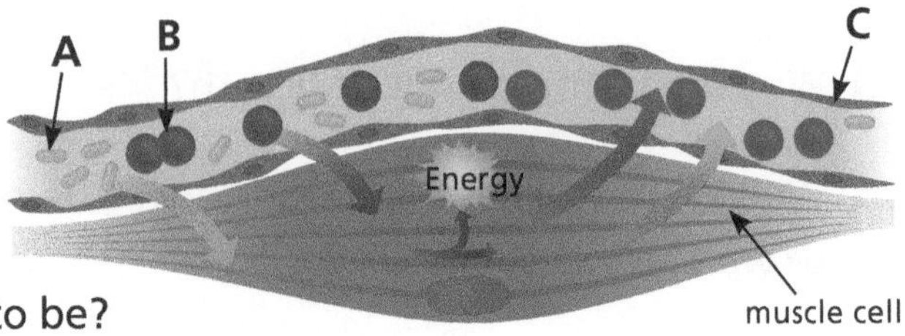

   a) Suggest what substance molecule **A** is most likely to be. [1]

   b) What is the name of cell **B**? [1]

   c) What type of blood vessel is **C** most likely to be? [1]

   A   B   C

   Energy

   muscle cell

3. Which of the following statements are true about stem cells in plants. Put ticks (✓) in the boxes next to the correct answers.

Stem cells are found in meristematic tissues	
Stem cells are found in the tips of shoots and roots	
Stem cells do not have cell walls	
Stem cells are packed full of chloroplasts	

   [2]

   > Total Marks _____ / 6

The Challenges of Size

1. The pictures show a unicellular organism called a *Paramecium* and a monkey, which is a multicellular organism.

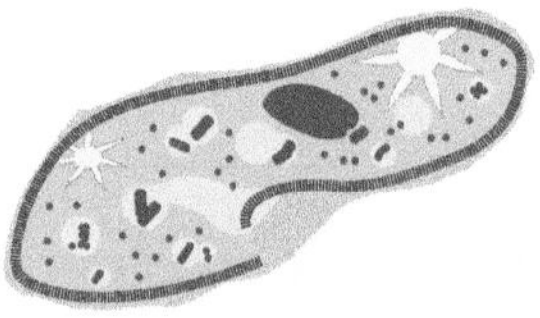

   Oxygen is able to pass into the *Paramecium* through the cell membrane.

   a) By what process will this happen? [1]

   b) Explain why the monkey has a specialised transport system that delivers oxygen to the lungs, but the *Paramecium* does not. [3]

   > Total Marks _____ / 4

The Heart and Blood Cells

1 The diagram below shows two blood vessels.

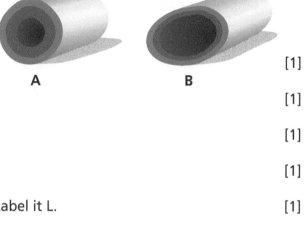

A   **B**

a) Which blood vessel will carry blood at high pressure? [1]

b) Which blood vessel has thick muscular walls? [1]

c) Which blood vessel has valves? [1]

d) Which blood vessel has the largest lumen? [1]

e) Draw an arrow on vessel **A** to show the lumen. Label it L. [1]

Total Marks .... / 5

Plants, Water and Minerals

1 Through which process do plants lose water?
Put a tick (✓) in the box next to the correct answer.

Translocation	
Transpiration	
Transfusion	

[1]

2 The diagram shows a cross-section through the stem of a plant.

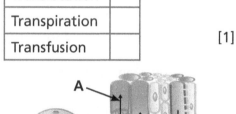

a) What is the name of the tubes labelled **A**? [1]

b) Name **one** substance transported in the tubes labelled **A**. [1]

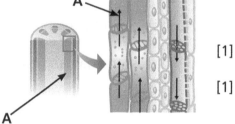

c) Which statement below best describes the tubes labelled **A**?
Put a tick (✓) in the box next to the best answer.

Made of living cells with large channels in end walls	
Made of dead cells with end walls removed	
Made of living cells packed with mitochondria	

[1]

3 What is the name of the process by which minerals move into plant roots against a concentration gradient? [1]

Total Marks .... / 5

Coordination and Control

1 Complete the table below about sense organs and the stimuli that they detect.

Sense Organ	Stimulus Detected
Eye	a) ....
b)	Sound
Nose	c) ....
d)	Taste
Skin	e) .... and f) ....

[6]

2 The passage below is about nerve impulses.

Choose the correct word from each pair.

Messages are carried by neurones as **electrical / chemical** impulses.

The junction between two neurones is called a **synapse / intersection**.

When the impulse reaches this junction, **hormones / neurotransmitters** are released.

These cross the gap and bind to **antigens / receptors** on the next neurone. [4]

3 The diagram below shows the stages of a reflex arc.

What do **A**, **B** and **C** on the diagram represent?

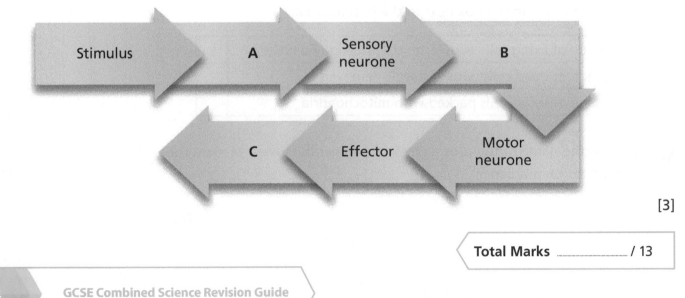

[3]

Total Marks .... / 13

The Endocrine System

1. Match each gland to the hormone it produces by drawing a line between them.

Gland	Hormone
Thyroid	Insulin
Pancreas	Thyroxine
Testes	Testosterone
Pituitary	Anti-diuretic hormone

[3]

2. HT The paragraph below is about control of thyroxine levels.

Complete the paragraph choosing the correct words from the selection below. You do **not** have to use all the words.

thyroid-stimulating hormone   **adrenaline**   **positive**   **glucose**
     metabolism   **thyroid**   **pituitary**   **negative**

Thyroxine is produced by the _____ gland. Its job is to control _____.

When levels of thyroxine in the blood fall, the decrease is detected by the _____

gland.

This gland produces the hormone _____, which causes an increase in production of

thyroxine. This is an example of _____ feedback.

[5]

Total Marks _____ / 8

Hormones and Their Uses

1. Oral contraceptives may contain oestrogen and progesterone, or progesterone on its own.

    a) Give **one** disadvantage of the progesterone-only pill.   [1]

    b) How does progesterone help to prevent pregnancy?   [1]

    c) Where in the body are oestrogen and progesterone produced?   [1]

    d) Give **one** disadvantage of using oral contraceptives.   [1]

Total Marks _____ / 4

Practice Questions

Maintaining Internal Environments

1. Insulin increases the permeability of cells to glucose, so glucose can move from the blood into the cells.

   a) By what process does the glucose move from the blood into the cells? [1]

   b) For what process is the glucose required? [1]

2. **HT** Another chemical, glucagon, is involved in the control of blood glucose levels.

   Which of the following statements about glucagon are **true**?

   A Glucagon is a hormone.
   B Glucagon is produced by the liver.
   C Glucagon is an enzyme.
   D Glucagon is produced by the pancreas. [2]

3. The graph below shows the effect three different sandwiches on a person's blood glucose levels when eaten.

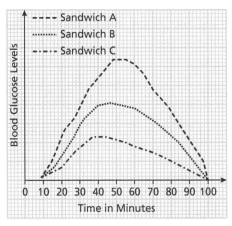

   Which sandwich would be most suitable for someone with diabetes? Explain why. [4]

4. Which of the following is not a risk factor for Type 2 diabetes?

   A aged over 40
   B smoking
   C obesity [1]

Total Marks _____ / 9

Recycling

1 The diagram shows the carbon cycle.

a) What letter represents each of the following processes?

   i) Photosynthesis

   ii) Animal respiration

   iii) Consumption

   iv) Microbial respiration

   v) Plant respiration   [5]

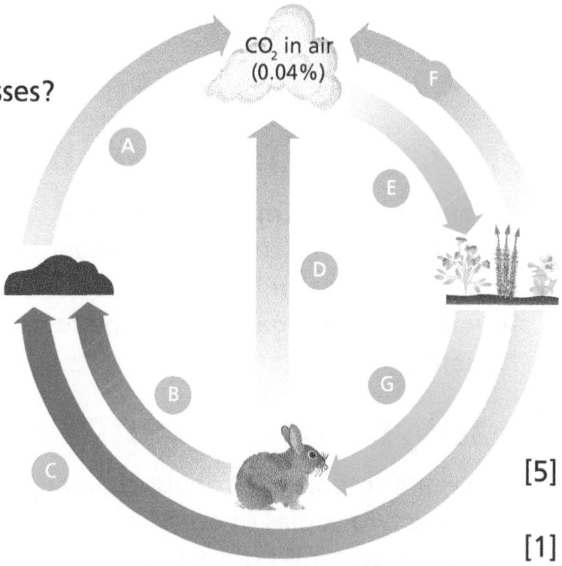

b) What process is shown by the letter **B**?   [1]

c) Name a process, **not** shown on the diagram, that releases carbon dioxide into the atmosphere. [1]

Total Marks _____ / 7

Interdependence

1 Divide the words below into **biotic** and **abiotic** components of an ecosystem.

animals   bacteria   rivers   trees   soil   sea   detritivores   [7]

2 In the Scottish island of Orkney, voles are eaten by birds of prey called hen harriers.
Voles are also eaten by owls.

a) Name **two** organisms that are in competition with each other for food.   [1]

b) What is the relationship between vole and owl?   [1]

c) Several years ago, stoats were introduced to the island. Stoats also eat voles.
   Local conservation groups are worried that the stoat numbers are growing rapidly.

   How is rapid growth of the stoat population likely to affect the numbers of hen harriers?
   Give a reason for your answer.   [2]

d) Suggest a factor, other than food, that two of the organisms may be in competition for.   [1]

Total Marks _____ / 12

Genes

You must be able to:

- Explain the relationship between genes, chromosomes and the genome
- Describe how genes and the environment influence phenotype.

Genes and Chromosomes

- **Chromosomes** are made of **DNA** and are found inside the nucleus.
- Each chromosome contains a number of **genes**.
- A gene is section of DNA that codes for a protein.
- A **genome** is an organism's complete set of DNA, including all of its genes.

Genotype and Phenotype

- Genes are responsible for our characteristics.
- We have two copies of every chromosome in our cells, therefore, we have two copies of each gene.
- The different forms of each gene are called **alleles**.
- We use capital and lower case letters to show if alleles are **dominant** or **recessive**.
- If two dominant alleles are present, e.g. BB, the dominant characteristic is seen, e.g. brown eyes.
- If two recessive alleles are present, e.g. bb, the recessive characteristic is seen, e.g. blue eyes.
- If one of each allele is present, e.g. Bb, the dominant character is seen, e.g. brown eyes.
- In the case of eye colour for example, brown eyes are dominant and would be shown as BB or Bb.
- Blue eyes on the other hand, are recessive and would be shown as bb.
- The **phenotype** is the characteristic that is seen, e.g. blue eyes.
- The **genotype** is the genes that are present, e.g. bb.

A Cell

Chromosomes

A Section of Chromosome

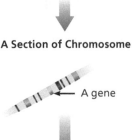
A gene

A Section of DNA

A Section of Uncoiled DNA

	A person who is **heterozygous** has two different alleles.	A person who is **homozygous** has both alleles the same.
Genes on Chromosomes		
Alleles	One allele for blue eyes, one for brown eyes	Both alleles are for blue eyes
Heterozygous or Homozygous	Heterozygous	Homozygous
Genotype	Bb	bb
Phenotype	Brown eyes	Blue eyes

> **Key Point**
>
> Capital letters represent dominant alleles.
>
> Lower case letters represent recessive alleles.

> **Key Point**
>
> A human body cell has 46 chromosomes, arranged in 23 pairs.

Environment and Phenotype

- Inherited variation is caused by the genes inherited from parents.
- Environmental variation is caused by environmental factors, e.g. diet.
- The phenotype of an organism is often the result of both genetic and environmental variation.
- For example:
  - a person who inherits light skin may, through prolonged exposure to the sun, develop darker skin
  - a person who inherits 'tall' genes may not grow tall if poorly nourished.
- Some characteristics show **continuous** variation, for example height. A person can be tall or short or anywhere in between the two.
- Characteristics that show **discontinuous** variation have a limited number of possible values. For example, blood groups are either A, B, AB or O.

> **Key Point**
>
> Height and weight are both examples of continuous variation.

Characteristics Controlled Solely by Inheritance
Eye colour
Hair colour
Blood group
Inherited diseases

Factors that are Influenced by Environment and Inheritance
Height and weight
Intelligence
Artistic or sporting ability
Skin colour

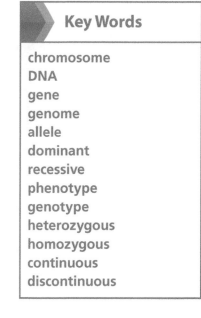

> **Key Words**
>
> chromosome
> DNA
> gene
> genome
> allele
> dominant
> recessive
> phenotype
> genotype
> heterozygous
> homozygous
> continuous
> discontinuous

Quick Test

1. What is a genome?
2. How do we show whether an allele is dominant or recessive?
3. Suggest **two** characteristics that may be affected by the environment.
4. Give an example of:
   a) continuous variation
   b) discontinuous variation.

Genetics and Reproduction

You must be able to:

- Describe meiosis and explain its role in genetic variation
- Use genetic crosses to determine the outcomes for sex and other characteristics
- Describe the work of Mendel in developing our understanding of genetics.

Meiosis

- Meiosis is a type of cell division that produces **gametes**.
- The cell divides twice to produce four gametes with genetically different sets of chromosomes.
- Gametes have half the number of chromosomes as body cells. This is called a **haploid** number.

| Cell with two pairs of chromosomes (diploid cell). | Each chromosome replicates itself. | Chromosomes part company and move to opposite poles. | Cell divides for the first time. | Copies now separate and the second cell division takes place. | Four haploid cells (gametes), each with half the number of chromosomes of the parent cell. |

- When the gametes fuse in fertilisation, the normal **diploid** number of chromosomes is restored.
- Fusion of gametes is a source of genetic variation.

Determination of Sex

- Sex inheritance is controlled by a whole chromosome rather than a gene.
- In humans, the 23rd pair of chromosomes in males contains an X and a Y chromosome; females have two X chromosomes.

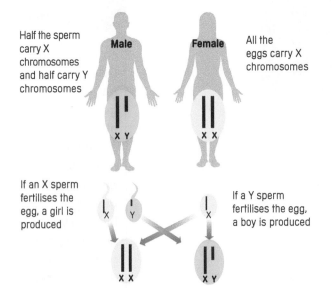

Half the sperm carry X chromosomes and half carry Y chromosomes

All the eggs carry X chromosomes

If an X sperm fertilises the egg, a girl is produced

If a Y sperm fertilises the egg, a boy is produced

Key Point

The chance of having a male child is always 50%, as is the chance of having a female child.

Genetic Crosses

- A genetic cross looks at the possible outcomes for offspring with parents of the same or different genotypes.
- All the offspring of a cross between a homozygous dominant and a homozygous recessive will appear to have the dominant characteristic.
- A cross between two heterozygous parents will give a ratio of three offspring with the dominant characteristic to one with the recessive.

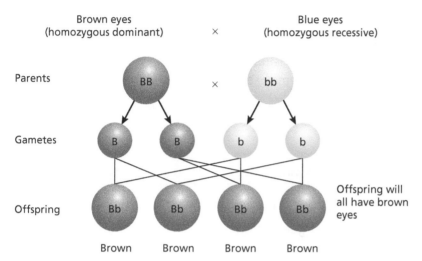

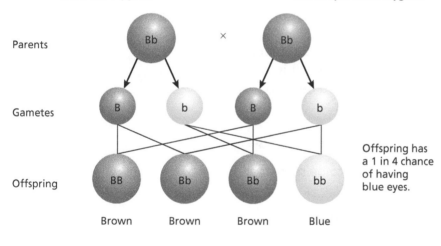

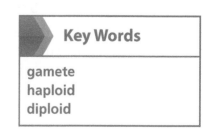

Key Point

Most of the characteristics we exhibit are not controlled by a single gene, but are the result of multiple gene inheritance.

Quick Test

1. What is the difference between a haploid cell and a diploid cell? Give an example of each.
2. Explain the role of meiosis in genetic variation.
3. Explain why there is a 50/50 chance of having a male or female baby.

Key Words

gamete
haploid
diploid

Natural Selection and Evolution

You must be able to:

- Understand there is extensive genetic variation within a population and relate this to evolution
- Explain how the development of ideas on evolution have impacted on modern biology, including the impact on classification systems.

Evolution and Natural Selection

- Evolution is the gradual change in the inherited characteristics of a population over a large number of generations, which may result in the formation of a new species.
- Charles Darwin's theory of evolution through natural selection suggested:
  - There is much variation within a species.
  - There is a struggle for survival and organisms have more offspring than can survive.
  - Those that are best adapted to their environment are likely to survive, breed and pass on their genes to the next generation (survival of the fittest).
  - Those that are least well adapted are likely to die.

Mutations

- Variation can arise because of **mutations** in a gene.
- If the mutation results in a characteristic that gives the organism an advantage over organisms without the characteristic, it is more likely to survive, breed and pass on the mutated gene to its offspring.

> **Key Point**
>
> Evolution by natural selection relies on random mutations.

Using Darwin's idea of survival of the fittest, you might hypothesise:

'Animals that are better camouflaged are less likely to be seen by predators.'

A model can be used to test this hypothesis using blue, green, red and brown strands of wool:

- Thirty 5cm strands of each colour of wool are scattered randomly in a field.
- A group of students are given 1 minute to find as many strands as they can.

Using the model, you might predict that:

- More red and blue strands of wool will be found because they are not camouflaged.
- Fewer green and brown strands will be found because they are better camouflaged.

> **Key Point**
>
> Scientific models are used to explain and predict the behaviour of real objects or systems that are difficult to observe directly.

Evidence for Evolution

- The fossil record gives us evidence of change over a long period of time, however there are gaps in this record.
- **Antibiotic resistance** provides evidence for evolution – bacteria divide very rapidly, so evolution by natural selection can be seen in a very short period of time:
  - A mutation may cause a bacterium to be resistant to an antibiotic.
  - Most of the bacteria in the population are killed by the antibiotic.
  - The antibiotic-resistant bacteria multiply rapidly to form a colony of resistant bacteria.

Developments in Classification Systems

- Developments in biology have had a significant impact on classification systems.
- The binomial classification system places organisms into groups based on a large number of characteristics such as anatomy, physiology, biochemistry and reproduction.
- This helps us to understand how organisms are related to each other.
- Darwin's theory of evolution provided a new explanation for how to group organisms:
  - nearness of descent
  - phylogeny (the sequence of events involved in the evolution of a species).
- **Phylogenetic** systems of classification help us understand the evolutionary history of organisms:
  - DNA sequencing can be used to show if organisms share common ancestors.
  - DNA sequencing has led to some organisms being reclassified.

Natural Classification of Sumatran Orang-utan

Level	Example
Kingdom	Animalia
Phylum	Chordata
Class	Mammalia
Order	Primates
Family	Hominidae
Genus	*Pongo*
Species	*Pongo abelii*

The Bornean orang-utan, *Pongo pygmaeus*, shares the same genus, which shows they are closely related. They developed as two separate species following the separation of Borneo and Sumatra.

Phylogenetic Tree

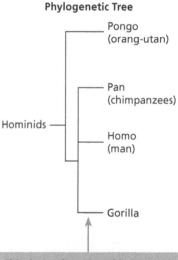

This shows that man, chimps, gorillas and orang-utans all share a common ancestor. It is thought that man is more closely related to gorillas than orang-utans.

Quick Test

1. Explain the process of natural selection.
2. Explain how antibiotic resistance develops.
3. How does DNA sequencing help to classify organisms?

Key Words

mutation
antibiotic resistance
phylogenetic

Monitoring and Maintaining the Environment

You must be able to:

- Explain how to carry out field investigations into the distribution and abundance of organisms in a habitat
- Describe how human interactions affect ecosystems
- Explain the benefits and challenges of maintaining biodiversity.

Distribution and Abundance

- Scientists studying the environment often want to investigate:
    - the distribution of an organism (where it is found)
    - the numbers of organisms in a given area or **habitat**.
- The following apparatus can be used for sampling:

Apparatus	What it is Used For
Pooter	Catching small insects on the ground
Pitfall traps	Catching small crawling insects
Pond net	Catching water invertebrates, small fish
Sweep net	Catching flying insects
Quadrat	Sampling the number of plant species
Line transects	Sampling the distribution of plant species

- A key can be used to identify the organisms found.

> **Key Point**
>
> It is important to gain a better understanding of the environment by collecting and interpreting data. This information is essential if we want to maintain the environment and variety of life.

A Quadrat

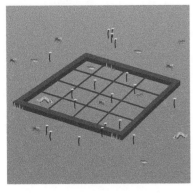

A Pooter

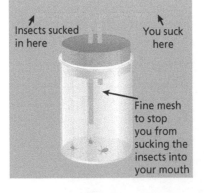

Insects sucked in here

You suck here

Fine mesh to stop you from sucking the insects into your mouth

A Pitfall Trap

A Sweep Net

The Capture–Mark–Recapture Method

- Follow the steps below to estimate the number of animals in an area using the capture–mark–recapture method:
  1. Count the animals caught in the trap.
  2. Mark them and release then.
  3. Count the animals caught a few days later, noting how many are marked / not marked.

$$\text{population size} = \frac{\text{number in 1st sample} \times \text{number in 2nd sample}}{\text{number in 2nd sample that are marked}}$$

Using Quadrats

- Follow the steps below to estimate the number of plants in a field:
  1. Measure the area of the field.
  2. Randomly place a number of quadrats (1 square metre) in the field.
  3. Count the number of plants in each quadrat.
  4. Work out the mean number of plants per quadrat.
  5. Multiply by the number of square metres in the field.

Human Interactions and Ecosystems

- Humans need to obtain and produce resources to survive.
- Their interactions have a huge impact on ecosystems.
- Deforestation, hunting and pesticides impact negatively on ecosystems.
- Captive breeding programmes, creating nature reserves, **sustainable** fishing and passing laws to protect animals have a positive impact on ecosystems.
- **Ecotourism** aims to reduce the impact of tourism on environments by not interfering with wildlife, leaving a low carbon footprint and supporting the local community. It is an example of sustainable development.

Biodiversity

- The greater the **biodiversity**, the greater the stability of an ecosystem and the greater the opportunity for medical discoveries.
- Biodiversity boosts the economy, e.g. a greater variety of plants means a greater variety of crops.
- The challenges of maintaining diversity arise due to:
  - **Political issues** – conservation strategies are often politically sensitive and there may be difficulty in gaining agreements between local, national and global partners.
  - **Economic issues** – conservation is often expensive, for example, trying to monitor conservation schemes.

> ## Quick Test
>
> 1. Why is biodiversity important?
> 2. Suggest **two** ways in which human activity impacts:
>   a) negatively on biodiversity
>   b) positively on biodiversity.
> 3. What is ecotourism?

> ### Key Point
>
> Activities such as farming, fishing, hunting and building can often have a negative impact on ecosystems.

> ### Key Words
>
> habitat
> sustainable
> ecotourism
> biodiversity

Investigations

You must be able to:

- Explain how to determine the number of organisms in a given area
- Plan and explain investigations
- Select appropriate apparatus and recognise when to apply sampling techniques
- Translate data and carry out statistical analysis, identifying potential sources of error
- Recognise the importance of peer review of results.

Planning and Carrying Out Investigations

- When planning an investigation many factors must be considered and certain steps should always be followed.

At the edge of the school field are some large trees that shade part of the field for much of the day.
A group of students wanted to find out if the shade from the trees affected the number of dandelions growing in the field.

Investigation:
How is the distribution of dandelions affected by light and shade?

The rationale for the investigation can be incorporated into the title.

Hypothesis:
There will be more dandelions growing the further you get from the trees because there will be more light.

The hypothesis should always be based on scientific knowledge.

Method:
1. Measure a transect from the trees to the edge of the field.
2. At 5-metre intervals along the transect, place a quadrat on the ground.

Apparatus should always be appropriate, e.g. you would not use a ruler to measure 5-metre intervals.

3. Count the number of dandelions in each quadrat.

Scientists use sampling and look at several small areas to get a good representation of the whole area.

4. Carry out two more transects, parallel to the first one, and count the dandelions in each quadrat. This will improve the reliability of the results.

Any data recorded should be reliable, which means readings should be repeated.

5. Work out the average number of dandelions at each distance from the trees.

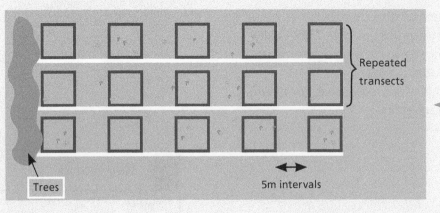

Repeated transects

A diagram often helps to clarify the method.

Trees

5m intervals

Results:

Distance from Trees (m)	0	5	10	15	20
Number of Dandelions in Each Transect	8	9	7	6	5
	8	8	7	2	6
	9	6	5	6	4
Mean Number	8.3	7.6	6.3	4.7	5.0

Data needs to be presented in an organised way, e.g. tables are useful for presenting data clearly.

Data sometimes needs to be analysed statistically by calculating the mean, mode or median.

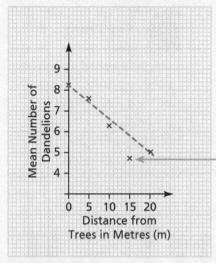

Line graphs are often useful for spotting anomalous results, which are often due to human errors, e.g. misreading the tape measure, or may occur randomly by chance.

Conclusion:

- The results show that the further you go from the trees, the fewer dandelions are growing, i.e. the opposite of what was predicted.

Evaluation:

- It was difficult to control variables such as the moisture content of the soil and the ground temperature, which could have affected the number of dandelions.
- To improve the experiment, a digital light meter could have been used to give some light readings for each quadrat.
- The experiment could be repeated at a different time of the year.

The evaluation should consider the method and results and what improvements could be made.

If the conclusion is different to what was predicted, the question 'why' must be asked and this may generate a new hypothesis, which can then be tested.

Key Point

Systemic / systematic errors occur when the same error is made every time. Systematic errors are only likely to be discovered if someone else tries to repeat the experiment and does not achieve the same results.

Key Point

A **peer review** is when one scientist evaluates another's experiment. This is valuable for scrutinising the design of the experiment and the validity of the data, and providing credibility.

Key Words

rationale
hypothesis
sampling
reliable
anomalous
systemic / systematic error
peer review

Quick Test

1. How do you ensure results from an investigation are reliable?
2. What is an anomalous result?
3. What are systemic / systematic errors?
4. Why is peer review important?

Feeding the Human Race

You must be able to:

- Explain the impact of the selective breeding of food plants and domesticated animals
- Describe the process of genetic engineering and the benefits and risks of using gene technology in food production.

Selective Breeding

- **Selective breeding** is the process of finding plants or animals with the best characteristics and breeding them.
- The process is repeated many times until the desired characteristic is present in all offspring.
- Selective breeding has been used to produce:
  - disease-resistant wheat
  - dairy cattle that give high milk yields
  - wheat that grows in areas of high salt levels.

> ### Key Point
>
> Selective breeding can be used to increase the yields of food crops, meat and milk, but it takes place slowly over many generations.

Genetic Engineering

- Genetic engineering involves altering the genome of an organism by adding a gene, or genes, from a different organism.
- It uses enzymes to 'cut and paste' genes.

Benefits of GM Foods	Risks of GM Foods
Higher crop yields	Once out in the wild it is impossible to recall genetically modified organisms
More nutritious crops	Genetically modified organisms may breed with other non-GM organisms passing on these new foreign genes into the wild population, for example, spread of herbicide-resistant genes might lead to super weeds
Crops can grow in harsh environments	We do not know the long-term effects of eating GM food; GM foods may harm the consumer, for example, causing allergic reactions
Crops resistant to pests and disease	
Better flavour food	
Food with longer shelf life	

HT The process of genetic engineering:

1 A strand of DNA is taken from the organism that has the useful characteristic.

2 The gene for the useful characteristic is isolated and cut from the DNA using **restriction enzymes**. Some enzymes produce DNA with short, single-stranded pieces at the ends – these are called sticky ends.

3 The desired gene is 'pasted' into the DNA of the organism that requires the useful characteristic using **ligase enzymes**.

Steps of Genetic Engineering

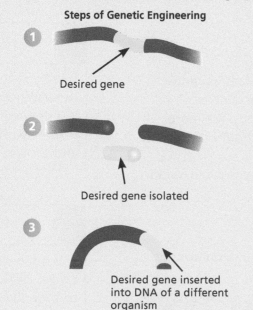

1 Desired gene

2 Desired gene isolated

3 Desired gene inserted into DNA of a different organism

HT A **plasmid** vector is a small loop of DNA often containing 'marker' genes for antibiotic resistance:

- Protein-making genes can be inserted into plasmids.
- The plasmids are mixed with bacteria.
- Only the bacteria that take up the plasmid will grow on medium containing antibiotics, due to the marker genes.
- These **host bacteria** are encouraged to multiply, producing large amounts of the desired protein.

> **Key Point**
>
> Genetic engineering modifies the genome of an organism, introducing desirable characteristics.

> **Key Words**
>
> selective breeding
> HT restriction enzymes
> HT ligase enzymes
> HT plasmid vector
> HT host bacteria

Quick Test

1. Explain the risks associated with GM crops.
2. HT What are restriction enzymes used for?
3. HT What is a plasmid?

Monitoring and Maintaining Health

You must be able to:

- Describe different types of diseases
- Describe how diseases spread and how this spread may be reduced
- Describe defence responses in animals to disease
- Explain how white blood cells and platelets are adapted to their function.

Diseases

- **Communicable** diseases are easily transmitted.
- They are caused by bacteria, viruses, fungi, protoctista or parasites.
- Sometimes diseases interact with each other, for example:
  - People with HIV (human immunodeficiency virus) are more likely to catch tuberculosis (TB) than those without HIV because of a weakened immune system.
  - Infection with some types of human papillomavirus can lead to the development of cervical cancer.

Spread of Disease

- Communicable diseases can be spread in humans in a number of ways, as shown in the table below:

Method of Spread	Examples
In the air through droplets, when people sneeze or cough	Chicken pox, tuberculosis, colds
By eating contaminated food	*Salmonella* food poisoning
By drinking contaminated water	Cholera
Contact – person to person or person to object	Athlete's foot
Animals – through bites or scratches	Malaria

- Spread of disease in humans can be reduced by good hygiene:
  - handle and prepare food safely
  - wash hands frequently
  - sneeze into a tissue then bin it
  - do not share personal items, e.g. a toothbrush
  - get vaccinated
  - do not touch wild animals.
- Communicable diseases can spread rapidly and infect millions of people globally.
- Statistics on the incidence and spread of communicable diseases are collected by Communicable Disease Centres.

Key Point

Health is more than just the absence of disease. It is defined as a state of complete physical, mental and social wellbeing.

Key Point

HIV is the virus that causes AIDS. It is possible to have the virus but not have AIDS. The virus is usually sexually transmitted and attacks the immune system making sufferers vulnerable to infection. There is no cure, but there are treatments to help manage the disease.

Key Point

Scientific quantities (i.e. statistics) allow us to predict likely trends in the number of cases of diseases and how the disease will spread nationally and globally. This is very important when planning for a country's future health needs.

Plants and Disease

- Plant diseases are caused by bacteria, viruses or fungi, and are spread by contact, insects, wind or water.
- Examples of plant diseases include: virus tobacco mosaic virus (TMV), fungal Erysiphe graminis (barley powdery mildew) and crown gall disease.
- Plants have cell walls and waxy epidermal cuticles, which form a barrier against plant pathogens.
- Plants can also produce toxic chemicals and pathogen-degrading enzymes as a response to infection.
- Spread of disease in plants can be reduced by:
  - using **insecticides** to kill pests which may carry disease
  - allowing space between plants
  - crop rotation
  - spraying crops with **fungicide** or **bactericide**.

Human Defences Against Disease

- A human's first line of defence is to stop microorganisms entering the body:
  - the skin acts as a physical barrier
  - **platelets** help the blood to clot and seal wounds to prevent pathogens entering
  - mucous membranes in the lungs produce mucus, which traps microorganisms
  - acid in the stomach kills microorganisms in food.

White Blood Cells

- White blood cells engulf and destroy microbes by a process called **phagocytosis**.
- Some white blood cells, called B lymphocytes, also produce **antibodies**.
- The lymphocyte recognises **antigens** on the surface of the invading pathogen.
- It produces antibodies that lock onto the antigen.
- The antibodies are specific for that antigen.

Antibody Production by White Blood Cells

Microorganisms (antigens) invade the body.

The white blood cell forms antibodies.

The antibodies cause the microorganisms to clump.

The white blood cell destroys the microorganisms.

Phagocytosis

White blood cell (phagocyte)

Microorganisms invade the body.

The white blood cell surrounds and ingests the microorganisms.

The white blood cell starts to digest the microorganisms.

The microorganisms have been digested by the white blood cell.

Key Point

There are two types of white blood cell involved in defence:

- Phagocytes destroy microorganisms by ingestion.
- Lymphocytes produce antibodies and form memory cells.

Key Words

communicable
platelets
phagocytosis
antibody
antigen

Quick Test

1. Name **four** groups of organism which cause disease.
2. Suggest **three** ways diseases can be spread in humans.
3. Give **three** human defences against disease.

Prevention and Treatment of Disease

You must be able to:

- Explain the role of vaccination and medicines in the prevention and treatment of disease
- Describe the process of developing new medicines.

Vaccination

- Vaccination (or immunisation) prevents people from getting a disease.
- The vaccine contains a dead or weakened version of the disease-causing organism.
- White blood cells (B lymphocytes) recognise the antigens present in the vaccine and produce antibodies.
- **Memory cells** are also produced that will recognise the disease-causing organism should the body come into contact with it again.
- If this happens, they will make lots of antibodies very quickly to prevent the person catching the disease.

Memory Lymphocytes and Antibody Production

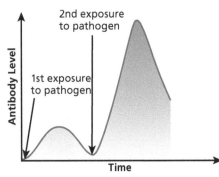

- If a large percentage of the population have been vaccinated, only a small percentage of the population will be at risk from catching the disease and passing it on to others.
- Vaccination is important in controlling the spread of disease.
- It is often possible to have one injection that vaccinates against several different diseases.

Common Vaccinations Available in the UK

Vaccination	Age Offered
5-in-1 (diphtheria, tetanus, whooping cough, polio and Hib (Haemophilus Influenzae Type B)	8 weeks
Pneumococcal	8 weeks
Meningitis B	8 weeks
Rotavirus	8 weeks
Measles, mumps and rubella (MMR) vaccine	1 year
HPV (girls only)	13 years
Flu vaccine	65 years

Antibiotics, Antivirals and Antiseptics

- **Antibiotics**, e.g. penicillin, are used to treat bacterial infections.
- They cannot be used for viral infections because viruses are found inside cells and the antibiotic would damage the cell.
- **Antivirals** are drugs that treat viral infections.
- **Antiseptics** are chemicals that kill microorganisms outside the body. They can be used on skin, on surfaces and on wounds.

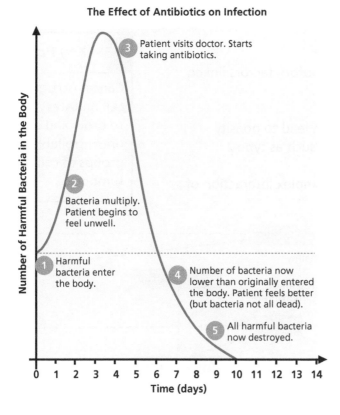

The Effect of Antibiotics on Infection

3 Patient visits doctor. Starts taking antibiotics.

2 Bacteria multiply. Patient begins to feel unwell.

1 Harmful bacteria enter the body.

4 Number of bacteria now lower than originally entered the body. Patient feels better (but bacteria not all dead).

5 All harmful bacteria now destroyed.

Number of Harmful Bacteria in the Body

Time (days)

> ### Key Point
> It is important to finish a course of antibiotics. Even when a person is feeling better, there are still infectious organisms in the body.

Developing New Medicines

- New medicines must be tested for toxicity, efficacy (effectiveness) and dosage before they are released to the public.
- There are a number of stages in developing a new drug:
  1. Tested on computer models or cells grown in the laboratory.
  2. Tested on animals, e.g. mice.
  3. Tested on small numbers of healthy volunteers in clinical trials (low doses often used).
  4. Further clinical trials on people with the disease, using control groups who are given a **placebo**.

Quick Test

1. How does vaccination work?
2. Distinguish between antibiotics, antivirals and antiseptics.
3. What are the **three** diseases that the MMR vaccination offers protection against?

> ### Key Words
> **memory cells**
> **antibiotic**
> **antiviral**
> **antiseptic**
> **placebo**

Non-Communicable Diseases

You must be able to:

- Recall some non-communicable diseases, suggest factors that contribute to them and evaluate data on them
- Evaluate some of the treatments for cardiovascular disease
- Discuss benefits and risks associated with using stem cells and gene technology in medicine
- Suggest why understanding the human genome is important in medicine.

Non-Communicable Diseases

- Many non-communicable diseases have contributory factors linked to lifestyle factors.
- Many of these factors interact with each other.
- For example, poor diet and lack of exercise can lead to obesity, which in turn is a risk factor for many diseases such as type 2 diabetes and **cardiovascular** disease.
- The arrows on the diagram below show the complex interaction of these factors.

> **Key Point**
>
> Cancer occurs when a cell mutates and begins to grow and divide uncontrollably. Such groups of cells are called tumours.

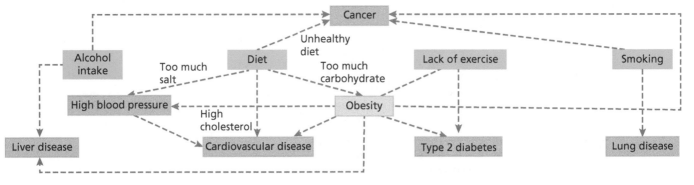

- Heart disease can be treated by:
  - **changes to lifestyle** – stopping smoking, eating healthily, exercising more
  - **medication** – there are a variety of medicines that can be taken to reduce high blood pressure, which is linked to heart disease, e.g. **statins** can be prescribed to lower cholesterol levels and aspirin can be taken to reduce the risk of further heart attacks
  - **surgery** – **stents** can be placed in narrowed arteries and heart transplants can replace damaged or diseased hearts.

> **Key Point**
>
> Cardiovascular disease is disease of the heart and / or blood vessels.

Evaluating Data

- Smoking and lung cancer in the UK:

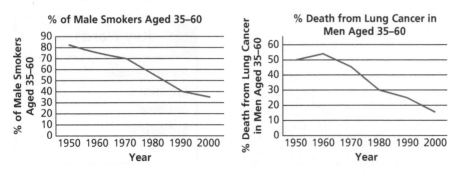

> **Key Point**
>
> You will often be asked to evaluate data to suggest the link between two variables.

- While the two graphs on smoking (on page 74) cannot prove that lung cancer is caused by smoking, they do suggest that as the percentage of men smoking decreased, so too did deaths from lung cancer.
- It is, therefore, likely that the two variables are linked.

Use of Stem Cells in Medicine

- Stem cells are cells which can differentiate to become any cell type found in the body (see page 25).
- They can be used to make new tissue and to replace tissues damaged by disease, e.g. to grow new nerve tissue to treat paralysis, or new heart valves to replace faulty ones.
- There are benefits and risks associated with using stem cells, as shown in the table below.

Potential Uses for Stem Cells

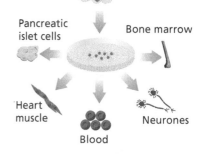

Benefits	Risks
Personal benefits to the person undergoing treatment	The stem cells may form tumours in the patient so may worsen the disease
Benefits to society since the process will provide knowledge that could lead to better future treatments	Stem cells may be rejected by the patient, which could lead to organ failure
Stem cells can be used to test new drugs	

- There are many ethical issues associated with using stem cells, e.g. some say that the use of stem cells sourced from human embryos is unethical and violates human rights as the embryos have 'no choice'.

Use of Gene Technology in Medicine

- Gene technology could be used in medicine to replace faulty genes, offering a cure for inherited conditions such as cystic fibrosis or diabetes.
- Replacing a faulty gene in the body has proved difficult to do.
- A virus is used to deliver the new gene and there are risks that the virus could harm the patient or deliver the gene to the wrong cell.

The Human Genome

- Genes can affect our chances of getting certain diseases.
- By studying the human genome, scientists hope to be able to predict the likelihood of a person getting a particular disease.
- Preventative action can then be taken.
- This is an example of personalised medicine.

Key Point

Some people are worried about the speed of developments in gene therapy, and are concerned that society does not fully understand the implications of these developments.

Quick Test

1. How does diet impact on cardiovascular disease?
2. What is cancer?
3. What are the **three** options for treatment of cardiovascular disease?
4. Give **one** risk of using stem cell technology in medicine.

Key Words

cardiovascular
statin
stent

Review Questions

Coordination and Control

1 Bryn runs into the road without looking.
There is a car coming towards him.
The driver of the car has to brake suddenly.

   a) What sense organ does the driver use to see Bryn? [1]

   b) Was braking a reflex action or a voluntary response for the driver? [1]

   c) The response was to push down on the brake pedal – what was the effector that
   carried out the response? [1]

2 The diagram below shows a neurone.

   a) What type of neurone is shown? [1]

Axon

Muscle cells

   b) Put an arrow on the diagram to show the direction of an impulse through the neurone. [1]

Total Marks _____ / 5

The Endocrine System

1 The graph to the right shows the chance of a woman
conceiving a baby naturally at different ages.

   a) Between what ages is there a significant drop
   in the chances of conceiving? [1]

   b) Use the graph to predict the chance of a woman
   aged 37 conceiving. [1]

   c) HT Women who have difficulty conceiving may be
   given hormone treatments to stimulate the
   release of eggs.

   i) Which hormones are commonly present in
   these drugs? [2]

   ii) Suggest why a woman aged over 45 is
   unlikely to be offered such treatment. [1]

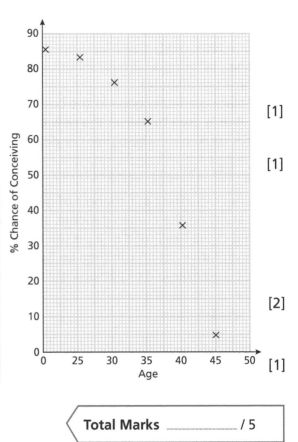

Total Marks _____ / 5

Hormones and Their Uses

1 Complete the passage below about oral contraceptives by choosing the correct words from the list. You do **not** have to use all the words.

> side effects   sexually transmitted diseases   FSH   more
> less   oestrogen   progesterone

Oral contraceptives are taken to prevent pregnancy. The combined pill contains the hormones

_____ and _____ . It is _____ reliable than the progesterone-

only pill. One of the disadvantages of using the contraceptive pill is it can cause _____ . [4]

> **Total Marks** _____ / 4

Maintaining Internal Environments

1 The flow chart shows the control of blood glucose levels by the hormone insulin.

a) By what process is glucose absorbed into the blood?

b) Complete the fourth and fifth level in the flow chart.

c) **HT** During exercise, blood glucose levels may drop.

 i) Why might glucose levels drop?

 ii) Explain how the body responds to the drop in glucose levels.

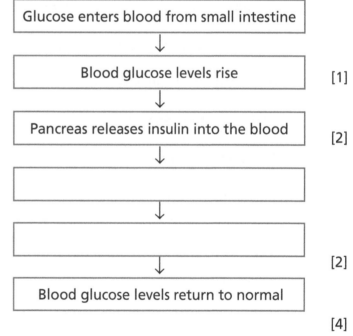

Glucose enters blood from small intestine

↓

Blood glucose levels rise   [1]

↓

Pancreas releases insulin into the blood   [2]

↓

[ ]

↓

[ ]   [2]

↓

Blood glucose levels return to normal   [4]

2 Hormones play an important part in regulating the body's internal conditions.

a) What term is given to organs that produce hormones?   [1]

b) How do hormones travel to their site of action?   [1]

> **Total Marks** _____ / 11

Review Questions

Recycling

1 The diagram below shows a bean plant.

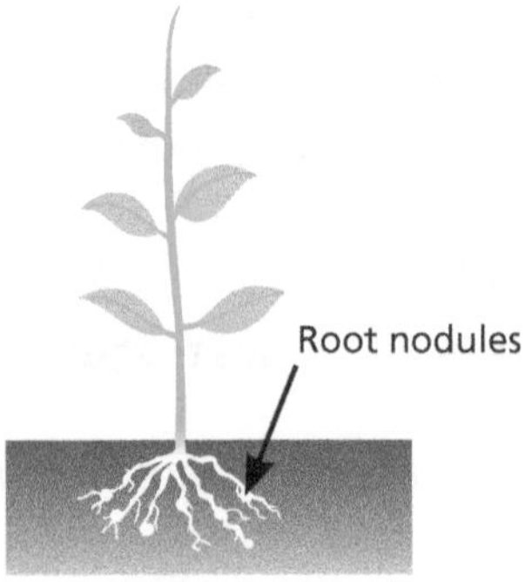

a) What type of bacteria will be present in the root nodules? [1]

b) Why are these bacteria important to the nitrogen cycle? [2]

Root nodules

2 George has an allotment. He grows peas, beans, cabbages and carrots.
Pea plants have root nodules.
Each year George plants his cabbages and carrots where peas and beans
were grown the year before.

a) Why does he do this? [1]

b) If the beans are not picked from the plant, they fall off and begin to decay.

Give the names of **two** types of organisms that cause decay. [2]

c) Worms in the soil help speed up the decay process.

Explain how they do this. [2]

3 The diagram below shows the nitrogen cycle.

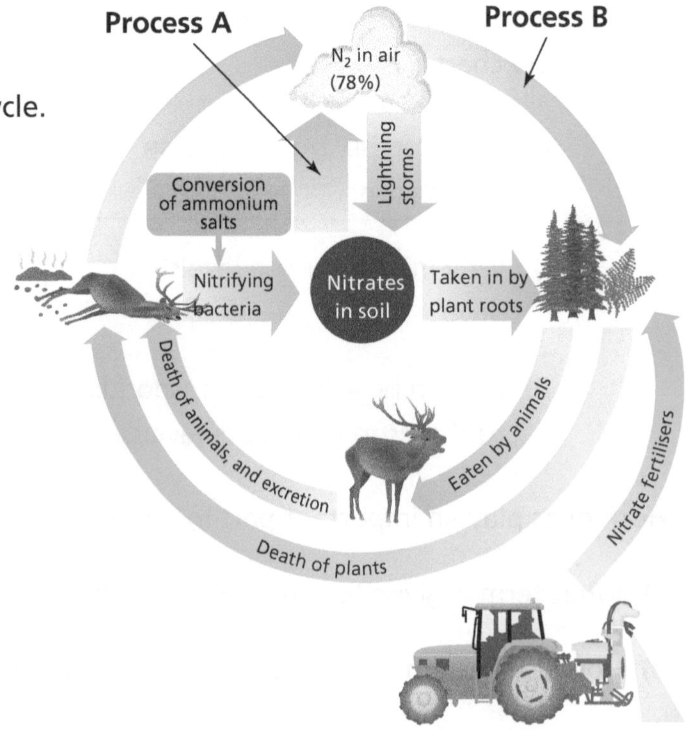

a) What is the name of the bacteria
involved in Process **A**? [1]

b) What is the name of the bacteria
involved in Process **B**? [1]

Process A   Process B

N₂ in air
(78%)

Lightning storms

Conversion
of ammonium
salts

Nitrifying
bacteria

Nitrates
in soil

Taken in by
plant roots

Death of animals and excretion

Eaten by animals

Nitrate fertilisers

Death of plants

Total Marks _____ / 10

Interdependence

1 In a community, the numbers of animals stay fairly constant.
This is because of limiting factors, which stop any one population from becoming too large.

Suggest **two** limiting factors. [2]

2 The food chain below is from a rock pool on a beach.

seaweed ➡ mussel ➡ crab ➡ gull

a) Name a producer in the food chain. [1]

b) Name a consumer in the food chain. [1]

c) Finish the sentences about the rock pool using words from the box.
You do **not** have to use all the words.

habitat   artificial   community   natural   population   niche

A rock pool is an example of a _____ ecosystem. The plants, fish and water

invertebrates make up the _____. The _____ of mussels in the rock

pool is likely to be larger than that of the crabs. [3]

d) While examining the above rock pool some students noticed lots of seaweed on
the beach.

i) Below is a list of factors that affect the distribution of seaweed.

Decide whether each of the five factors are **biotic** or **abiotic**.

The aspect of the beach (which direction it faces)	**Biotic / Abiotic**
The temperature	**Biotic / Abiotic**
Accessibility of the beach to humans	**Biotic / Abiotic**
The amount of fishing in the area	**Biotic / Abiotic**
The slope of the beach	**Biotic / Abiotic** [5]

ii) There is an oil spill near to the beach.

Suggest why, in the following weeks, there is a lot more seaweed on the beach. [1]

Total Marks _____ / 13

Genes

1. Put the following structures in order of size, starting with the largest.

   nucleus   gene   cell   chromosome [1]

2. The diagram below shows the pairs of alleles for genes that code for tongue rolling, eye colour and attached earlobes.

   Use the diagram to answer the following questions.

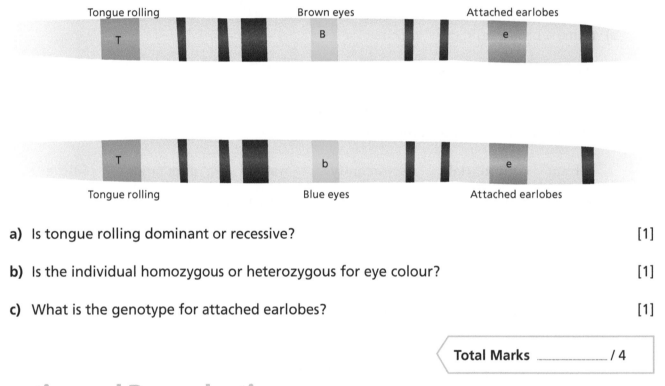

   a) Is tongue rolling dominant or recessive? [1]

   b) Is the individual homozygous or heterozygous for eye colour? [1]

   c) What is the genotype for attached earlobes? [1]

   Total Marks _____ / 4

Genetics and Reproduction

1. Fill in the spaces on the diagram below to show how many chromosomes are present in each of these human cells.

   [3]

2. In meiosis, how many gametes are produced from one parent cell? [1]

3. Cystic fibrosis is an inherited disease caused by a recessive gene.
People who have one dominant and one recessive allele are known as carriers.
People who have two recessive alleles will have cystic fibrosis.

The diagram shows a cross between two carriers of cystic fibrosis.

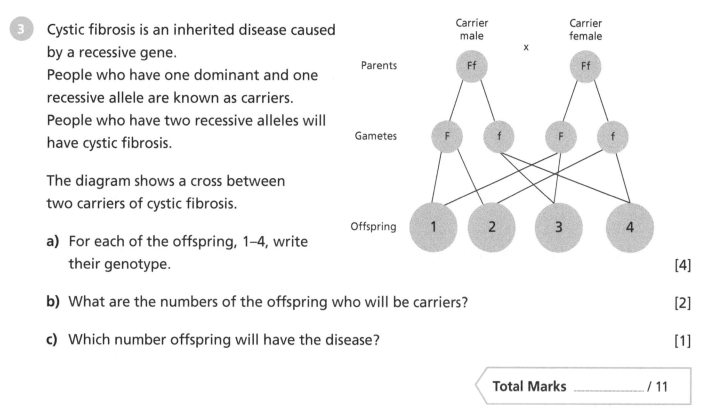

a) For each of the offspring, 1–4, write their genotype. [4]

b) What are the numbers of the offspring who will be carriers? [2]

c) Which number offspring will have the disease? [1]

Total Marks .... / 11

Natural Selection and Evolution

1. Evolution is a slow process but can be seen quickly when bacteria develop antibiotic resistance.

Explain how antibiotic resistance develops. [3]

2. The diagram below shows a phylogenetic tree.

Use the diagram to decide which pair of organisms is most closely related in each question below.

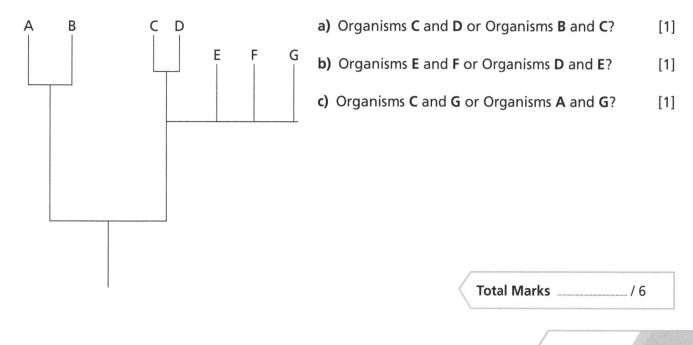

a) Organisms **C** and **D** or Organisms **B** and **C**? [1]

b) Organisms **E** and **F** or Organisms **D** and **E**? [1]

c) Organisms **C** and **G** or Organisms **A** and **G**? [1]

Total Marks .... / 6

Monitoring and Maintaining the Environment

1   Some students wanted to survey the variety of organisms in an area by a canal.

Draw a line from each survey to the best piece of apparatus for the students to use.

Survey	Apparatus

Survey
The variety of water invertebrates in the canal
The variety of plants growing by the side of the canal
The variety of flying insects in the long grass by the side of the canal
The variety of invertebrates found under the hedges along the side of the canal

Apparatus
Quadrat
Pitfall trap
Sweep net
Pond net

[3]

2   Some students used the capture–mark–recapture method to estimate the number of slugs in a garden.
In their first sample they caught 12 slugs, which they marked.
In their second sample, which was collected a few days later, there were 10 slugs of which four were marked.

What is the estimated population size of slugs in the garden? Show your working.   [2]

Total Marks _____ / 5

Investigations

1   Look at the diagram of a pea, before and after germination.

If provided with water and a suitable temperature, germination should take about three days.

A group of students want to find out the optimum temperature for germination of the pea.

Before   After

Describe an investigation the students could do. Your description should include how to ensure the investigation is a fair test.   [5]

Total Marks _____ / 5

Feeding the Human Race

1 HT Below are the stages of genetically engineering cabbages to produce a toxin that harms insects.
The toxin is produced naturally by bacteria.

    A   The gene is pasted into the DNA of a cabbage at an early stage of development.

    B   The gene for toxin production is cut out of the DNA.

    C   A strand of DNA is isolated from the bacteria.

  a) Put the stages in the correct order. [1]

  b) What is the name of the enzymes used in stages **A** and **B**? [2]

2 The following statements are about selectively breeding cows that produce good quantities of meat. They are not in the correct order.

A Offspring that produce a lot of meat are chosen.

B A cow and a bull that provide good quantities of meat are chosen as the parents.

C The offspring are bred together.

D The process is repeated over many generations.

E The parents are bred with each other.

Put the statements in the correct order. [1]

3 HT Scientists often use plasmids in genetic engineering.

  a) What is a plasmid? [1]

  b) Why do scientists use plasmids in genetic engineering? [1]

Total Marks _____ / 6

Monitoring and Maintaining Health

1 Draw a line from each disease to the method by which it is spread.

Disease	Method of Spread
Malaria	Contact
Cholera	By air
Tuberculosis	By water
Athlete's foot	Animal vector

[3]

2. Which of the following statements about HIV and AIDS are **true**?

   A   AIDS is caused by HIV.
   B   AIDS can be cured if diagnosed early.
   C   The HIV virus can be treated with antivirals to prolong life expectancy.
   D   The virus can be spread by droplets in the air.   [2]

3. White blood cells recognise antigens on the surface of microorganisms and produce antibodies to attack them.

   a) Look at the diagram. Which letter corresponds to the microorganism, the antigen and the antibody?   [2]

   b) What is the name of the white blood cell that produces antibodies? Circle the correct answer.

   phagocyte   **lymphocyte**   **blastocyte**   [1]

   Total Marks _____ / 8

Prevention and Treatment of Disease

1. Paul goes to the doctor with earache.
   The doctor prescribes antibiotics and tells Paul that he must be sure to finish the course even if he feels better.

   a) What type of organism may be responsible for Paul's earache?   [1]

   b) Why did the doctor tell Paul to finish the course even if he is feeling better?   [2]

2. Put the stages below, about developing new medicines, in order.

   A   Tests on animals
   B   Tests on healthy volunteers
   C   Clinical trials
   D   Tests on computer models or cells grown in the laboratory   [3]

3. Explain why it is important to test new drugs before they are released to the public.   [3]

4. In the early 1800s, a doctor called Semmelweiss suggested that 'something' on the hands of doctors and surgeons caused infections and could be spread from patient to patient. By insisting that doctors washed their hands, he reduced patient deaths on hospital wards from 12% to 1%.

   a) What was the 'something' on the hands of doctors and surgeons? [1]

   b) In modern hospitals, staff, patients and visitors are encouraged to wash their hands regularly.

      i) What type of substance is used in hand wash in hospitals? [1]

      ii) Suggest **one** precaution, other than hand washing, that surgeons take to reduce the spread of infection. [1]

   Total Marks _____ / 12

Non-Communicable Diseases

1. Beatrice has been told that her arteries are coated with fatty deposits and that her cholesterol levels are above normal.
   The doctor wants to treat her with a drug to reduce her cholesterol levels.
   The doctor tells her she may need an operation to make her arteries wider.

   a) Suggest what drug the doctor may want to prescribe. [1]

   b) What could be placed inside Beatrice's arteries to make them wider? [1]

2. Many lifestyle factors influence how likely it is that someone will suffer from certain diseases.

   Explain the link between diet and heart disease. [3]

3. Ben goes to the doctor for an annual check up.
   He drinks about 30 units of alcohol a week and is considerably overweight.
   The doctor takes Ben's blood pressure and finds it is high.

   What advice should the doctor give Ben? [4]

4. What are **two** conditions for which obesity is a risk factor? [2]

5. Which of the following statements best describes cancer?

   A   Cells begin to grow in the wrong place.
   B   Cells become infected with cancer chemicals.
   C   Cells begin to grow and divide uncontrollably. [1]

   Total Marks _____ / 12

Particle Model and Atomic Structure

You must be able to:

- Use the particle model to explain states of matter and changes in state
- Describe the structure of the atom
- Explain how the atomic model has changed over time
- Calculate the number of electrons, protons and neutrons in atoms, ions and isotopes.

Particle Theory of Matter

- The particle theory of matter is based on the idea that all matter is made up of small particles.
- It is used to explain the structure and properties of solids, liquids and gases.

States of Matter

- When particles are heated (given energy), the energy causes them to move more.
- Even in a solid, the particles vibrate.
- At 0°C, water changes **state** from solid to liquid. This is called the melting point.
- The water **molecules** move faster as the water is heated.
- At 100°C, the water molecules change from liquid to gas. This is called the boiling point.
- When a substance is changing state, the temperature remains constant.
- If gaseous water is cooled, the molecules slow down as they lose kinetic energy and liquid water forms.
- Changes in state are physical changes – it is relatively easy to go back to a previous state.
- Chemical changes are permanent changes where an atom or molecule chemically joins to another atom or molecule.
- The particle model uses circles or spheres to represent particles.

HT The particle model has some limitations.

HT It fails to address the forces of attraction between particles, the size of the particles and that there is space between the particles.

The Atom

- The **atom** is the smallest part of an element that retains the properties of that element.
- Atoms are very small, with a radius of about 1^{-10}m and a mass of about 10^{-23}g.
- An atom has a central, positively charged nucleus, surrounded by shells of negatively charged electrons.
- The radius (width) of the nucleus is far smaller than that of the whole atom.
- The nucleus is made up of protons and neutrons.
- An atom has no overall charge because the charges of the (positive) protons and (negative) electrons cancel each other out.

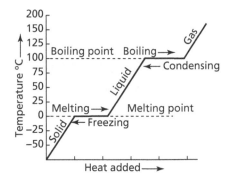

Key Point

The particles within a substance only have no movement at absolute zero (–273.15°C). At all other temperatures the particles will vibrate.

Atomic Particle	Relative Charge	Relative Mass
Proton	+1	1
Neutron	0	1
Electron	–1	0.0005 (zero)

Key Point

No matter how many electrons or neutrons there are in an atom, it is the number of protons that defines an element and its properties.

The History of the Atom

- The model of atomic structure has changed over time.
- In the early 1800s, John Dalton proposed that all atoms of the same element were identical.
- In 1897, J. J. Thomson discovered the electron and put forward the 'plum pudding model' with charges spread evenly throughout the atom.
- In 1911, Ernest Rutherford with his assistants, Geiger and Marsden, discovered that the atom had a dense centre – the nucleus – by firing **alpha particles** at gold foil.
- In 1913, Niels Bohr predicted that electrons occupy energy levels or orbitals.

> **Key Point**
>
> It has taken centuries for scientists, working together, to arrive at the current atomic model. It was experimental evidence that led to changes in the model.

Mass Number and Atomic Number

- The **mass number** is the total number of protons and neutrons in an atom.
- The **atomic number** is the number of protons in an atom.
- An **ion** is formed when an element gains or loses electrons.

> number of protons = atomic number
>
> number of electrons = number of protons (in an atom)
>
> number of neutrons = mass number – atomic number

Element	Symbol	Mass Number	Atomic Number	Protons	Neutrons	Electrons
Hydrogen	$_1^1H$	1	1	1	0	1
Helium	$_2^4He$	4	2	2	2	2
Sodium	$_{11}^{23}Na$	23	11	11	12	11
HT Oxide	$_8^{16}O^{2-}$	16	8	8	16 – 8 = 8	8 + 2 = 10

> The oxide ion is formed when an oxygen atom gains two extra electrons.

- **Isotopes** are atoms of the same element that have:
  - the same atomic number
  - a different mass number, due to a different number of neutrons.
- For example, carbon has three main isotopes:

Isotope	Symbol	Mass Number	Atomic Number	Protons	Neutrons	Electrons
Carbon-12	$_6^{12}C$	12	6	6	6	6
Carbon-13	$_6^{13}C$	13	6	6	7	6
Carbon-14	$_6^{14}C$	14	6	6	8	6

Chlorine has two isotopes:

$_{17}^{35}Cl$ Mass number = 35
Atomic number = 17

$_{17}^{37}Cl$ Mass number = 37
Atomic number = 17

35 – 17 = 18 neutrons 37 – 17 = 20 neutrons

> Taking away the atomic number from the mass number gives the number of neutrons. So, there are two extra neutrons in chlorine-37 compared to chlorine-35.

1. What are isotopes?
2. How many neutrons does an element have with a mass number of 14 and an atomic number of 6?
3. Why have different models of the atom been proposed over time?

> **Key Words**
>
> state
> molecule
> atom
> alpha particle
> mass number
> atomic number
> ion
> isotopes

Purity and Separating Mixtures

You must be able to:

- Suggest appropriate methods to separate substances
- Work out empirical formulae using relative molecular masses and relative formula masses
- Calculate the R_f values of different substances that have been separated using chromatography.

Purity

- In chemistry something is pure if all of the particles that make up that substance are the same, e.g. pure gold only contains gold atoms and pure water only contains water molecules.
- All substances have a specific melting point at room temperature and pressure.
- Comparing the actual melting point to this known value is a way of checking the purity of a substance.
- Any impurities cause the substance to melt at a different temperature.
- **Formulations** are mixtures that have been carefully designed to have specific properties, e.g. alloys.

Relative Atomic, Formula and Molecular Mass

- Every element has its own **relative atomic mass** (A_r).
- This is the ratio of the average mass of one atom of the element to one-twelfth of the mass of an atom of carbon-12.
- The **relative molecular mass** (M_r) is the sum of the relative atomic masses of each atom making up a molecule.
- For example, the M_r of O_2 is $2 \times 16 = 32$.
- The **relative formula mass** (M_r) is the sum of the relative atomic masses of all the atoms that make up a compound.

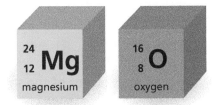

The relative atomic mass of magnesium is 24 and of oxygen is 16.

Calculate the relative formula mass of H_2O.

H:	$2 \times 1 = 2$
O:	$1 \times 16 = 16$
H_2O:	$2 + 16 = 18$

Multiply the number of atoms of each element in the molecule by the relative atomic mass.

Add them all up to calculate the M_r.

Empirical Formula

- The empirical formula is the simplest whole number ratio of each type of atom in a compound.
- It can be calculated from the numbers of atoms present or by converting the mass of the element or compound.

What is the empirical formula of a compound with the formula $C_6H_{12}O_6$?

$$C = \frac{6}{6} = 1 \qquad H = \frac{12}{6} = 2 \qquad O = \frac{6}{6} = 1$$

The empirical formula is written as CH_2O.

Work out the smallest ratio of whole numbers by dividing each by the smallest number. This would be $C_1H_2O_1$.

Remember, the 1 is not written.

- For example, all alkenes have the empirical formula C_1H_2 although the 1 is not written, so it would appear as CH_2.

> What is the empirical formula of a compound containing 24g of carbon, 8g of hydrogen and 32g of oxygen?

Elements	Carbon	:	Hydrogen	:	Oxygen
Mass of element	24	:	8	:	32
A_r of element	12	:	1	:	16
$\dfrac{\text{Mass of element}}{A_r}$	2	:	8	:	2
Divide by smallest number	÷ 2		÷ 2		÷ 2
Ratio of atoms in empirical formula	1	:	4	:	1

List all the elements in the compound.

To find the number of moles, divide the mass of each element by its relative atomic mass.

Divide each answer by the smallest number in step 2 to obtain a ratio.

The ratio may have to be scaled up to give whole numbers.

The empirical formula is therefore CH_4O.

Remember, it is incorrect to write the 1 for an element.

Separation Techniques

- Techniques that can be used to separate mixtures include:
- **Filtration** – a solid is separated from a liquid (e.g. copper oxide solid in copper sulfate solution).
- **Crystallisation** – a solvent is evaporated off to leave behind a solute in crystal form (e.g. salt in water).
- **Distillation** – two liquids with significantly different boiling points are separated, i.e. when heated, the liquid with the lowest boiling point evaporates first and the vapour is condensed and collected.
- **Fractional distillation** – a mixture of liquids with different boiling points are separated (e.g. petrol from crude oil).
- **Chromatography** – substances in a mixture are separated using a **stationary phase** and a **mobile phase**.
  - **Paper chromatography** – this is useful for separating mixtures of dyes in solution (e.g. dyes in ink).
  - **Thin layer chromatography (TLC)** – this is more accurate than paper chromatography and uses a thin layer of an inert solid for the stationary phase.
  - **Gas chromatography** – this separates gas mixtures by passing them through a solid stationary phase.
- Substances separated by chromatography can be identified by calculating their R_f **values**.

Key Point

Substances move up the stationary phase at different rates depending upon their properties. The rate will remain the same as long as the conditions are the same.

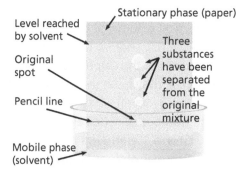

Level reached by solvent

Stationary phase (paper)

Three substances have been separated from the original mixture

Original spot

Pencil line

Mobile phase (solvent)

$$R_f = \frac{\text{distance moved by the compound}}{\text{distance moved by the solvent}}$$

LEARN

- Separated substances can be identified by comparing the results to known R_f values.

Key Words

relative atomic mass (A_r)
relative molecular mass (M_r)
relative formula mass (M_r)
chromatography
stationary phase
mobile phase
R_f value

Quick Test

1. What is the relative formula mass of $Mg(OH)_2$?
2. What is paper chromatography used to separate?
3. What is the empirical formula of a compound with the formula C_2H_6?

Bonding

You must be able to:

- Explain how metals and non-metals are positioned in the periodic table
- Describe the electronic structure of an atom
- Draw dot and cross diagrams for ions and simple covalent molecules.

The Periodic Table

- An element contains one type of atom.
- Elements cannot be chemically broken down into simpler substances.
- There are about 100 naturally occurring elements.
- The design of the modern periodic table was first developed by Mendeleev.
- Elements in Mendeleev's table were placed into groups based on their **atomic mass**.
- Mendeleev's method was testable and predicted elements not yet discovered.
- However, some elements were put in the wrong place because the values used for their atomic masses were incorrect.
- The modern periodic table is a modified version of Mendeleev's table.
- It takes into account the arrangement of electrons, the number of electrons in the outermost shell, and atomic number.

Groups

- A vertical column of elements in the periodic table is a **group**.
- Lithium (Li), sodium (Na) and potassium (K) are in Group 1.
- Elements in the same group have similar chemical properties because they have the same number of electrons in their **outer shell** (or energy level).
- The number of outer electrons is the same as the group number:
  - Group 1 elements have one electron in their outer shell.
  - Group 7 elements have seven electrons in their outer shell.
  - Group 0 elements have a full outer shell.

Periods

- A horizontal row of elements in the periodic table is a **period**.
- Lithium (Li), carbon (C) and neon (Ne) are in Period 2.
- The period for an element is related to the number of occupied electron shells it has.
- For example, sodium (Na), aluminium (Al) and chlorine (Cl) have three shells of electrons so they are in the Period 3.

> **Key Point**
>
> The number of protons in a nucleus of an element never changes. That's why the periodic table shows the atomic number.

Metals and Non-Metals

- The majority of the elements in the periodic table are metals.
- Metals are very useful materials because of their properties:
  - They are lustrous, e.g. gold is used in jewellery.
  - They are hard and have a high density, e.g. titanium is used to make steel for drill parts.
  - They have high tensile strength (are able to bear loads), e.g. steel is used to make bridge girders.
  - They have high melting and boiling points, e.g. tungsten is used to make light-bulb filaments.
  - They are good conductors of heat and electricity, e.g. copper is used to make pans and wiring.
- Metals can react with non-metals to form ionic compounds.
- For example, metals react with oxygen to form metal oxides.

Electronic Structure

- An element's position in the periodic table can be worked out from its **electronic structure**.
- For example, sodium's electronic structure is 2.8.1 (atomic number = 11):
  - It has three orbital shells, so it can be found in Period 3.
  - It has one electron in its outer shell, so it can be found in Group 1.
- The electronic structure can also be shown using a dot and cross diagram, in which each cross represents an electron.

Chemical Bonds

- Chemical bonds are *not* physical structures.
- They are the transfer or sharing of electrons, which leads to the atoms involved becoming more stable.
- An **ionic bond** is formed when one or more electrons are donated by one atom or molecule and received by another atom or molecule.
- When an ionic compound is in solution, or in a molten state, the ions move freely.
- When an ionic compound is solid, ions are arranged in a way to cancel out the charges.
- A **covalent bond** is formed when atoms share electrons to complete their outermost shell.

Sodium atom, Na
2.8.1

Sodium ion, Na$^+$
[2.8]$^+$

+ 1e$^-$

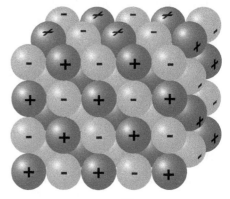

➕ Positively charged ion   ⚪ Negatively charged ion

atomic mass
group
outer shell
period
electronic structure
ionic bond
covalent bond

Quick Test

1. Give two ways that elements are arranged in the modern periodic table.
2. Draw a dot and cross diagram to show the electronic structure of magnesium.
3. Write the electronic structure for chlorine.

Models of Bonding

You must be able to:

- Describe and compare the type of bonds in different substances and their arrangement
- Use a variety of models to represent molecules
- Identify the limitations of different models.

Models of Bonding

- **Models** can be used to show how atoms are bonded together.
- Dot and cross diagrams can show:
  - each shell of electrons or just the outer shell
  - how electrons are donated or shared.
- Methane is a **covalent** compound. Each molecule is made up of a carbon atom joined to four hydrogens (CH_4).

Methane, CH_4

Methane, CH_4

$$H\!-\!\!\underset{\displaystyle H}{\overset{\displaystyle H}{\underset{|}{\overset{|}{C}}}}\!\!-\!H$$

Each line or shared pair of electrons shows a covalent bond.

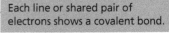
Methane, CH_4

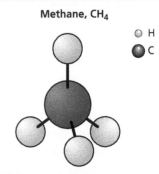

○ H
● C

- Ball and stick models give an idea of the 3D shape of a molecule or compound.
- Each model has limitations:
  - The scale of the nucleus to the electrons is wrong in most models.
  - Models show bonds as physical structures.
  - Most models do not give an accurate idea of the 3D shape of a molecule.
  - The bond lengths are not in proportion to the size of the atoms.
  - Models aid our understanding about molecules, but they are not the real thing.

Key Point

Scientists use models to help solve problems. As atoms are too small to be seen with the naked eye, models are a helpful way of visualising them.

Ion Formation

- Metals give away electrons to become positive ions:

Sodium atom, Na
2.8.1

Sodium ion, Na$^+$
[2.8]$^+$

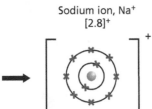

+ 1e$^-$

Sodium gives away a single electron to become a Na$^+$ ion.

- Non-metals gain electrons to become negative ions:

Chlorine atom, Cl
2.8.7

Chloride ion, Cl$^-$
[2.8.8]$^-$

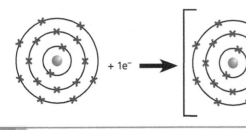

Chlorine gains an electron to become a Cl$^-$ ion.

- Ionic bonds are the electrostatic forces of attraction that hold the ions together.

Simple Molecules

- When non-metals or non-ionic molecules join together, the atoms share electrons and form a covalent bond. These are called **simple molecules**.
- Hydrogen gas, H_2, is a covalent molecule.

Hydrogen Atoms   **A Hydrogen Molecule**

Covalent bond

Outermost shells overlap

Each hydrogen atom now has a full outermost shell, with two electrons.

Giant Covalent Structures

- **Giant covalent structures** are formed when the atoms of a substance form repeated covalent bonds.
- Silicon dioxide is a compound made up of repeating silicon and oxygen atoms joined by single covalent bonds.

Polymers

- A **polymer** is formed when repeated units are covalently bonded together.
- For example, when lots of ethene molecules are joined together they form poly(ethene).

- Silicon atom
- Oxygen atom

Covalent bond

Ethene Monomers   **Poly(ethene) Polymer**

Atoms are shown by their element symbol. Bonds are shown with lines. Two lines together indicate a double bond (two covalent bonds between atoms).

Metals

- Metal atoms are held together by strong metallic bonds.
- The metal atoms lose their outermost electrons and become positively charged.
- The electrons can move freely from one metal ion to another.
- This causes a sea of **delocalised** (free) electrons to be formed.

Free electron

Positive metal ion

Quick Test

1. What is meant by the term 'sea of delocalised electrons'?
2. Give two limitations of a dot and cross model of a covalent compound.
3. What is meant by the term 'giant covalent structure'?

Key Words

model
covalent
ionic bond
simple molecule
giant covalent structure
polymer
delocalised

Properties of Materials

You must be able to:

- Describe how carbon can form a wide variety of different molecules
- Explain the properties of diamond, graphite, fullerenes and graphene
- Explain how the bulk properties of materials are related to the bonds they contain.

Carbon

- Carbon is the sixth element in the periodic table and has an atomic mass of 12.
- Carbon is in Group 4 because it has four electrons in its outer shell.
- This means that it can make up to four covalent bonds with other atoms.
- It can also form long chains of atoms and rings.
- There is a vast variety of naturally occurring and synthetic (man-made) carbon-based compounds, called **organic** compounds.

> **Key Point**
>
> There are a few carbon compounds that are non-organic. They include the oxides of carbon, cyanides, carbonates and carbides.

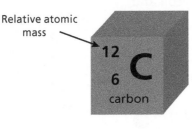

Relative atomic mass

12
6 **C**
carbon

Allotropes of Carbon

- Each carbon atom can bond with up to four other carbon atoms.
- Different structures are formed depending on how many carbon atoms bond together.
- These different forms are called **allotropes** of carbon. They do not contain any other elements.
- Graphite is formed when each carbon atom bonds with three other carbon atoms:
  - Graphite has free electrons so it can conduct electricity, e.g. in electrolysis.
  - The layers are held together by weak bonds, so they can break off easily, e.g. in drawing pencils and as a dry lubricant.
- **Graphene** is a single layer of graphite:
  - In this form, the carbon is 207 times stronger than steel.
  - Graphene has free electrons so it can conduct electricity.
  - It is used in electronics and solar panels.
- Diamond is formed when each carbon atom bonds with four other carbon atoms:
  - Diamond cannot conduct electricity, as all its outermost electrons are involved in bonding.
  - Diamonds are very hard. They are used in drill bits and polished diamonds are used in jewellery.
  - Diamond is extremely strong because each atom forms the full number of covalent bonds.

Graphite

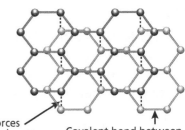

Weak forces between layers

Covalent bond between two carbon atoms within a layer

Graphene

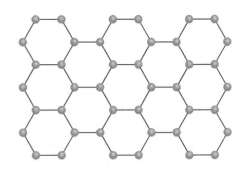

Diamond

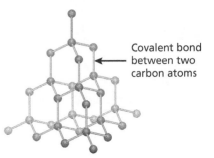

Covalent bond between two carbon atoms

- **Fullerenes** are tubes and spherical structures formed using only carbon atoms:
  - They are used as superconductors, for reinforcing carbon-fibre structures, and as containers for drugs being introduced into the body.

Structure of Buckminsterfullerene

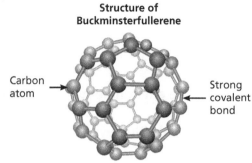

Carbon atom

Strong covalent bond

Bonding and Changing State

- Bonding is an attraction between atoms in elements and compounds.
- If the attraction is weak, then it is easy to separate the atoms compared to those with a stronger attraction.
- The ions in ionic substances are more easily separated when they are in solution or molten, as they can move about freely.
- When an ionic substance is in its crystal (solid) form, i.e. when the distance between ions is at its smallest, it is very difficult to separate the ions due to the strong electrostatic forces.
- They form a giant lattice structure.
- The melting point of ionic substances is, therefore, very high.
- For example, the melting point of NaCl is 801°C.
- Covalent bonds are very strong.
- If there are many bonds, e.g. in a giant covalent compound, the melting point will be very high (higher than for ionic compounds).
- For example, graphite melts at 3600°C.
- Simple covalent molecules have very low boiling points.
- The simplest gas, hydrogen, melts at −259°C and has a boiling point of −252.87°C.
- This is because the **intermolecular forces** that hold all the molecules together are weak and, therefore, easily broken.

Key Point

Don't confuse intermolecular forces (the forces between molecules) with the intramolecular forces (e.g. the covalent bonds between the atoms in the molecules).

Bulk Properties

- The properties of materials are related to:
  - the different types of bonds the molecules contain
  - the ways in which the bonds are arranged
  - the strength of the bonds in relation to the intermolecular forces.
- Ionic and covalent bonds are always stronger than intermolecular forces.
- The individual atoms do not have the same properties as the bulk material.

Quick Test

1. What is meant by the term 'organic compound'?
2. Why is diamond so strong?
3. Explain, giving an example, what allotropes are.

Key Words

organic
allotropes
graphene
fullerenes
intermolecular force

Review Questions

Genes

1 Corey describes himself to his friend, listing the features below.

For each feature, state whether it is caused by **genetics**, the **environment** or a **combination** of both.

a) 1.6 metres tall   b) Blue eyes   **[2]**

c) Pierced eyebrow   d) Weight 90kg   **[2]**

e) Scar on left cheek   **[1]**

2 Sickle cell anaemia is a serious inherited blood disorder. The red blood cells, which carry oxygen around the body, develop abnormally.
It is caused by a recessive gene and a person with sickle cell anaemia must have two recessive alleles.

Use the letters **A** = no sickle cell anaemia and **a** = sickle cell anaemia.

a) What would be the phenotype of someone with the following alleles?

   i) AA   ii) Aa   iii) aa   **[3]**

b) What would be the genotype of someone who was a carrier of sickle cell anaemia?   **[1]**

3 Lesley was studying variation within her class.
She collected information on her friends' height, weight, eye colour, shoe size and whether they had freckles.

a) Divide the list above into continuous and discontinuous variations.   **[5]**

b) Lesley plotted a bar chart of her results for shoe size.

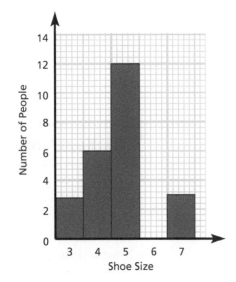

i) Seven people in Lesley's class wear a size 6 shoe. Plot this information on the chart.   **[1]**

ii) Suggest how many people in Lesley's class might wear size 8 shoes.   **[1]**

Total Marks _____ / 16

Genetics and Reproduction

1 What is the name of the type of cell division that leads to the formation of gametes? [1]

2 In humans, brown eyes is the dominant trait and blue eyes is the recessive trait.

Brown Eyes   Brown Eyes

Parental   Bb   Bb

Gametes

Offspring

a) Complete the genetic cross shown. [2]

b) Are the parents homozygous or heterozygous for eye colour? [1]

c) What is the ratio of brown eyes to blue eyes in the offspring? [1]

d) Circle the correct option in the sentence below.

If both parents had blue eyes, there would be a **0% / 25% / 50% / 75% / 100%** chance that their offspring will have blue eyes. [1]

Total Marks _____ / 6

Natural Selection and Evolution

1 In 1753, Carl Linnaeus classified the grey wolf as *Canis lupus,* the domestic dog as *Canis canis* and the coyote as *Canis latrans.*

In 1993, analysis of mitochondrial DNA from all three animals showed:
* the grey wolf and domestic dog share 99.8% of their DNA
* the grey wolf and coyote share 96% of their DNA.

Following this discovery, the domestic dog was reclassified as *Canis lupus familiaris.*

a) Which of the words in the table describes the word '*Canis*'? Put a tick (✓) in the box next to the correct option.

Family	
Genus	
Species	

[1]

b) Suggest why the domestic dog was reclassified. [1]

c) Apart from results of DNA sequencing, suggest **one** other reason why organisms may need to be reclassified. [1]

d) There are many species of dogs. What is meant by the word 'species'? [1]

Total Marks _____ / 4

Review Questions

Monitoring and Maintaining the Environment

1 Neonicotinoid pesticides are new nicotine-like chemicals that act on the nervous systems of insects. They do not affect the nervous system of mammals like some previous pesticides.

These pesticides are water soluble, which means they can be applied to the soil and taken up by the whole plant, which then becomes toxic to any insects that try to eat it.

Neonicotinoids are often applied as seed treatments, which means coating the seeds before planting.

Dutch scientists are concerned that their use is responsible for the decline in the numbers of swallows, starlings and thrushes over the past 10 years.

The scientists have linked decreasing numbers of birds to areas where there are high levels of neonicotinoids in the surface water on fields.

The pesticide can remain in some soil types for up to three years.

a) Suggest **one** advantage and **one** disadvantage of the pesticide being water soluble. [2]

b) Why are neonicotinoids less harmful than some previous pesticides? [1]

c) Scientists have data to link decreasing bird numbers with pesticide levels, but they have yet to discover how the pesticide is causing this decrease.

Suggest **two** possible ways in which the pesticide could be responsible for decreasing bird numbers. [2]

d) The table below shows some data on bird numbers.

Average Concentration of Neonicotinoid in Surface Water in ng/ml	0	10	20	30
Number of Visiting Birds	12 000	11 988	11 650	11 600

Calculate the percentage decrease in the bird population when levels of neonicotinoids reach 30ng/ml.

Show your working and give your answer to two decimal places. [2]

Total Marks .... / 7

Investigations

1 Some students wanted to investigate the distribution of the meadow buttercup plant in the area around a local river. They had read that the meadow buttercup prefers damp areas to dry areas.

The equipment they used is shown below.

The students' hypothesis was: 'Meadow buttercups prefer damp areas to dry areas'.

Tape measure   Quadrat   Moisture meter

a) Describe how the students could use the equipment to test their hypothesis.
In your answer you should make reference to how each piece of equipment would be used. **[5]**

b) Another group of students were investigating the distribution of water snails in the river. They used a pond net to sweep through the water and counted the number of snails in the net. They did this 10 times in total before repeating the investigation further upstream. Their results are shown below.

	Number of Snails in the Net at Each Sweep									
Downstream	3	6	5	4	6	2	2	3	4	4
Upstream	6	1	8	7	10	8	7	6	9	9

i) Calculate the mean, mode and median number of snails in one sweep of the downstream sample. **[3]**

ii) Identify the anomalous result in the upstream data and suggest **two** reasons for this anomalous result. **[3]**

> **Total Marks** _____ / 11

Feeding the Human Race

1 The graph shows the number of hectares used globally for genetically modified (GM) crops since 1998.

a) What are genetically modified crops? **[1]**

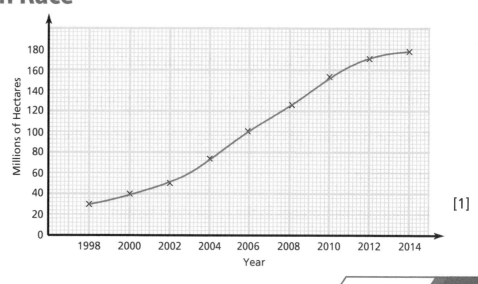

b) Between which years was there the highest increase in the use of land for GM crops? [1]

c) What percentage increase was there in use of land for GM crops between 1998 and 2014?
Show your working. [2]

d) Give **two** reasons why scientists may want to genetically modify crop plants. [2]

2 HT Genetic engineering can be used to produce rice which contains genes to combat vitamin A deficiency. These genes originally come from maize.

Explain how enzymes would be used in the transfer of genes from the maize to the rice. [4]

Total Marks _____ / 10

Monitoring and Maintaining Health

1 Use the diseases listed below to answer the questions.

malaria   flu   athlete's foot   tuberculosis

a) Which disease is caused by a fungus? [1]

b) Which disease is caused by a virus? [1]

c) Which disease would be treated with antibiotics? [1]

2 The diagram below shows one way in which the body deals with invading microorganisms.

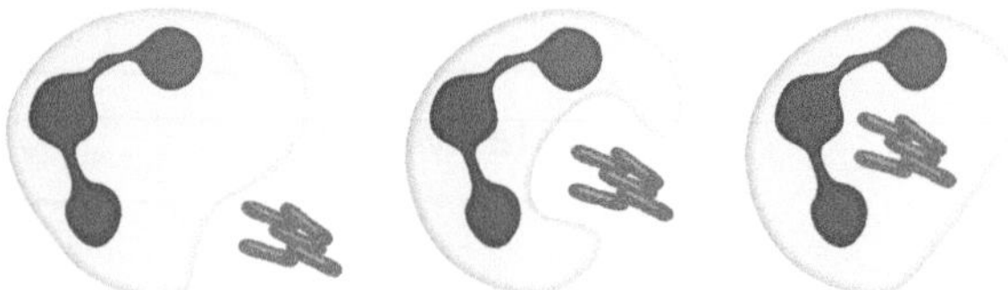

a) What is the name of this process? [1]

b) What type of cell is involved in this process? [1]

3 The human body has a number of mechanisms to prevent microorganisms from gaining entry.

Describe how each of the following helps to defend the body.

a) Platelets [2]

b) Mucous membranes in the respiratory system [1]

Total Marks _____ / 8

Prevention and Treatment of Disease

1 Tanya and Charlie are best friends.
Tanya has been immunised against measles, but Charlie has not.
They come into contact with someone who has measles.
Charlie catches measles, but Tanya does not.

a) What was in the measles vaccine that Tanya was given? [1]

b) Explain why Tanya does **not** catch measles, but Charlie does. [3]

c) How is measles spread from one person to another? [1]

d) Charlie goes to the doctor. The doctor advises plenty of rest and painkillers if necessary.

Why does the doctor **not** prescribe antibiotics for Charlie? [2]

2 When new drugs are developed in the laboratory they are eventually tested in clinical trials.

State **two** ways that drugs are tested before clinical trials. [2]

3 In clinical trials, a control group of patients are often given a placebo.

What is the purpose of the control group? [1]

Total Marks .... / 10

Non-Communicable Diseases

1 The table below shows how the percentage of adults with obesity in the United States changed over a 50-year period.

Year	1962	1974	1980	1994	2002	2008	2012
% Adults with Obesity	13	13	15	23	31	35	36

a) Plot a line graph of these results. [3]

b) Use the graph to predict what the percentage of people with obesity would have been in 2015. [1]

c) Name **two** diseases that people with obesity are more likely to suffer from than people of normal weight. [2]

Total Marks .... / 6

Particle Model and Atomic Structure

1   Look at the heating curve for water.

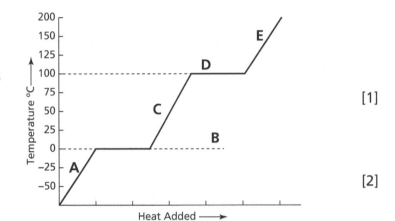

a) Which point, **A**, **B**, **C**, **D** or **E**, indicates the boiling point?   [1]

b) Describe what will happen to the particles in water as the temperature changes from −50°C to −25°C.   [2]

2   HT Look at the diagram of water molecules in distilled water.

What is in the gaps between the water molecules in the diagram?   [1]

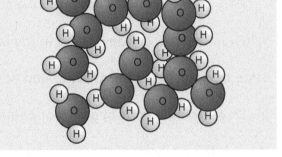

3   Look at the following elements on your periodic table and answer the questions below:
lithium, **oxygen**, **neon**, **silicon** and **calcium**.

a) Which element has a mass number of 20?   [1]

b) Which element has an atomic number of 20?   [1]

c) What would the mass number of an isotope of Si be if it had two extra neutrons?   [1]

4   Water vapour at 140°C was cooled to −40°C.

a) Draw a labelled diagram showing the changes that take place during the cooling process.   [4]

b) Describe what happens to the particles in the water as it cools.   [2]

5   What are the charges and relative masses of:

a) An electron?   b) A proton?   c) A neutron?   [6]

Total Marks _____ / 19

Purity and Separating Mixtures

1 a) What does the term **pure** mean in chemistry? [1]

b) Describe how melting points can be used to help identify a pure substance. [2]

2 Athina is separating food colourings using chromatography.

a) Calculate the R$_f$ value for the two colours in **X**. Show your working. [3]

b) Which of the food colourings, **A, B, C, D** or **E**, matches **X**? [1]

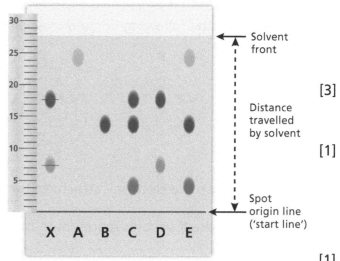

3 A compound has the formula C$_8$H$_8$S$_2$.

What is its empirical formula? [1]

4 What is the empirical formula of a compound containing 84g of carbon, 16g of hydrogen and 64g of oxygen? Show your working. [3]

5 What is the best method for separating water from an ink solution?

A Distillation   **C** Evaporation

B Filtration   **D** Chromatography [1]

Total Marks _____ / 12

Bonding

1 Look at the following chemical symbols from the periodic table.

a) Write down the atomic number for each element. [2]

b) Potassium oxide has the formula: K$_2$O.
Work out the **relative formula mass** for K$_2$O. [1]

c) Calculate the **relative molecular mass** for oxygen. [1]

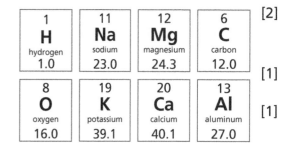

2 a) Why do elements in the same group of the periodic table have the same properties? [1]

   b) Lithium, carbon and neon are in the same period.

      What feature do these elements share? [1]

3 The electronic structure of potassium, K, is written as 2.8.8.1.

   a) Which of the following dot and cross diagrams represents K?

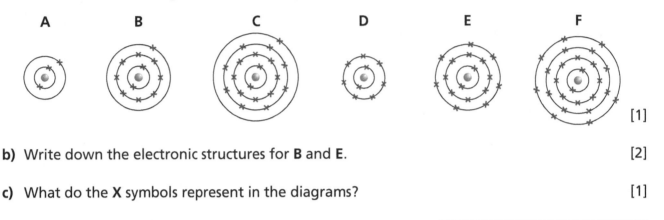

   A   B   C   D   E   F

   [1]

   b) Write down the electronic structures for **B** and **E**. [2]

   c) What do the **X** symbols represent in the diagrams? [1]

   Total Marks _____ / 10

Models of Bonding

1 a) Describe what is meant by the term **ion**. [1]

   b) Draw the dot and cross diagram for a sodium ion, Na^+, and a chloride ion, Cl^-. [2]

2 a) Describe what is meant by the term **covalent bond**. [2]

   b) Chlorine gas, Cl_2, is a covalent molecule.

      Use a dot and cross diagram to show the covalent bond between the chlorine atoms. [2]

3 Silicon dioxide is a compound that forms a giant covalent structure.

   a) What is meant by the term **giant covalent structure**? [2]

   b) Water freezes into ice.
      Although it can form giant structures, ice is not a **giant covalent structure**.

      Suggest why this is the case. [2]

   Total Marks _____ / 11

Properties of Materials

1 **a)** Explain how carbon can form a variety of different molecules. [3]

b) Describe what is meant by the term **allotrope**. [1]

c) Give the names of **three** allotropes of carbon. [3]

2 Graphene is used in electronics and solar panels.

Graphene

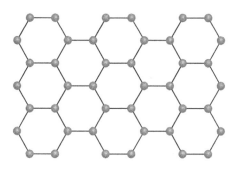

a) Explain why graphene is used for these purposes. [2]

b) Other than cost, explain why diamond is **not** used for these purposes. [1]

c) Draw the structure of diamond. [2]

3 Ionic compounds can conduct electricity.

a) Describe the conditions required for an ionic compound to conduct electricity. [2]

b) Why do ionic compounds in their crystalline form typically have very high melting points? [2]

4 Simple covalent compounds typically have very low melting points.

Which of the following would have the lowest melting point?

A CH_4

B H_2O

C H_2

D O_2 [1]

Total Marks _____ / 17

Introducing Chemical Reactions

You must be able to:

- Use names and symbols to write formulae and balanced chemical equations
- Describe the states of reactants and products in a chemical reaction.

Law of Conservation of Mass

- The law of conservation of mass means that no atoms are created or destroyed.
- This means that, in a chemical reaction, the mass of the **products** will always equal the mass of the **reactants**.
- The atoms in a reaction can recombine with other atoms, but there will be no change in the overall number of atoms.
- This allows chemists to make predictions about chemical reactions. For example:
  - What might be formed when chemicals react together?
  - How much of the chemical or chemicals will be made?

> **Key Point**
>
> Chemicals are not 'used up' in a reaction. The atoms are rearranged into different chemicals.

Formulae and State Symbols

- Compounds can be represented using **formulae**, which use symbols and numbers to show:
  - the different elements in the compound
  - the number of atoms of each element in a molecule of the compound.
- A small subscript number following a symbol is a multiplier – it tells you how many of those atoms are present in a molecule.
- If there are brackets around part of the formula, everything inside the brackets is multiplied by the number on the outside.

Sulfuric acid has the formula H_2SO_4.
This means that there are two hydrogen atoms, one sulfur atom and four oxygen atoms.

The ratio of the number of atoms of each element in sulfuric acid is 2H : 1S : 4O.

$Ca(NO_3)_2$
This means that there is one calcium atom and two nitrate (NO_3) groups.
In total there are one calcium, two nitrogen and six oxygen atoms present in this compound.

- There are four state symbols, which are written in brackets after the formula symbols and numbers:
  - (s) = **solid**
  - (l) = **liquid**
  - (g) = **gas**
  - (aq) = **aqueous** (dissolved in water).

$$H_2O(l) \qquad CO_2(g) \qquad H_2SO_4(aq) \qquad S_8(s)$$

Balancing Equations

- Equations show what happens during a chemical reaction.
- The reactants are on the left-hand side of the equation and the products are on the right.
- Remember, no atoms are lost or gained during a chemical reaction so the equation must be balanced.
- There must always be the same number of each type of atom on both sides of the equation.
- A large number written before a molecule is a **coefficient** – it is a multiplier that tells you how many copies of that whole molecule there are.

$2H_2SO_4(aq)$ means there are two molecules of $H_2SO_4(aq)$ present.

- To balance an equation:

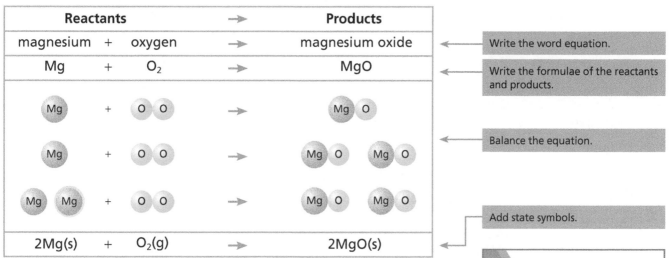

Reactants	→	Products	
magnesium + oxygen	→	magnesium oxide	← Write the word equation.
Mg + O_2	→	MgO	← Write the formulae of the reactants and products.
			← Balance the equation.
			Add state symbols.
$2Mg(s)$ + $O_2(g)$	→	$2MgO(s)$	←

- You should be able to balance equations by looking at the formulae, without drawing the atoms. For example:

calcium carbonate	+	nitric acid	→	calcium nitrate	+	carbon dioxide	+	water
$CaCO_3$	+	HNO_3	→	$Ca(NO_3)_2$	+	CO_2	+	H_2O
$CaCO_3$	+	$2HNO_3$	→	$Ca(NO_3)_2$	+	CO_2	+	H_2O
$CaCO_3(s)$	+	$2HNO_3(aq)$	→	$Ca(NO_3)_2(aq)$	+	$CO_2(g)$	+	$H_2O(l)$

- Equations can also be written using displayed formulae. These must be balanced too.

Key Point

If you find the numbers keep on increasing on both sides of an equation you are trying to balance, it is likely you have made a mistake. Restart by checking the formulae and then rebalancing the equation.

Key Words

products
reactants
formulae
solid
liquid
gas
aqueous
coefficient

Quick Test

1. What is the formula of calcium hydroxide?
2. Write the balanced symbol equation for the reaction: sodium + chlorine → sodium chloride.
3. How many of each atom are present in this formula: $2MgSO_4$?

Chemical Equations

You must be able to:

- Recall the formulae of common ions and use them to deduce the formula of a compound
- HT Use names and symbols to write balanced half equations
- HT Construct balanced ionic equations.

Formulae of Common Ions

- Positive ions are called cations.
- Negative ions are called anions.
- There are a number of common ions that have a set charge.
- It is useful to learn these, although the charges can be worked out if necessary.

Positive Ions (Cations)		Negative Ions (Anions)	
Name	Formula	Name	Formula
Hydrogen	H^+	Chloride	Cl^-
Sodium	Na^+	Bromide	Br^-
Silver	Ag^+	Fluoride	F^-
Potassium	K^+	Iodide	I^-
Lithium	Li^+	Hydroxide	OH^-
Ammonium	NH_4^+	Nitrate	NO_3^-
Barium	Ba^{2+}	Oxide	O^{2-}
Calcium	Ca^{2+}	Sulfide	S^{2-}
Copper(II)	Cu^{2+}	Sulfate	SO_4^{2-}
Magnesium	Mg^{2+}	Carbonate	CO_3^{2-}
Zinc	Zn^{2+}	Hydrogen carbonate	HCO_3^-
Lead	Pb^{2+}		
Iron(II)	Fe^{2+}		
Iron(III)	Fe^{3+}		
Aluminium	Al^{3+}		

- The roman numerals after a transition metal's name tell you its charge, e.g. iron(II) will have the charge Fe^{2+} and iron(III) will have the charge Fe^{3+}.
- When combining ions to make an ionic compound, it is important that the charges cancel each other out so the overall charge is neutral.

$Cu^{2+} + Cl^-$ ←

The formula is: $CuCl_2$

> **Key Point**
>
> Although ionic compounds are written as a formula (e.g. $CuCl_2$), they are actually dissociated when in solution, i.e. the ions separate from each other.

> Two negative charges are needed to cancel the charge on the copper cation. These will come from having two chloride ions.

HT Stoichiometry

- **Stoichiometry** is the measurement of the relative amounts of reactants and products in chemical reactions.
- It is based on the conservation of mass, so knowing quantities or masses on one side of an equation enables you to work out the quantities or masses on the other side of the equation.
- When looking at the stoichiometry of a chemical reaction it is common to look at the ratios of the molecules and compounds.

HT Half Equations

- **Half equations** can be written to show the changes that occur to the individual ions in a reaction:
  1. Write the formulae of the reactants and the products.
  2. Balance the number of atoms.
  3. Add the charges present.
  4. Add electrons (e^-) so that the charges on each side balance.

Hydrogen ions to hydrogen gas:
1. Write formulae: $H^+ \rightarrow H_2$
2. Balance numbers: $2H^+ \rightarrow H_2$
3. Identify charges: 2^+ 0
4. Add electrons: $2H^+ + 2e^- \rightarrow H_2$

Chloride ions to chlorine gas:
1. $Cl^- \rightarrow Cl_2$
2. $2Cl^- \rightarrow Cl_2$
3. 2^- 0
4. $2Cl^- \rightarrow Cl_2 + 2e^-$

> **HT Key Point**
>
> It is convention to show added electrons only; the electrons being taken away are not shown.

HT Balanced Ionic Equations

- When writing a balanced ionic equation, only the **species** that actually change form, i.e. gain or lose electrons, are written.
- The species that stay the same, the **spectator ions**, are ignored.
  1. Write the full balanced equation with state symbols.
  2. Write out all the soluble ionic compounds as separate ions.
  3. Delete everything that appears on both sides of the equation (the spectator ions) to leave the **net ionic equation**.

lead nitrate + potassium chloride $\longrightarrow$ lead chloride + potassium nitrate

1. $Pb(NO_3)_2(aq) + 2KCl(aq) \longrightarrow PbCl_2(s) + 2KNO_3(aq)$

2. $Pb^{2+}(aq) + 2NO_3^-(aq) + 2K^+(aq) + 2Cl^-(aq) \longrightarrow$
   $PbCl_2(s) + 2K^+(aq) + 2NO_3^-(aq)$

3. $Pb^{2+}(aq) + 2Cl^-(aq) \longrightarrow PbCl_2(s)$

> The spectator ions, NO_3^-(aq) and K^+(aq), are removed.

> This is the net ionic equation.

> **Key Words**
>
> cations
> anions
> charge
> HT stoichiometry
> HT half equation
> HT species
> HT spectator ions
> HT net ionic equation

Quick Test

1. What are the formulae of barium oxide, copper fluoride and aluminium chloride?
2. Aluminium ions have a charge of 3^+ and oxide ions have a charge of 2^-. What is the formula of aluminium oxide?
3. HT What is the net ionic equation for the reaction of $Na_2CO_3(aq) + BaCl_2(aq)$?

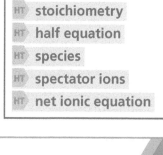

Moles and Mass

You must be able to:

HT Explain what a mole is

HT Calculate the relative molecular mass, mass and number of moles of substances from equations and experimental results.

Moles

- In chemistry it is important to accurately measure how much of a chemical is present.
- Atoms are very small and there would be too many to count in even 1g of substance.
- Instead a measurement is used that represents a known, precise number of atoms – a **mole**.
- A mole represents a set amount of substance – the amount of substance that contains the same number of atoms as 12g of the element **carbon-12**.
- The number of atoms in 1 mole of carbon-12 is a very large number: 6.022×10^{23} atoms.
- This number is known as **Avogadro's constant**.

Calculations Using Moles

- Every element in the periodic table has an atomic mass.
- This means that the mass of one mole of an element will be equivalent to that element's **relative atomic mass** in grams (g).
- The mass of one mole of any compound is its relative formula mass (M_r) in g.
- The **relative molecular mass** of a compound is numerically the same as the relative formula mass. Its units are g/mol.
- You can use the following formulae to calculate the number of moles of an element or compound:

$$\text{number of moles} = \frac{\text{mass}}{\text{relative molecular mass}}$$

$$\text{relative molecular mass} = \frac{\text{mass}}{\text{number of moles}}$$

What is the relative molecular mass of magnesium hydroxide, $Mg(OH)_2$?

Mg: 1×24 =   24
O: 2×16 =   32
H: 2×1 =   2
M_r: $24 + 32 + 2$ =   58

The relative formula mass of $Mg(OH)_2$ is 58, so the relative molecular mass of $Mg(OH)_2$ is 58g/mol.

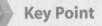

HT **Key Point**

Carbon-12 is the pure isotope of carbon, which has the atomic mass of precisely 12.

HT **Key Point**

6.022×10^{23}g is written in standard form notation because writing 0.000 000 000 000 000 000 000 06022g is extremely awkward.

The formula has been given: $Mg(OH)_2$

How many moles of ethanol are there in 230g of ethanol?
(The relative formula mass of ethanol is 46.)

$$\text{number of moles} = \frac{\text{mass}}{\text{relative molecular mass}}$$

$$= \frac{230\text{g}}{46\text{g/mol}} = 5\text{mol}$$

- If the mass of one mole of a chemical is known, then the mass of one atom or molecule can be worked out.

One mole of sulfur has a mass of 32g.
What is the mass of one sulfur atom?

$$\frac{\text{atomic mass of element}}{\text{Avogadro's constant}} = \frac{32\text{g}}{6.022 \times 10^{23}} = 5.3 \times 10^{-23}\text{g}$$

Calculating Masses of Reactants or Products

- The ratio of the experimental mass to the atomic mass of the constituent atoms can be used to predict the amount of product in a reaction or vice versa.

How much water will be produced when 2 moles of hydrogen is completely combusted in air?

$$2H_2(g) + O_2(g) \rightarrow 2H_2O(l)$$

relative molar mass of water = (2 × 1) + 16 = 18g/mol
mass of water produced = 2 × 18 = 36g

72g of water is produced in the same reaction, how much oxygen was reacted?

$$2H_2(g) + O_2(g) \rightarrow 2H_2O(l)$$

relative molecular mass of water = (2 × 1) + 16 = 18g/mol
relative molecular mass of oxygen = 2 × 16 = 32g/mol

moles of water produced = $\frac{72}{18}$ = 4mol

moles of oxygen used = 2mol
mass of oxygen used = 2 × 32 = 64g

> **Key Point**
>
> Showing the units in an equation helps because they cancel out. If the final unit matches what you are trying to find out, you have done the calculation correctly.

2 moles of hydrogen produce 2 moles of water.

Since 2 moles of water are formed from 1 mole of oxygen, divide by 2.

Quick Test

1. Write the equation that you would use to work out mass from the relative molecular mass and number of moles.
2. 16g of oxygen reacts fully with hydrogen. How much water is produced?
3. The relative atomic mass of caesium (Cs) is 133. What is the mass of a single atom?

> **Key Words**
>
> mole
> carbon-12
> Avogadro's constant
> relative atomic mass
> relative molecular mass

Energetics

You must be able to:

- Explain the difference between endothermic and exothermic reactions
- Draw and label reaction profiles for an endothermic and an exothermic reaction
- HT Calculate energy changes in a chemical reaction considering bond energies.

Reactions and Temperature

- In a chemical reaction, energy is taken in from or given out to the surroundings.
- **Exothermic** reactions release energy to the surroundings causing a temperature rise, e.g. when wood burns through combustion.
- The energy given out by exothermic chemical reactions can be used for heating or to produce electricity, sound or light.
- **Endothermic** reactions absorb energy from the surroundings and cause a temperature drop.
- For example, when ethanoic acid (vinegar) and calcium carbonate react, the temperature of the surroundings decreases.
- Endothermic reactions can be used to make cold packs, which are used for sports injuries.

Key Point

Energy is never lost or used up, it is just transferred.

Activation Energy

- Most of the time chemicals do not spontaneously react.
- A minimum amount of energy is needed to start the reaction. This is called the **activation energy**.
- For example, paper does not normally burn at room temperature.
- To start the combustion reaction, energy has to be added in the form of heat from a match. This provides enough energy to start the reaction.
- As the reaction is exothermic, it will produce enough energy to continue the reaction until all the paper has reacted (burned).

Reaction Profiles

- A graph called a **reaction profile** can be drawn to show the energy changes that take place in exothermic and endothermic reactions.

HT Energy Change Calculations

- In a chemical reaction:
  - making bonds is an exothermic process (releases energy)
  - breaking bonds is an endothermic process (requires energy).
- Chemical reactions that release more energy by making bonds than breaking them are exothermic reactions.

Reaction Profile for an Exothermic Reaction

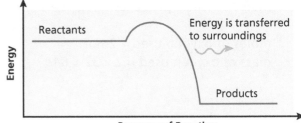

Reaction Profile for an Endothermic Reaction

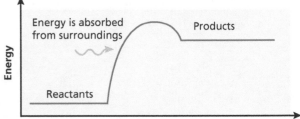

- That energy was originally stored in the bonds between atoms in the reactants.
- Chemical reactions that need more energy to break bonds than is released when new bonds are made are endothermic reactions.
- The energy taken in from the environment is converted to bond energy between the atoms in the products.
- To work out whether a reaction is exothermic or endothermic, calculations can be carried out using information about how much energy is released when a bond forms and how much energy is needed to break a bond.
- The steps to follow are:
  1. Write out the balanced equation and look at the bonds.
  2. Add up the energies associated with breaking bonds in the reactant(s).
  3. Add up the energies associated with making bonds in the product(s).
  4. Calculate the energy change using the equation below:

> **LEARN** **HT**
>
> **energy change = energy used to break bonds – energy released when new bonds are made**

- If the energy change is negative, the reaction is exothermic (more energy is released making bonds than is used breaking them).
- If the energy change is positive, the reaction is endothermic (less energy is released making bonds than is used breaking them).

> **HT** **Key Point**
>
> In the exam, you will be given the bond energy values. You do not have to memorise them.

Hydrogen reacts with iodine to form hydrogen iodide. Calculate the energy change for this reaction.

Bond	Bond Energy (kJ/mol)
H–H	436
I–I	151
H–I	297

$H_2(g) + I_2(g) \rightarrow 2HI(g)$

Total energy needed to break the bonds in
the reactants = 436 + 151
               = 587kJ/mol

Total energy released making the bonds in
the product   = 2 × 297
               = 594kJ/mol

Energy change = 587 – 594
               = –7kJ/mol

> The reactants contain one H–H bond and one I–I bond. The products contain two H–I bonds.

> Energy change is negative, so the reaction is exothermic.

> **Key Words**
>
> exothermic
> endothermic
> activation energy
> reaction profile
> **HT** environment
> **HT** bond energy

Types of Chemical Reactions

You must be able to:

- Explain whether a substance is oxidised or reduced in a reaction
- **HT** Explain oxidation and reduction in terms of loss and gain of electrons
- Predict the products of reactions between metals or metal compounds and acids.

Oxidation and Reduction

- When oxygen is added to a substance, it is **oxidised**.
- When oxygen is removed from a substance, it is **reduced**.
- The substance that gives away the oxygen is called the **oxidising agent**.
- The substance that receives the oxygen is the **reducing agent**.

> **copper oxide + hydrogen** ⟶ **copper + water**

Copper oxide is the oxidising agent (it loses the oxygen). Hydrogen is the reducing agent (it gains the oxygen to form water).

HT Loss and Gain of Electrons

- Chemists modified the definition of oxidation and reduction when they realised that substances could be oxidised and reduced without oxygen being present.
- The definition now focuses on the loss or gain of electrons in a reaction:
  - If a substance gains electrons, it is reduced.
  - If a substance loses electrons, it is oxidised.

> $2Na(s) + Cl_2(g) \longrightarrow 2NaCl(s)$

Sodium gives away the single electron in its outermost shell, so it has been oxidised. Chlorine receives the electrons from the two sodium atoms, so it has been reduced.

> **HT** **Key Point**
>
> **OILRIG:** **O**xidation **I**s **L**oss (of electrons), **R**eduction **I**s **G**ain (of electrons).

Acids and Alkalis

- When an acid or alkali is dissolved in water, the ions that make up the substance move freely.
  - An **acid** produces hydrogen ions, $H^+(aq)$.
  - An **alkali** produces hydroxide / hydroxyl ions, $OH^-(aq)$.
- For example, a solution of hydrochloric acid, HCl, will dissociate into $H^+(aq)$ and $Cl^-(aq)$ ions.
- A solution of sodium hydroxide, NaOH, will dissociate into $Na^+(aq)$ and $OH^-(aq)$ ions.

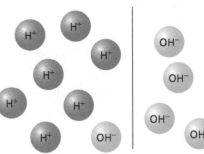

low pH = lots of H^+ lots of OH^- = high pH

Neutralisation

- **Neutralisation** occurs when an acid reacts with an alkali or a **base**, to form a **salt** and water.

> **acid + base** ⟶ **salt + water**

- For example, hydrochloric acid reacts with sodium hydroxide to produce sodium chloride and water:

$$HCl(aq) + NaOH(aq) \longrightarrow NaCl(aq) + H_2O(l)$$

- The reaction can be rewritten to only show the species that change:

$$H^+(aq) + OH^-(aq) \longrightarrow H_2O(l)$$

Reacting Metals with Acid

- Many metals will react in the presence of an acid to form a salt and hydrogen gas.

metal + acid $\longrightarrow$ salt + hydrogen

- The reactivity of a metal determines whether it will react with an acid and how vigorously it reacts.
- Metals can be arranged in order of reactivity in a **reactivity series**.
- If there is a reaction, then the name of the salt produced is based on the acid used:
  - Hydrochloric acid forms chlorides.
  - Nitric acid forms nitrates.
  - Sulfuric acid forms sulfates.

magnesium + hydrochloric acid $\longrightarrow$
magnesium chloride + hydrogen
$$Mg(s) + 2HCl(aq) \longrightarrow MgCl_2(aq) + H_2(g)$$

Reacting Metal Carbonates with Acid

- Metal carbonates also react with acids to form a metal salt, plus water and carbon dioxide gas.

metal carbonate + acid $\longrightarrow$
salt + water + carbon dioxide

- The salts produced are named in the same way as for metals reacting with acids.

magnesium carbonate + sulfuric acid $\longrightarrow$
magnesium sulfate + water + carbon dioxide
$$MgCO_3(s) + H_2SO_4(aq) \longrightarrow MgSO_4(aq) + H_2O(l) + CO_2(aq)$$

Quick Test

1. What gas is made when metal carbonates react with acid?
2. What salt is made when zinc oxide is reacted with nitric acid?
3. Write the word equation for the reaction between copper oxide and sulfuric acid.

> ## Key Point
>
> Remember, ionic substances separate from each other when dissolved or molten. The ions move freely and are not joined together.

> ## Key Point
>
> Water is not an ionic compound. It is a polar molecule (it has positively charged hydrogen and negatively charged oxygen), which means that ionic substances can dissolve easily into it.

Reactivity Series

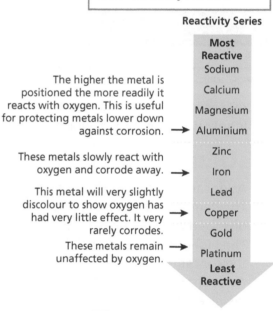

Most Reactive

The higher the metal is positioned the more readily it reacts with oxygen. This is useful for protecting metals lower down against corrosion. $\rightarrow$

Sodium
Calcium
Magnesium
Aluminium

Zinc

These metals slowly react with oxygen and corrode away. $\rightarrow$ Iron

Lead

This metal will very slightly discolour to show oxygen has had very little effect. It very rarely corrodes. $\rightarrow$ Copper

Gold

These metals remain unaffected by oxygen. $\rightarrow$ Platinum

Least Reactive

> ## Key Words
>
> oxidised
> reduced
> oxidising agent
> reducing agent
> acid
> alkali
> neutralisation
> base
> salt
> reactivity series

pH, Acids and Neutralisation

You must be able to:

- Describe techniques to measure pH
- HT Explain the terms dilute, concentrated, weak and strong in relation to acids
- HT Explain pH in terms of dissociation of ions.

Measuring pH

- Indicators change colour depending on whether they are in acidic or alkaline solutions.
- Single indicators, such as litmus, produce a sudden colour change when there is a change from acid to alkali or vice versa.
- pH is a scale from 0 to 14 that provides a measure of how acidic or alkaline a solution is.
- Universal indicator is a mixture of different indicators, which gives a continuous range of colours.
- The pH of a solution can be estimated by comparing the colour of the indicator in solution to a pH colour chart.

0 1 2 3 4 5 6 7 8 9 10 11 12 13 14
Neutral
Strongly acidic   Weakly acidic   Weakly alkaline   Strongly alkaline

- pH can also be measured electronically using an electronic data logger with a pH probe, which gives the numerical value of the pH.

HT Dilute and Concentrated Acids

- Acids can be dilute or concentrated.
- The degree of dilution depends upon the amount of acid dissolved in a volume of water.
- The higher the ratio of acid to water in a solution, the higher the concentration.
- Acids dissociate (split apart) into their component ions when dissolved in solution.
- The concentration is measured as the number of moles of acid per cubic decimetre of water (mol/dm^3).
- For example, $1 mol/dm^3$ is less concentrated than $2 mol/dm^3$ of the same acid.

HT Strong and Weak Acids

- The terms weak acid and strong acid refer to how well an acid dissociates into ions in solution.
- Strong acids easily form H^+ ions.

$$HCl(aq) \longrightarrow H^+(aq) + Cl^-(aq)$$
$$HNO_3(aq) \longrightarrow H^+(aq) + NO_3^-(aq)$$
$$H_2SO_4(aq) \longrightarrow 2H^+(aq) + SO_4^{2-}(aq)$$

> **Key Point**
> Judging something using the eye is a qualitative measurement and has more variation than a quantitative measurement, such as a pH reading from a pH probe.

> **Key Point**
> Don't confuse the term 'concentrated' with how 'strong' an acid or alkali is.

- These strong acids fully ionise.
- Acids that do not fully ionise form an **equilibrium mixture**.
- This means that the ions that are formed can recombine into the original acid. For example:

$$\text{ethanoic acid} \rightleftharpoons \text{ethanoate ions} + \text{hydrogen ions}$$
$$CH_3COOH(aq) \rightleftharpoons CH_3COO^-(aq) + H^+(aq)$$

HT Changing pH

- pH is a measure of how many hydrogen ions are in solution.
- Changing the concentration of an acid leads to a change in pH.
- The more concentrated the acid, the lower the pH and vice versa.
- The concentration of hydrogen ions will be greater in a strong acid compared to a weak acid.
- The pH of a strong acid will therefore be lower than the equivalent concentration of weak acid.
- As the concentration of H^+ ions increases by a factor of 10, the pH decreases by one unit.
- A solution of an acid with a pH of 4 has 10 times more H^+ ions than a solution with a pH of 5.
- A solution of an acid with a pH of 3 has 100 times more H^+ ions than a solution with a pH of 5.

Concentrated weak acid – a lot of acid present, but little dissociation of acid

Concentrated strong acid – a lot of acid present with a lot of dissociation to form many hydrogen ions

$$\text{acid (HA)} \rightleftharpoons \text{hydrogen ion } (H^+) + \text{anion } (A^-)$$

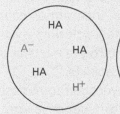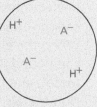

Dilute weak acid – little acid present with little dissociation of acid

Dilute strong acid – little acid present but with a high degree of dissociation

HT Neutralisation and pH

- For neutralisation to occur, the number of H^+ ions must exactly cancel the number of OH^- ions.
- pH curves can be drawn to show what happens to the pH in a neutralisation reaction:
  - An acid has a low pH – when an alkali is added to it, the pH increases.
  - An alkali has a high pH – when an acid is added to it, the pH decreases.
- You should be able to read and interpret pH curves (like the one opposite) to work out:
  - the volume of acid needed to neutralise the alkali
  - the pH after a certain amount of acid has been added.

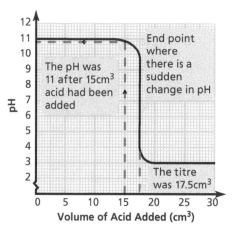

The pH was 11 after 15cm³ acid had been added

End point where there is a sudden change in pH

The titre was 17.5cm³

Volume of Acid Added (cm³)

Key Words

pH
HT dilute
HT concentrated
HT dissociate
HT weak acid
HT strong acid
HT equilibrium mixture

Electrolysis

You must be able to:

- Predict the products of electrolysis of simple ionic compounds
- Describe electrolysis in terms of the ions present
- Describe the competing reactions taking place during electrolysis.

Simple Electrolysis

- Ionic compounds can be broken down into simpler substances using an electric current.
- In **electrolysis**, the solution containing the ionic compound (e.g. copper(II) sulfate solution) is called an **electrolyte**.
- When the electrolyte melts or dissolves, the ions in the compound dissociate.
- Metals and hydrogen form positive ions, called **cations**.
- Non-metals form negative ions, called **anions**.
- The negative electrode is called a **cathode** and the positive electrode is called the **anode**.
- When electricity is passed through the electrolyte, the ions move.
- The cations will move to the cathode and the anions will move to the anode.
- At the cathode, the metal ions gain electrons to become metal atoms.
- At the anode, the non-metal ions lose electrons to become non-metal atoms.

> **Key Point**
>
> Unless the ions can move (i.e. the substance is in solution or molten) electrolysis will not occur.

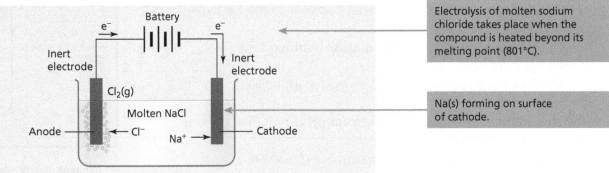

Electrolysis of molten sodium chloride takes place when the compound is heated beyond its melting point (801°C).

Na(s) forming on surface of cathode.

Species Involved

- The atoms, molecules or ions involved in a chemical reaction are called **species**.
- The reactions that take place at each electrode during electrolysis can be written as half equations, which show what happens to each species involved.
- For example, the half equations for the electrolysis of molten sodium chloride are:

At the cathode:
$$Na^+ + e^- \longrightarrow Na$$

This is a reduction process as electrons are gained.

At the anode:
$$2Cl^- \longrightarrow Cl_2 + 2e^-$$

This is an oxidation process as electrons are lost.

- When a substance lower in the reactivity series is in solution, e.g. copper sulfate solution, the reaction is a little different:

> **At the cathode:**
> $Cu^{2+} + 2e^- \longrightarrow Cu$
> **At the anode:**
> $2H_2O \longrightarrow O_2 + 4H^+ + 4e^-$ ←

The sulfate ions stay in solution. They move to the anode but are not discharged.

Types of Electrodes

- **Inert electrodes** do not react during electrolysis.
- Typically they are made from carbon.
- Electrodes can be made out of inert metals instead, such as platinum, which will not react with the products of electrolysis.
- However, platinum electrodes are very expensive.
- Non-inert or **active electrodes** can be used for processes such as electroplating.
- For example, using copper electrodes with copper sulfate solution:

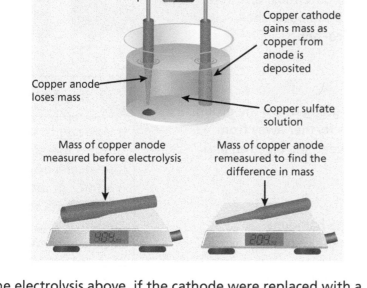

Active Electrodes

Copper cathode gains mass as copper from anode is deposited

Copper anode loses mass

Copper sulfate solution

Mass of copper anode measured before electrolysis

Mass of copper anode remeasured to find the difference in mass

- In the electrolysis above, if the cathode were replaced with a metal object it would become covered in copper metal, i.e. it will be copper-plated.

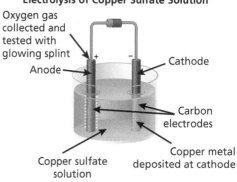

Electrolysis of Copper Sulfate Solution

Oxygen gas collected and tested with glowing splint

Anode

Cathode

Carbon electrodes

Copper metal deposited at cathode

Copper sulfate solution

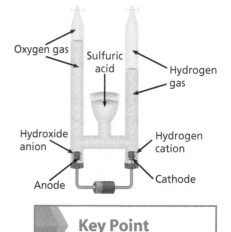

Electrolysis of Dilute Sulfuric Acid

Oxygen gas

Sulfuric acid

Hydrogen gas

Hydroxide anion

Hydrogen cation

Anode

Cathode

> **Key Point**
>
> Ionic solutions conduct electricity because the ions that make up the solution move to the electrodes, *not* because electrons move through the solution.

> **Key Words**
>
> electrolysis
> electrolyte
> cations
> anions
> cathode
> anode
> species
> inert electrode
> active electrode

> **Quick Test**
>
> 1. In what state must an ionic substance be for electrolysis to occur?
> 2. At which electrode will oxygen form during electrolysis?
> 3. Describe how you could silver-plate a key using electrolysis.

Predicting Chemical Reactions

You must be able to:

- Describe the properties of elements in Group 1, Group 7 and Group 0
- Predict possible reactions of elements from their position in the periodic table
- Work out the order of reactivity of different metals.
- Describe the tests to identify gases produced in a chemical reaction.

Group 1

- Group 1 metals all have one **electron** in their outer shell.
- The outer shell can take a maximum of eight electrons (for elements 3–20) or two for H and He.
- Elements are **reactive** because atoms will gain or lose electrons until they have a full outer shell.
- The first three elements in the group are lithium (Li), sodium (Na) and potassium (K).
- It is easier for a Group 1 metal to donate its outermost electron than gain seven more electrons to achieve a full outer shell.
- The elements in Group 1 have similar physical and chemical properties.
- Density increases going down the group.
- The first three metals float on water.
- The Group 1 metals become more reactive going down the group.
- This is because the outermost electron gets further away from the nucleus, so the force of attraction between the electron and nucleus is weaker.

Reactivity increases, and melting and boiling points decrease as you go down the group.

> ### Key Point
>
> It is wrong to say that a Group 1 metal 'wants' to lose an electron. There is a force of attraction between the electrons (negative) and the nucleus (positive). The atom cannot hold onto its outermost electron if there is a stronger force of attraction present.

Group 7

- The non-metals in Group 7 are known as the **halogens**.
- They all have seven electrons in their outer shell, so they have similar chemical properties.
- It is easier for a Group 7 halogen to gain a single electron than donate seven electrons to achieve a full outer shell.
- Fluorine (F), chlorine (Cl), bromine (Br) and iodine (I) are halogens.
- The elements in Group 7 have similar physical and chemical properties.
- Their melting and boiling points are very low and they are not very dense.
- Unlike the alkali metals, halogens become less reactive going down the group.
- This is because the closer the outer shell is to the nucleus, the stronger the force of attraction, and the easier it is to gain an electron.

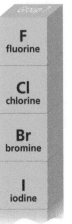

Reactivity decreases, and melting and boiling points increase as you go down the group.

Group 0

- The non-metals in Group 0 are known as the **noble gases**.
- These elements all have a full outer shell and do not react with other elements.

Predicting Reactivity

- In general, for metals:
  - the lower down the group, the more reactive the element
  - the fewer the electrons in the outer shell, the more reactive it will be.
- For example, potassium (K) and calcium (Ca) are in the same period on the periodic table.
- K is more reactive than Ca because Ca has two electrons in its outer shell and K only has one.
- A metal higher in the **reactivity series** can displace another metal from a compound, e.g. copper is lower than magnesium in the reactivity series, so it will be displaced by magnesium.

$$CuSO_4(aq) + Mg(s) \longrightarrow MgSO_4(aq) + Cu(s)$$

- In general, for non-metals:
  - the higher up the group, the more reactive the element
  - the greater the number of electrons in the outer shell, the more reactive it will be.

Metal Reactions with Acids and Water

- Very reactive metals will react with acid to form metal salts + hydrogen:

$$metal + acid \longrightarrow metal\ salt + hydrogen$$

- Reactive metals will react with water to form metal hydroxides + hydrogen.
- The more reactive the metal, the quicker it will donate its outer electron(s). Positive metal ions are formed.

$$magnesium + water \longrightarrow magnesium\ hydroxide + hydrogen$$
$$Mg(s) + 2H_2O(l) \longrightarrow Mg(OH)_2(aq) + H_2(g)$$

> **Key Point**
>
> The reactivity series also includes hydrogen and carbon. Although non-metals, these elements also displace metals.

$Mg(OH)_2$ is an ionic compound so it will dissociate into $Mg^{2+}(aq)$ and $OH^-(aq)$ ions.

Identifying Gases

Gas	Properties	Test for Gas
Oxygen, O_2	A colourless gas that helps fuels burn more readily in air.	Relights a glowing splint.
Carbon dioxide, CO_2	A colourless gas produced when fuels are burned in a sufficient supply of oxygen.	Turns limewater milky.
Hydrogen, H_2	A colourless gas. It combines violently with oxygen when ignited.	When mixed with air, burns with a squeaky pop.
Chlorine, Cl_2	A green poisonous gas that bleaches dyes.	Turns damp indicator paper white.

> **Quick Test**
>
> 1. Which of these halogens would be the least reactive: bromine, chlorine, fluorine or iodine?
> 2. Explain why caesium is more reactive than lithium.
> 3. Which of the following metals will displace copper from $CuSO_4$: silver, platinum, lithium?

> **Key Words**
>
> electron
> reactive
> halogen
> noble gas
> reactivity series

Review Questions

Particle Model and Atomic Structure

1 Tad places a cup of water on a windowsill.
When he returns four days later, he notices that all of the water has disappeared.

   a) What has happened to the water?   [1]

   b) Tad decides to place another cup on the windowsill.

      What would Tad need to do to increase the rate of water loss?   [1]

   c) Draw two boxes:

      i) In one box show the arrangement of the water particles in the cup.   [1]

      ii) In the other box, draw the arrangement of water particles in the air.   [1]

2 Which of the following is the approximate **mass** of an atom?

   A 1×10^{-1}g   **B** 1×10^{-10}g   **C** 1×10^{-15}g   **D** 1×10^{-23}g   [1]

3 Complete the table showing the structure of an atom.   [5]

Subatomic Particle	Relative Mass	Relative Charge
	1	
Neutron		0 (neutral)
Electron		

4 In 1911, Geiger and Marsden fired alpha particles at gold foil.
They expected all of the alpha particles to pass straight through the atoms.
Instead, a substantial number bounced straight back.

   a) What subatomic particle must have been present in the middle of the atom for this to happen?   [1]

   b) Suggest why only some of the alpha particles bounced back.   [2]

5 For each of the following elements, work out how many **neutrons** are present:

2		11		23
He		**Na**		**V**
helium		sodium		vanadium
4.0		23.0		50.9

       [3]

Total Marks .... / 16

Purity and Separating Mixtures

1 Sanjit is testing a mystery pure substance.
He heats the substance until it boils.

    a) How could he confirm the identity of the substance? [2]

    b) Sanjit now takes a sample of water and boils it.
He finds that the water boils at 101°C.

       Assuming that the thermometer is working correctly, why does the water boil at
this temperature? [2]

2 What is the empirical formula of each of the following substances?

    a) $C_6H_{12}O_6$ [1]

    b) CH_3COOH [1]

    c) $CH_3CH_2CH_2COOH$ [1]

3 Calculate the relative formula masses of the following compounds:

    a) $C_6H_{12}O_6$ b) CH_3COOH [2]

    c) CO_2 d) H_2SO_4 [2]

4 What is the scientific term for a substance created by mixing two or more elements, at least
one of which is a metal? [1]

5 Three substances are separated using thin layer chromatography.

Substance	R_f Value
A	0.6
B	0.3
C	0.7

Which substance travelled the furthest in the solid stationary phase? [1]

Total Marks .... / 13

Review Questions

Bonding

1 Which of the following scientists first devised the structure of the modern periodic table? Circle the correct answer.

Bohr   Mendel   Mendeleev   Marsden   Rutherford   Thomson   [1]

2 Here are the electronic structures of five elements:

Element	Electronic Structure
A	2.1
B	2.8.1
C	2.8.3
D	2.8.8
E	2.2

a) Which elements are in Period 2 of the periodic table? [2]

b) Which elements are in the same group? [2]

c) Which is the most reactive metal? [1]

d) What is the atomic number of each element? [5]

3 The majority of elements in the periodic table are metals.

a) Give **three** properties of metals. [3]

b) What is the general name given to the product formed when a metal reacts with oxygen? [1]

c) What type of compound is formed when metals react with non-metals? [1]

Total Marks _____ / 16

Models of Bonding

1 Which of the following diagrams correctly shows the bonding for fluorine gas (F_2)?

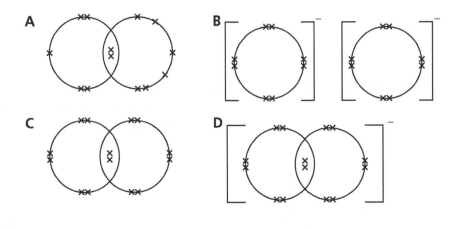

A   B   C   D

[1]

2 Draw a dot and cross diagram to show the covalent bonds in **methane** (CH_4). [2]

3 What advantage do ball and stick models have over dot and cross diagrams? [2]

4 Show, using dot and cross diagrams, how magnesium becomes a magnesium ion. [2]

Total Marks _____ / 7

Properties of Materials

1 Carbon has a number of different **allotropes**.

a) What is meant by the term **allotrope**? [1]

b) Graphite is used as a dry lubricant in air compressors.

Referring to the structure of graphite, explain why it is a good dry lubricant. [4]

c) Diamonds are often used in drill bits.

Referring to the structure of diamond, explain why diamonds are used in drill bits. [3]

d) Give **one** use of graphene. [1]

2 Draw **one** line from each material to its property.

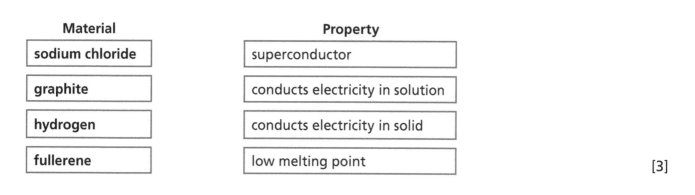

Material	Property
sodium chloride	superconductor
graphite	conducts electricity in solution
hydrogen	conducts electricity in solid
fullerene	low melting point

[3]

Total Marks _____ / 12

Introducing Chemical Reactions

1 What is the **law of conservation of mass**? [1]

2 a) What do the **subscript** numbers that appear after an element symbol mean, e.g. Cl_2? [1]

   b) Write the number of atoms of each element shown in each formula below:

   i) $C_6H_{12}O_6$ iii) H_2O_2 [2]

   ii) CH_3CH_2COOH iv) $Ca(NO_3)_2$ [2]

3 Write down the **four** state symbols. [1]

4 Write the **balanced symbol equations** for the following reactions, including state symbols:

   a) magnesium + oxygen $\rightarrow$ magnesium oxide [2]

   b) lithium + oxygen $\rightarrow$ lithium oxide [2]

   c) calcium carbonate + hydrochloric acid $\rightarrow$ calcium chloride + carbon dioxide + water [2]

   d) aluminium + oxygen $\rightarrow$ aluminium oxide [2]

Total Marks _____ / 15

Chemical Equations

1 What are the charges on these common ions?

   a) copper(II) b) oxide c) iron(III) d) sulfide [4]

2 HT Write the half equation for each of the following reactions:

   a) Hydrogen ions to hydrogen gas [1]

   b) Iron(II) ions to iron solid [1]

   c) Copper(II) ions to copper solid [1]

   d) Zinc to zinc ions [1]

3 HT Write the ionic equation for the following reaction.
All the compounds involved are soluble, except for silver chloride.

silver nitrate + lithium chloride → lithium nitrate + silver chloride [2]

Total Marks _____ / 10

Moles and Mass

1 HT What does a **mole** represent in chemistry? [1]

2 HT Which of the following is Avogadro's constant?

A 6.022×10^{32} C 3.142×10^{32}

B 6.022×10^{23} D 3.142×10^{23} [1]

3 HT What unit is **molecular mass** measured in? [1]

4 HT Cyanobacteria are organisms that can convert atmospheric nitrogen into nitrates.
Abigail is preparing stock solutions containing different metals to investigate how they affect the growth of cyanobacteria.

42	23
Mo	**V**
molybdenum	vanadium
95.9	50.9

She weighs out 287.7g of the element molybdenum.

a) How many moles of molybdenum does she have?
   Show your working. [2]

b) Abigail needs to weigh out 5 moles of vanadium.

   What mass of vanadium should she use?
   Show your working. [2]

5 HT What is the relative molecular mass of glucose, $C_6H_{12}O_6$?
(Relative atomic mass of C = 12, H = 1 and O = 16.) [1]

6 HT Calculate the mass of one atom of each of the following elements.
Show your working. Give your answer to one decimal place.

a) Vanadium   c) Caesium   [4]

b) Molybdenum   d) Bismuth   [4]

7 HT Five moles of hydrogen are completely combusted in air.

How much water is produced in the reaction? [2]

Total Marks _____ / 18

Energetics

1 a) A reaction gives out energy to the environment.

What type of reaction is it? [1]

b) A reaction takes in energy from the environment.

What type of reaction is it? [1]

2 Give **two** ways in which the energy released from a reaction can be used. [2]

3 Explain what is meant by the term **activation energy**. [1]

4 Draw a reaction profile for an exothermic reaction. [1]

5 Draw a reaction profile for an endothermic reaction. [1]

6 HT Which of the following is an **exothermic** process?

making chemical bonds   breaking chemical bonds [1]

7 HT Peter reacted hydrogen gas with fluorine gas to form hydrogen fluoride gas.

The equation for the reaction is: $H_2(g) + F_2(g) \rightarrow 2HF(g)$

Bond	Bond Energy (kJ/mol)
H–F	565
H–H	432
F–F	155

Calculate the energy change for this reaction and state whether the reaction is **exothermic** or **endothermic**. [4]

8 HT A series of reactions was carried out and the energy changes were recorded.

For each energy change, state whether it was **exothermic** or **endothermic**.

a) +90kJ/mol   b) −181kJ/mol   c) +20kJ/mol   d) +8kJ/mol   [4]

9 HT Look at the reaction profile for $H_2(g) + F_2(g) \rightarrow 2HF(g)$

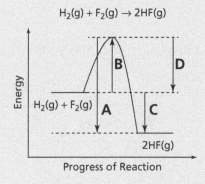

Which letter on the diagram shows the energy change for the reaction?   [1]

Types of Chemical Reactions

Total Marks .... / 17

1 Which reactions involve a reactant being oxidised?

A magnesium + oxygen → magnesium oxide

B water (solid) → water (liquid)

C copper + oxygen → copper oxide

D barium carbonate + sodium sulfate → barium sulfate + sodium carbonate   [2]

2 HT Explain what **oxidation** and **reduction** mean in terms of electrons.   [2]

3 For each of the following reactions, write a balanced equation, including state symbols.
Then state which species has been oxidised and which has been reduced.

a) sodium + chlorine → sodium chloride   [3]

b) magnesium + oxygen → magnesium oxide   [3]

c) lithium + bromine → lithium bromide   [3]

d) copper(II) oxide + hydrogen → copper + water   [3]

4 What ions are produced by a) an acid and b) an alkali? [2]

5 What is the general equation for the neutralisation of a base by an acid? [1]

6 Dilute sulfuric acid and sodium hydroxide solution are reacted together.

   a) Write the balanced symbol equation for the reaction. [2]

   b) Which ions are not involved in the reaction? [2]

   c) HT Write the ionic equation for the reaction between dilute sulfuric acid and sodium hydroxide solution. [2]

   Total Marks _____ / 25

pH, Acids and Neutralisation

1 HT What is meant by the term **weak acid**? [1]

2 HT Look at the concentrations below. For each pair, which is more concentrated? [1]

   a) $1mol/dm^3$ H_2SO_4 OR $2mol/dm^3$ H_2SO_4 b) $3mol/dm^3$ HNO_3 OR $2mol/dm^3$ HNO_3 [1]

3 HT How many more times concentrated are the H^+ ions in a solution with a pH of 6 compared to a solution with a pH of 3? [1]

   Total Marks _____ / 4

Electrolysis

1 What are the ions of a) metals and b) non-metals called? [2]

2 Why is it not possible to carry out electrolysis on crystals of table salt (sodium chloride) at room temperature and pressure? [1]

3 Describe how you could copper-plate a nail using copper(II) sulfate solution. [3]

4 Why are inert electrodes often used in electrolysis? [1]

   Total Marks _____ / 7

Predicting Chemical Reactions

1   Which of the following statements is **correct** about a Group 7 element?

   A   It conducts electricity.   B   It is malleable.

   C   It dissolves in water to make bleach.   D   It reacts with water to form hydrogen.   [1]

2   How many electrons are in the outermost shell of a Group 1 element?   [1]

3   Sort the following elements into order of reactivity, with the most reactive first:

   bromine   **chlorine**   **fluorine**   **iodine**   [1]

4   a) Draw the electronic structure of sodium (atomic number 11) and potassium
       (atomic number 19).   [2]

   b) Sodium is placed into a container of water.
       It moves around rapidly on top of the water and burns with a yellow flame.

       Describe what happens when potassium is placed into a container of water.   [2]

   c) Which of the following best describes why potassium reacts differently to sodium?

       A   Sodium is denser than potassium.

       B   Potassium is denser than sodium.

       C   Potassium's outermost electron is closer to the nucleus.

       D   Potassium's outermost electron is further away from the nucleus.   [1]

5   Look at the periodic table on page 275.

   Which is the **least** reactive Group 2 metal?   [1]

6   Zinc is reacted with sulfuric acid: $Zn(s) + H_2SO_4(aq) \rightarrow ZnSO_4(aq) + H_2(g)$

   What **ions** will be formed in this reaction?   [1]

7   Describe the test for hydrogen gas.   [2]

Total Marks .... / 12

Controlling Chemical Reactions

You must be able to:

- Suggest methods for working out the rate of reaction
- Describe factors that affect the rate of reaction
- Explain how surface area to volume ratios change in solids.

Rates of Reaction

- The **rate of reaction** is a measure of how much product is made in a specific time.
- It can be measured in:
  - g/s or g/min for mass changes
  - cm^3/s or cm^3/min for volume changes.
- Chemical reactions stop when one of the reactants is used up.
- The amount of product produced depends on the amount of reactant(s) used.
- Often there is an excess of one of the reactants.
- The one that is used up first is the limiting reactant.

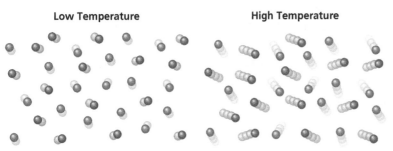

Low Temperature   High Temperature

Changing the Temperature

- Chemical reactions happen when particles collide with enough energy for the reactants to bond together or split apart.
- The more successful **collisions** there are between particles, the faster the reaction.
- At low temperatures, the particles move slowly as they have less **kinetic energy**.
- This means that the particles collide less often, and with lower energy, so fewer collisions will be successful – the rate of reaction will be slow.
- At high temperatures, the particles move faster as they have more kinetic energy.
- This means that particles successfully collide more often, and with higher energy, giving a faster rate of reaction.

Changing the Concentration of Liquids

- In a reaction where one or both reactants are in low concentrations, the particles will be spread out in solution.
- The particles will collide with each other less often, so there will be fewer successful collisions.

- If there is a high concentration of one or both reactants, the particles will be crowded close together.
- The particles will collide with each other more often, so there will be many more successful collisions.
- To measure the effect of concentration on rate of reaction, create a series of different concentrations of a reactant by diluting a stock solution of known concentration.
- Once a range of at least five different concentrations has been created, the reactants can be reacted together.

Low Concentration

High Concentration

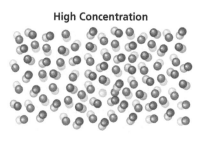

Changing the Pressure of Gases

- When a gas is under a low pressure, the particles are spread out.
- The particles will collide with each other less often, so there will be fewer successful collisions.
- When the pressure is high, the particles are crowded more closely together.
- The particles therefore collide more often, resulting in many more successful collisions.
- It is difficult to carry out an experiment where gas pressure is changed in a school laboratory.
- Specialised equipment is needed to ensure that the pressure can be maintained and measured.

Low Pressure

High Pressure

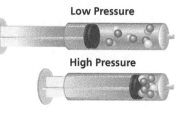

Changing the Surface Area of Solids

- The larger the surface area of a solid reactant, the faster the reaction.
- Powdered solids have a large surface area compared to their volume – they have a high surface area to volume ratio.
- This means there are more particles exposed on the surface for the other reactants to collide with.
- The greater the number of particles exposed, the greater the chance of them colliding successfully, which increases the rate of the reaction.
- As a result, powders react much faster than a lump of the same reactant.
- To measure the effect of surface area on rate of reaction, take the reactant and create a set of different surface areas, e.g. by cutting the solid into progressively smaller sections.
- Once a suitable range of surface areas has been created, they can then be reacted with the other reactant.

Lump of Solid (Large Particles)   Powdered Solid (Small Particles)

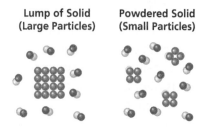

Quick Test

1. What is the effect of increasing the temperature of a reaction mixture?
2. How could the rate of reaction be increased in a reaction involving gaseous reactants?
3. Describe the surface area to volume ratio of a large block of calcium carbonate compared to calcium carbonate powder.

Key Words

rate of reaction
collisions
kinetic energy
pressure
surface area

Catalysts and Activation Energy

You must be able to:

- Analyse the rate of reaction using a graph
- Describe the characteristics of catalysts and their effect on the rate of reaction
- Explain the action of catalysts in terms of activation energy.

Analysing the Rate of Reaction

- From a graph you can find out the following:
  1. How long it takes to make the maximum amount of products:
     Draw a vertical line from the beginning of the flat line (which indicates the reaction has finished) down to the x-axis (time).
  2. How much product was made:
     Draw a horizontal line from the highest point on the graph across to the y-axis.
  3. Which reaction is the quicker:
     Compare the gradient (steepness) of graphs for the same reaction under different conditions.

- $\dfrac{1}{time}$ is **proportional** to the **rate of reaction**.

Effect of Catalysts

- A **catalyst** is a substance that speeds up the rate of a chemical reaction without being used up or changed in the reaction.
- Catalysts are very useful materials, as only a small amount is needed to speed up a reaction involving large amounts of reactant.
- You can see how a catalyst affects the rate of reaction by comparing graphs of reactions with and without the catalyst.
- The graph on the right shows two reactions that eventually produce the same amount of product.
- One reaction takes place much faster than the other because a catalyst is used.
- As catalysts are not used up in a reaction, they are not a reactant.
- When a catalyst is used in a reaction, the name of the catalyst is written above the arrow in the equation.
- For example, hydrogen peroxide is relatively stable at room temperature and pressure. It breaks down into water and oxygen very slowly.
- In the presence of the **inorganic** catalyst, manganese(IV) oxide, the breakdown is very rapid.

Volume of CO_2 Produced (cm^3)

The line is steeper for the 45°C line, so the rate is faster.

relative rate of reaction for 45°C = $\dfrac{1}{5min}$ = 0.2

relative rate of reaction for 30°C = $\dfrac{1}{10min}$ = 0.1

Key Point

When looking at a graph, it may be obvious which line is steeper but it pays to calculate the gradient as well.

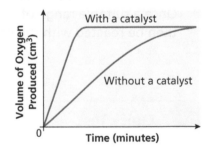

Without a catalyst: hydrogen peroxide ⟶ water + oxygen (very slow)

With a catalyst: hydrogen peroxide *manganese(IV) oxide* water + oxygen (very fast)

Catalysts and Activation Energy

- Catalysts work by lowering the **activation energy** for a reaction.

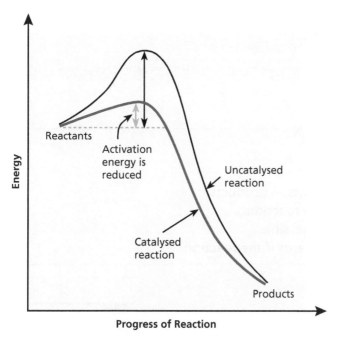

- This means that the reaction is more likely to happen, because less energy is needed to get it started.
- This may be achieved by providing a surface for the reactants to temporarily **adhere** to (stick to).
- Adhering to the catalyst's surface uses electrons, weakening the bonds in the reactants, lowering the activation energy.

Biological Catalysts

- **Enzymes** and certain pigments are biological catalysts.
- They are proteins and, as with inorganic catalysts, they lower the activation energy for a reaction.
- Examples include:
  - **amylase** – an enzyme that breaks down the starch (a polymer) into glucose
  - **chlorophyll** – a pigment that catalyses photosynthesis
  - **catalase** – an enzyme that, like manganese(IV) oxide, can break down hydrogen peroxide into water and oxygen.

> **Key Point**
>
> Don't forget, catalysts are not used up in a reaction.

> **Key Words**
>
> proportional
> rate of reaction
> catalyst
> inorganic
> activation energy
> adhere
> enzymes

Quick Test

1. What does the gradient (steepness) of a reaction graph tell you?
2. How do catalysts affect activation energy?
3. Which biological molecules can catalyse a reaction?

Equilibria

You must be able to:

- Recall that some reactions may be reversed by altering the reaction conditions
- Describe what a dynamic equilibrium is
- **HT** Make predictions about how the equilibrium point will change under different conditions.

Reversible Reactions

- Most chemical reactions are of the type: reactants → products
- The reaction is complete, i.e. it is an **irreversible** reaction.
- There are, however, some reactions that are **reversible**.
- A reversible reaction can go forwards or backwards if the reaction conditions are changed.

Dynamic Equilibrium

- A reversible reaction can reach **equilibrium** (a balance).
- This means that the rate of the forward reaction is equal to the rate of the reverse reaction.
- A closed system is one where the conditions of the reaction (such as pressure and temperature) are not changed and no substances are added or removed.
- At equilibrium in a closed system:
  - there is no change in the amounts and concentrations of reactants and products
  - reactants are reacting to form products and products are reacting to re-form reactants at the same rate.
- This is called a **dynamic equilibrium**.
- For example, heating a mixture of hydrogen and iodine gas:

> **Heating a mixture of hydrogen and iodine gas.**
> $$H_2(g) \; + \; I_2(g) \; \rightleftharpoons \; 2HI(g)$$

- Unlike an irreversible reaction, the reactants will *not* completely form hydrogen iodide.
- When the reaction starts, the forward reaction producing HI will dominate.
- This is because the concentration of HI will be very low.
- As the concentration of HI increases, the rate of the reverse reaction (making H_2 and I_2) will also increase.
- Eventually the rate of the forward reaction will equal the rate of the reverse reaction.
- The reaction is in dynamic equilibrium.

HT Changing the Equilibrium

- The position of the equilibrium can be changed by altering:
  - the temperature
  - the pressure
  - the concentration of the reactant(s) and/or the product(s).

Key Point

The $\rightleftharpoons$ symbol indicates a reversible reaction.

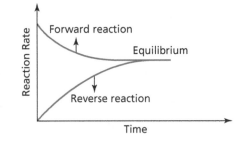

Key Point

Remember, at equilibrium in a closed system the concentrations of the reactants and products remain the same.

- If the equilibrium lies to the right of the reaction, the concentration of the products is greater than the concentration of the reactants.
- If the equilibrium lies to the left of the reaction, the concentration of the products is less than the concentration of the reactants.

HT Le Chatelier's Principle

- Le Chatelier's principle states that when the conditions of a system are altered, the position of the equilibrium changes to try and restore the original conditions.
- This means that predictions can be made about:
  - what will happen when certain conditions are changed
  - what conditions need to be changed to produce the highest yield of a substance.

HT Changing Conditions

- If a reaction is exothermic in the reactant → product direction, Le Chatelier's principle indicates that increasing the temperature will:
  - decrease the yield of products
  - increase the amount of reactants (to restore the original conditions).
- If a reaction is endothermic in the reactant → product direction, then an increase in temperature will:
  - increase the yield of products
  - decrease the amount of reactants (to restore the original conditions).
- Pressure changes only apply to reactions involving gases.
- Pressure changes will alter the equilibrium if there is a difference in the amount of gas in the products and the reactants.
- Applying Le Chatelier's principle, if there are more moles of gas reactants compared to products:
  - an increase in pressure will lead to more product
  - a decrease in pressure will cause the equilibrium to move towards the side of the equation with the greatest number of moles.
- If the concentration of one of the substances in a reaction increases, then Le Chatelier's principle means that the equilibrium will shift to restore the original conditions:
  - If the concentration of a reactant is increased, more product will be made.
  - If the concentration of a product is increased, more reactant will be made.

Key Point

With reversible reactions, you have to be clear which direction is exothermic. If the forward reaction is exothermic, the reverse reaction will be endothermic and vice versa.

Key Point

A catalyst does not affect the equilibrium position. It speeds up the rate of reaction, so equilibrium will be reached more rapidly.

Key Words

irreversible
reversible
equilibrium
dynamic equilibrium
HT Le Chatelier's principle
HT exothermic
HT endothermic

Quick Test

1. HT Name three factors that can change the position of the equilibrium for a reaction.
2. No reactions take place when a reversible reaction is at equilibrium. True or false?
3. HT If a forward reaction is exothermic, what would the effect of increasing the temperature be? Explain your answer.

Improving Processes and Products

You must be able to:

- Describe how to extract metals from their ores using carbon
- Explain the use of electrolysis to extract metals higher than carbon in the reactivity series
- HT Evaluate different types of biological metal extraction.

Industrial Metal Extraction Using the Reactivity Series

- It is only the least reactive metals that are found in their pure, elemental form (e.g. gold, platinum).
- All other metals react with other elements to form **minerals**, which are substances made up of metal compounds (typically oxides, sulfides or carbonates).
- An **ore** is a rock that contains minerals from which a metal can be extracted profitably.
- Carbon, when heated with a metal oxide, will **displace** a metal that is below it in the reactivity series.
- The products formed will be molten metal and carbon dioxide.
- An example is the extraction of copper from its ore.
- There are a number of different ores of copper. Each is a different copper compound.
- For example, cuprite is copper(I) oxide (Cu_2O) and chalcocite is copper(I) sulfide (Cu_2S).
- In the school laboratory, the displacement reaction can easily be demonstrated with pure copper(II) oxide.

> **copper(II) oxide + carbon ⟶ copper + carbon dioxide**

- The copper(II) oxide used in laboratories is free of impurities, which means displacement can take place more effectively.
- In industry, the reactions are scaled up to maximise the amount of metal that can be extracted.

Industrial Metal Extraction Using Electrolysis

- Metals that are higher than carbon in the reactivity series cannot be displaced from their compounds by carbon.
- Industrial chemists use **electrolysis** to extract metals instead.
- The ore is heated to high temperatures to make it molten so that the ions move freely.
- The mixture is poured into an electrolytic cell with electrodes made of carbon.
- A typical example is the extraction of aluminium from its ore:
  - The aluminium ore is bauxite, which consists mainly of aluminium oxide (Al_2O_3).
  - The reaction is: aluminium oxide → aluminium + oxygen

> **Key Point**
>
> Carbon is a non-metal. However, in displacement reactions, it can displace less reactive metals as if it were one.

Extracting Copper from Copper(II) Oxide

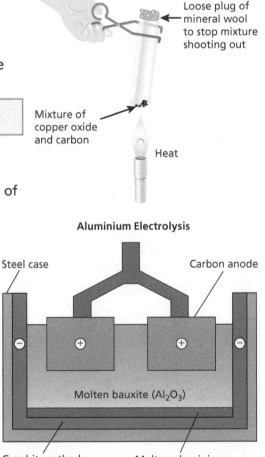

Loose plug of mineral wool to stop mixture shooting out

Mixture of copper oxide and carbon

Heat

Aluminium Electrolysis

Steel case

Carbon anode

Molten bauxite (Al_2O_3)

Graphite cathode

Molten aluminium

HT	At the cathode:	$Al^{3+}(l) + 3e^- \longrightarrow Al(l)$
	At the anode:	$2O^{2-}(l) \longrightarrow O_2(g) + 4e^-$

- The oxygen reacts with the carbon anode to form carbon dioxide.
- This wears the anode out, so it has to be replaced regularly.

HT Biological Extraction Methods

- In recent years, scientists have developed ways of using living organisms to extract metals.
- **Bacteria**:
  - Bacteria are single-celled organisms that reproduce rapidly.
  - They can be bred to survive in very high concentrations of metal ions.
  - As they feed, they accumulate the ions in their cells.
  - Chemists can then extract the ions from the bacteria.
  - As the bacteria accumulate a specific ion, they will ignore the waste material.
  - This means that this process is very efficient.
  - It is much cheaper than extracting metals in a furnace (by displacement).
  - However, once the ions are removed from the ore there is a risk of them leaking from the bacteria into the environment.
- Phytoextraction:
  - Heavy metals, such as lead, are dangerous to most living organisms in high concentrations.
  - Plants cannot normally tolerate high levels of metal ions.
  - However, plants can be bred, or genetically engineered, to accumulate heavy metal ions. These are called hyperaccumulators.
  - The plants take in the ions via the roots and store them in non-essential tissues.
  - The willow tree has been used in this way to remove the heavy metal cadmium from contaminated areas.
  - Advantages are that the process is environmentally friendly – the soil is not harmed in the process – and it is cheap.
  - The disadvantage is that it takes a while for the plants to grow, so it is a long-term project.

Key Point

As scientists learn how organisms grow and respond to the environment, they can start creating new technologies to address problems such as pollution.

Quick Test

1. Galena is an ore containing lead. Suggest an appropriate method for extracting the lead.
2. Why do the anodes have to be replaced regularly in the electrolysis of molten aluminium?
3. HT Give two reasons why bacteria are used for metal extraction.

Key Words

minerals
ore
displace
electrolysis
HT phytoextraction
HT hyperaccumulators

Life Cycle Assessments and Recycling

You must be able to:

- Describe the differences between industrial and laboratory scale production
- Describe a life cycle assessment for a material or a product
- Explain recycling and evaluate when it is worth undertaking.

Scales of Production

- Manufacturing chemicals on an industrial scale is very different to producing chemicals in a laboratory.

Laboratory Production	Industrial Production
Small quantities of chemicals used	Large quantities of chemicals used
Chemicals are expensive to buy	Bulk buying means chemical are cheaper
Energy usage is not an issue	There is a need to reduce costs and, therefore, energy usage
Equipment is usually glassware	Large-scale expensive factory equipment is needed
Chemists work on each reaction	Automation is common

Life Cycle Assessments

- A **life cycle assessment** is used to determine the environmental impact of a product throughout its life cycle.
- A product can be viewed as having four stages in its life cycle:
  1. Obtaining raw materials
  2. Manufacture
  3. Use
  4. Disposal.
- For each stage, the assessment looks at how much energy is needed, how **sustainable** the processes are, and what the environmental impact is.

PACKAGING   DISTRIBUTION

LIFE CYCLE ASSESSMENT

MANUFACTURING   USAGE

MATERIALS   DISPOSAL

Key Point

Manufacture includes the processes and materials used to make the product, and packaging and distribution.

Plastic Bags	Paper Bags
Plastic bags are made out of plastic.	Paper bags are made from wood pulp.
The plastic is made from crude oil, which is not **renewable**.	The wood pulp is made from trees, which are renewable.
Making the plastic requires a lot of energy (for fractional distillation and cracking) and is not sustainable.	This means that paper bags are sustainable (low environmental impact).
The environmental impact of making the bags is high, as oil is running out.	Paper bags are often only used once. Therefore, they are disposed of in large numbers.
Plastic bags are often only used once, are non-biodegradable and disposed of in landfill sites.	Paper bags can be recycled and repulped to make new bags.

Recycling

- Recycling is the process of taking materials and using them to make new products, e.g. making a vase from a glass bottle.
- Recycling materials means:
  - less quarrying for raw materials is required
  - less energy is used to extract metals from ores
  - the limited ore and crude oil reserves will last longer (saves natural resources)
  - disposal problems are reduced.
- It is important to evaluate whether an object should be recycled or not.
- Sometimes recycling could be more environmentally damaging than just disposal.
- For example, plastic objects that are placed in recycling bins may have contaminants – they would have to be deep cleaned, which the recycling companies can't do due to high costs.

> **Key Point**
>
> Being environmentally friendly is seen as an advantage when marketing and selling a product.

Computer circuit boards could end up in landfill, but once reusable components are removed, they can be used in a number of products, e.g. the covers for notebooks.

> **Quick Test**
>
> 1. Why is returning glass bottles to a store for refilling *not* an example of recycling?
> 2. Suggest why some plastics are not suitable for recycling.
> 3. What is the purpose of a life cycle assessment?

> **Key Words**
>
> life cycle assessment
> sustainable
> renewable

Crude Oil

You must be able to:

- Explain the processes of fractional distillation and cracking
- Describe alternatives to crude oil as a fuel source.

Crude Oil

- Crude oil is a fossil fuel that formed over millions of years in the Earth's crust.
- It has become crucially important to our modern way of life.
- It is the main source of chemicals for the petrochemical industry, for making plastics and fuels such as petrol and diesel.
- The hydrocarbons present in crude oil are used extensively throughout the chemical industry.
- Crude oil is a finite resource, which means it is being used up much faster than it is being replaced.

> **Key Point**
>
> It is important to recognise that the components of crude oil can be changed into many other chemicals, including fuel, medicines and plastics.

Fractional Distillation

- Crude oil is a mixture of many hydrocarbons that have different boiling points.
- The hydrocarbon chains generally all belong to the alkane homologous series (general formula C_nH_{2n+2}).

Alkane	Methane, CH_4	Ethane, C_2H_6	Propane, C_3H_8	Butane, C_4H_{10}
Displayed Formula	H $\vert$ H$-$C$-$H $\vert$ H	H H $\vert$ $\vert$ H$-$C$-$C$-$H $\vert$ $\vert$ H H	H H H $\vert$ $\vert$ $\vert$ H$-$C$-$C$-$C$-$H $\vert$ $\vert$ $\vert$ H H H	H H H H $\vert$ $\vert$ $\vert$ $\vert$ H$-$C$-$C$-$C$-$C$-$H $\vert$ $\vert$ $\vert$ $\vert$ H H H H

- The longer the chain, the greater the intermolecular forces, so the higher the boiling point.
- This means that crude oil can be separated into useful fractions (parts) that contain mixtures of hydrocarbons with similar boiling points.
- The process used is **fractional distillation**.
- The crude oil is heated in a fractionating column.
- The column has a temperature gradient – it is hotter at the bottom of the column than at the top:
  - Fractions with low boiling points leave towards the top of the fractionating column.
  - Fractions with high boiling points leave towards the bottom of the fractionating column.

Fractional Distillation Tower

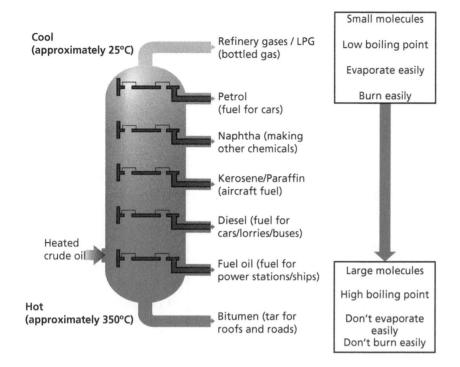

Cool (approximately 25°C)

Refinery gases / LPG (bottled gas)

Petrol (fuel for cars)

Naphtha (making other chemicals)

Kerosene/Paraffin (aircraft fuel)

Diesel (fuel for cars/lorries/buses)

Heated crude oil

Fuel oil (fuel for power stations/ships)

Hot (approximately 350°C)

Bitumen (tar for roofs and roads)

Small molecules

Low boiling point

Evaporate easily

Burn easily

Large molecules

High boiling point

Don't evaporate easily

Don't burn easily

Cracking

- There is not enough petrol in crude oil to meet demand.
- A process called **cracking** is used to change the parts of crude oil that cannot be used into smaller, more useful molecules.
- Cracking requires a catalyst, high temperature and high pressure.
- Petrol can be produced in this way, as can other short-chain hydrocarbons used to make polymers for plastics and fuels.
- For example, C_6H_{14} can be cracked to form C_2H_4 (ethene) and C_4H_{10} (butane).

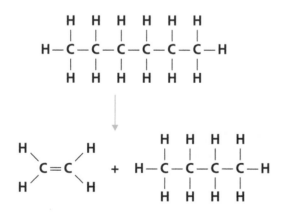

Quick Test

1. What is meant by the term 'cracking'?
2. Why do longer-chain hydrocarbons have higher boiling points?
3. Give **three** fractions of crude oil and **one** use for each.

Key Words

fractional distillation
cracking

Interpreting and Interacting with Earth's Systems

You must be able to:

- Describe how the Earth's atmosphere has developed over time
- Explain the greenhouse effect
- Describe the effects of greenhouse gases.

The Earth's Atmosphere

- The current composition of the Earth's **atmosphere** is 78% nitrogen, 21% oxygen and 1% other gases, including 0.040% (400ppm) carbon dioxide.
- The Earth's atmosphere has not always been the same as it is today – it has gradually changed over billions of years.
- A current theory for how the Earth's atmosphere evolved is:
  1. A hot, volcanic Earth released gases from the crust.
  2. The initial atmosphere contained ammonia, carbon dioxide and water vapour.
  3. As the Earth cooled, its surface temperature gradually fell below 100°C and the water vapour condensed into liquid water.
  4. Some carbon dioxide dissolved in the newly formed oceans, removing it from the atmosphere.
  5. The levels of nitrogen in the atmosphere increased as nitrifying bacteria released nitrogen.
  6. The development of primitive plants that could photosynthesise removed carbon dioxide from the atmosphere and added oxygen.

The Greenhouse Effect

- The Earth's atmosphere does a very good job at keeping the Earth warm due to the **greenhouse effect**.
- Greenhouse gases prevent **infrared** radiation from escaping into space.
- Carbon dioxide, methane and water vapour are powerful greenhouse gases.

Infrared radiation

> **Key Point**
>
> The atmosphere is very thin. If the Earth were the size of a basketball, the atmosphere would only be as thick as a layer of clingfilm wrapped around it.

> **Key Point**
>
> Scientists take a variety of samples from all over the planet and analyse them to build models of the gases that were present in the atmosphere at different times in Earth's history.

> **Key Point**
>
> Scientific consensus (over 97% of scientists globally) is that it is extremely likely humans are the main cause of climate change.

Climate Change and the Enhanced Greenhouse Effect

- Evidence strongly indicates that humans have contributed to increasing levels of carbon dioxide, creating an **enhanced greenhouse effect**.
- Three key factors have affected the balance of carbon dioxide in the atmosphere:
  1. Burning **fossil fuels**, which releases CO_2 into the atmosphere.
  2. Deforestation over large areas of the Earth's surface, which reduces photosynthesis, so less CO_2 is removed from the atmosphere.
  3. A growing global population, which directly and indirectly contributes to the above factors.
- Carbon dioxide levels have increased from approximately 0.028%, at the time when humans first started using industrial technology, to the 0.040% seen today.
- As a result, the average global temperatures are changing, bringing about changes in weather patterns (climate change) over the entire planet.
- The evidence points to humans being the main cause of global climate change. However, these are still controversial issues.
- There are natural sources of greenhouse gases, e.g. volcanic eruptions, carbon dioxide emissions from the oceans and respiration.

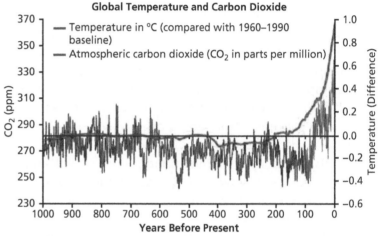

Global Temperature and Carbon Dioxide

— Temperature in °C (compared with 1960–1990 baseline)
— Atmospheric carbon dioxide (CO_2 in parts per million)

- Detailed global temperature and CO_2 measurements were not made in the past, so scientists have to take samples and model what the atmosphere was like at different times.
- There is still a lot of research being carried out to gather more evidence and improve the climate models.
- To slow or stop climate change, we need to:
  - Reduce our reliance on fossil fuels, and find and use alternative energy sources.
  - Be more aware of our energy use and not waste energy.
  - Implement **carbon capture** schemes, such as planting forests – the forests store carbon whilst they are growing.
  - Reduce waste sent to landfill – waste often produces methane gas, which has a greater greenhouse effect than carbon dioxide.

Quick Test

1. What was the main source of carbon dioxide, methane and ammonia in the Earth's early atmosphere?
2. What process led to the large amount of oxygen in the atmosphere?
3. Which greenhouse gas is produced by landfill sites?

Key Words

atmosphere
greenhouse effect
infrared
enhanced greenhouse
  effect
fossil fuels
carbon capture

Air Pollution and Potable Water

You must be able to:

- Describe the effects of CO, SO₂, NOₓ and particulates in the atmosphere
- Describe methods for increasing potable water.

Carbon Monoxide, Sulfur Dioxide and Oxides of Nitrogen (NO$_x$)

- When hydrocarbons are combusted with oxygen, they will produce carbon dioxide (CO_2) and water.
- If there is not enough oxygen, carbon monoxide (CO) is produced.
- Carbon monoxide is toxic.
- It bonds more tightly to haemoglobin than oxygen does, starving the blood of oxygen so that the body suffocates slowly.
- Gas appliances need to be checked regularly for this reason.
- Burning fossil fuels produces large amounts of carbon monoxide.
- Sulfur dioxide (SO_2), in small quantities, is used as a preservative in wine to prevent it from going off.
- Sulfur dioxide is produced when fossil fuels are burned.
- It is also produced in large quantities during volcanic eruptions.
- Sulfur dioxide dissolves in water vapour in the atmosphere causing acid rain, which kills plants and aquatic life, erodes stonework and corrodes ironwork.
- SO_2 also reacts with other chemicals in the air forming fine **particulates** (see below).
- Nitrogen and oxygen from the air react in a hot car engine to form nitrogen monoxide (NO) and nitrogen dioxide (NO_2).
- NO$_x$ is used to refer to the various nitrogen oxides formed.
- They can cause photochemical smog and acid rain.

> **Key Point**
>
> CO attaches to the haemoglobin because the shape of the CO molecule is so similar to the O_2 molecule.

Particulates

- Particulates are very small, solid particles that can cause lung problems and **respiratory** diseases.
- If a fuel is burned very inefficiently, carbon (soot) is produced.
- As well as causing respiratory problems, soot can coat buildings and trees in a black layer.
- Realising the dangers of **air pollutants**, governments have passed laws dictating the maximum levels that are allowed to be emitted, with the aim of improving air quality.

Potable Water

- Although the surface of the planet is covered in water, much of it is not safe to drink.
- Approximately 26% of the world's population only has access to unsafe water (e.g. contaminated by faeces).
- Providing potable water (water that is safe to drink) is an important goal for many countries.

Treating Waste Water

- When rain water goes down a drain, the toilet is flushed or a bath is emptied, the water enters the sewage system.
- This water is treated at a sewage treatment works before being piped to the tap:
  1. The water is filtered to remove large objects.
  2. The water is held in a **sedimentation tank** to remove small particulates.
  3. Bacteria are added to break down the particulates.
  4. Fine particles are removed by filtration.
  5. Chlorine is added to kill bacteria.
  6. The water is **ultra-filtered** through special membranes and stored in reservoirs.
  7. It is then treated before being piped back to the tap.

Key Point

All water molecules are identical, even if they have passed through other organisms.

Desalination

- The majority of the world's water is in the seas and oceans.
- The water contains dissolved salt, which means it is undrinkable.
- **Desalination** (removing the salt to make the water drinkable) is an expensive process.
- There are two ways this can be achieved:
- **Evaporation / condensation:**
  1. Heat the salt water to boiling point to evaporate the water off.
  2. Cool the water vapour to recondense it into liquid form.
- **Reverse osmosis:**
  1. The salt water is filtered and passed into an osmosis tank.
  2. The salt water is forced against a partially permeable membrane.
  3. The membrane only lets water pass through, separating out the salt molecules.
  4. The water has chemicals added (e.g. chlorine) to kill bacteria.
- Both processes are very expensive due to their high energy needs.

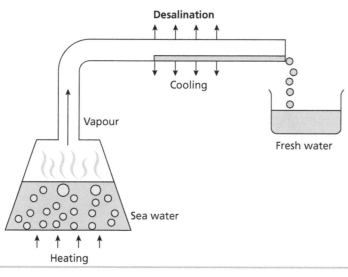

Quick Test

1. What is the biggest source of CO in the atmosphere?
2. Why is chlorine often added to drinking water?
3. What is meant by the term 'potable water'?

Key Words

particulates
respiratory
air pollutants
sedimentation tank
ultra-filtered
desalination
reverse osmosis

Introducing Chemical Reactions

1. Write down the number of atoms of each element in each of the following compounds.

    a) H_2SO_4 b) $Cu(NO_3)_2$ c) CH_3CH_2COOH d) C_2H_6 [4]

2. Balance the following equations:

    a) $CuO(s) + H_2SO_4(aq) \rightarrow CuSO_4(aq) + H_2O(l)$ [2]

    b) $Mg(s) + O_2(g) \rightarrow MgO(s)$ [2]

    c) $Mg(OH)_2(aq) + HCl(aq) \rightarrow MgCl_2(aq) + H_2O(l)$ [2]

    d) $CH_4(g) + O_2(g) \rightarrow CO_2(g) + H_2O(l)$ [2]

3. HT Write the half equation for each of the following reactions:

    a) Solid lead to lead ions   b) Aluminium ions to aluminium   [2]

    c) Bromine to bromide ions   d) Silver ions to solid silver   [2]

Total Marks _____ / 16

Chemical Equations

1. HT When writing a balanced ionic equation, which species appear in the equation?   [1]

2. HT Write the net ionic equation for:

    a) $AgNO_3(aq) + KCl(aq) \rightarrow AgCl(s) + KNO_3(aq)$ [1]

    b) magnesium nitrate (aq) + sodium carbonate (aq) →
                            magnesium carbonate (s) + sodium nitrate (aq)   [1]

Total Marks _____ / 3

Moles and Mass

1 HT Calculate the **number of moles** of each of the following elements:

    a) 6.9g of Li   **b)** 62g of P   [2]

2 HT Calculate the molar mass of ammonium chloride, NH_4Cl.
(The relative atomic mass of H = 1, Cl = 35.5 and N = 14.)   [1]

3 HT Barium chloride reacts with magnesium sulfate to produce barium sulfate and magnesium chloride.
What mass of barium sulfate will be produced if 5mol of barium chloride completely reacts? Show your working.   [2]

4 HT How many moles are there in 22g of butanoic acid, $C_4H_8O_2$?

    A 0.1   **B** 0.25   **C** 0.5   **D** 1   [1]

> **Total Marks** _____ / 6

Energetics

1 Atu pulls a muscle whilst playing rugby.
A cold pack is applied to his leg to help cool the muscle and prevent further injury.
The pack contains ammonium nitrate and water.
When the pack is crushed, the two chemicals mix and ammonium nitrate dissolves endothermically.

What is meant by the term **endothermic**?   [1]

2 HT Mark reacts hydrogen gas with chlorine gas: $H_2(g) + Cl_2(g) \rightarrow 2HCl(g)$

Bond	Bond Energies (kJ/mol)
H–Cl	431
H–H	436
Cl–Cl	243

    a) Calculate the energy change for the reaction and state whether the reaction is **endothermic** or **exothermic**.   [3]

    b) Draw the expected reaction profile for the reaction.   [1]

> **Total Marks** _____ / 5

Review Questions

Types of Chemical Reactions

1 HT Claudia places a copper wire into a solution of colourless silver nitrate solution. As time passes, Claudia notices that shiny crystals start developing on the surface of the copper wire. She also notices that the solution becomes a light blue colour.

    a) Write the balanced symbol equation for the reaction between copper and silver nitrate. [2]

    b) What are the shiny crystals on the wire? [1]

    c) What causes the blue coloration of the solution? [1]

> Total Marks _____ / 4

pH, Acids and Neutralisation

1 Nitric acid and sodium hydroxide are reacted together.

    a) Write the balanced symbol equation for the reaction. [2]

    b) Which ions are spectator ions? [2]

    c) Rewrite the equation you wrote for part **a)** showing only the reacting species. [2]

2 HT What is meant by the term **strong acid**? [1]

> Total Marks _____ / 7

Electrolysis

1 In what state(s) will ionic compounds conduct electricity? [1]

2 Masum is carrying out the electrolysis of water and sulfuric acid.

    a) Which of the following would be the most appropriate material for the electrodes?

        A wood   **B** copper   **C** carbon   **D** plastic [1]

    b) Write the names of the anions and cations involved in this electrolysis. [2]

    c) Write the reactions taking place at **i)** the anode and **ii)** the cathode. [2]

> Total Marks _____ / 6

Predicting Chemical Reactions

1 Which of the following properties does **not** apply to a Group 1 element?

    A conducts electricity

    B malleable

    C dissolves in water to make bleach

    D reacts with water to form hydrogen   [1]

2 Which of the following dot and cross diagrams could show a Group 0 element?

    A   **B**   **C**   **D**   [1]

3 Sort the following elements into order of reactivity, with the most reactive first:

    caesium   **lithium**   **potassium**   **rubidium**   **sodium**   [1]

4 **a)** Write the electronic configuration of each of the following elements:

    i) Li   **ii)** Mg   **iii)** Al   **iv)** O   [4]

    b) Pieces of lithium, magnesium and aluminium are placed into test tubes containing water. The results for Al and Mg are shown in the diagram below:

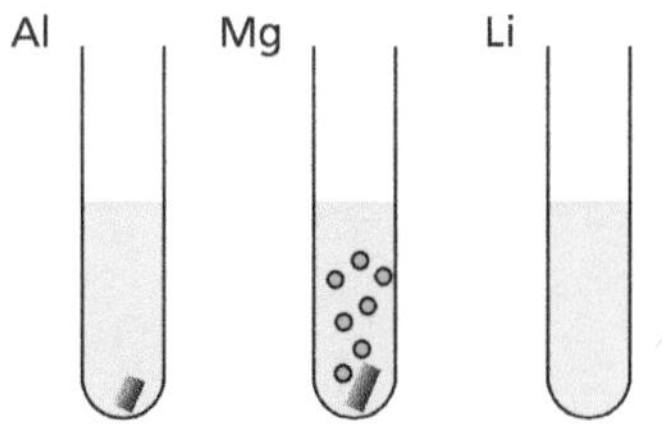

    Draw what would be seen in the tube containing the lithium.   [1]

    c) What gas is produced in the reaction?   [1]

5 The decomposition of hydrogen peroxide in the presence of a catalyst produces oxygen.

    Describe the test for oxygen gas.   [2]

Total Marks _____ / 11

Controlling Chemical Reactions

1   Niamh is investigating the effect of temperature on the breakdown of hydrogen peroxide in the presence of the catalyst manganese(IV) oxide.

In each experiment Niamh uses a different temperature. All other factors are kept the same.

Niamh's results are shown in the graph below.

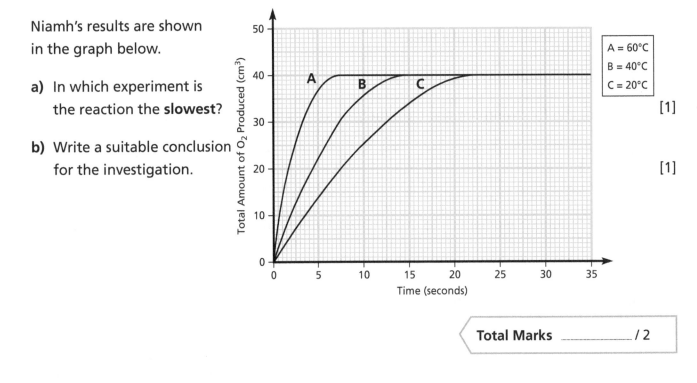

a) In which experiment is the reaction the **slowest**?   [1]

b) Write a suitable conclusion for the investigation.   [1]

Total Marks _____ / 2

Catalysts and Activation Energy

1   Explain, in terms of activation energy, how a catalyst affects the rate of reaction.   [2]

2   Body cells sometimes produce hydrogen peroxide, H_2O_2.
Hydrogen peroxide is dangerous in the human body.
The enzyme, catalase, speeds up the breakdown of hydrogen peroxide.

hydrogen peroxide $\xrightarrow{\textit{catalase}}$ water + oxygen

a) Why is catalase written above the reaction arrow?   [1]

b) In what way is catalase different to manganese(IV) oxide, which can also break down hydrogen peroxide?   [2]

3 Temperature affects particle movement.

   a) Draw two diagrams to show reacting particles at:

      i) a low temperature   ii) a high temperature.   [2]

   b) Explain why reactions are faster at a higher temperature.   [3]

> **Total Marks** _____ / 10

Equilibria

1 A dynamic equilibrium can be reached when there is a closed system.

   Explain what is meant by the term **closed system**.   [2]

2 Which of the following graphs shows a dynamic equilibrium?

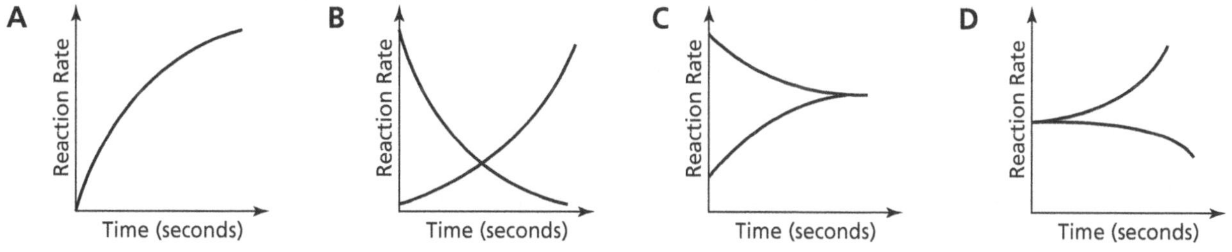

   A — Reaction Rate vs Time (seconds)   B — Reaction Rate vs Time (seconds)   C — Reaction Rate vs Time (seconds)   D — Reaction Rate vs Time (seconds)   [1]

3 HT A reversible reaction that reaches dynamic equilibrium is shown below:

   $$CH_3COOH(aq) \rightleftharpoons CH_3COO^-(aq) + H^+(aq)$$

   This is an endothermic reaction in the forward direction (reactant → product).

   a) What would happen to the amount of product if the temperature
      was increased?   [1]

   b) What would happen to the amount of product if the pressure were increased?   [1]

   c) What would happen to the reaction if the amount of $CH_3COO^-(aq)$
      was increased?   [1]

> **Total Marks** _____ / 6

Practice Questions

Improving Processes and Products

1 Which of the following metals **cannot** be displaced by carbon?

    A iron   **B** magnesium   **C** copper   **D** lead   [1]

2 Zac is extracting copper using carbon displacement.
He sets up the equipment as shown on the right:

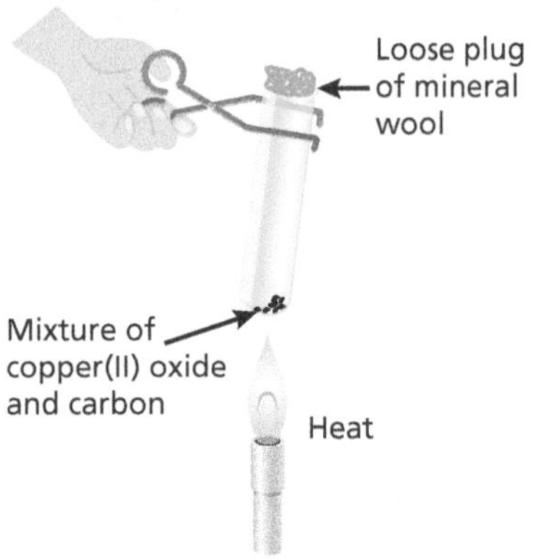

Loose plug of mineral wool

Mixture of copper(II) oxide and carbon

Heat

    a) Write the word equation for the displacement of copper
from copper(II) oxide by carbon.   [1]

    b) Suggest why there is a loose plug of mineral wool in the
top of the tube.   [1]

3 Which of the following is used to extract metals that are more
reactive than carbon from their ores?

    A magnetism   **B** chromatography   **C** explosives   **D** electrolysis   [1]

4 For an ionic compound, such as sodium chloride, to be electrolysed, which of the following
statements are true?

    A The amount of metal in the compound has to be greater than 90%.

    B The ionic compound needs to be molten or in solution.

    C A catalyst needs to be added to the ionic compound.

    D The ionic compound needs to be in a gaseous state.   [1]

5 **HT** Write the balanced symbol equation for the reaction that occurs at the cathode during
the electrolysis of copper(II) sulfate solution.   [1]

6 **HT** Molten aluminium oxide is being electrolysed.

    a) What ions are present in aluminium oxide?   [1]

    b) Write the balanced symbol equation for the reaction that takes place at:

        i) the cathode   **ii)** the anode.   [2]

7 Give **three** advantages of using bacteria to extract metals from contaminated soils.   [3]

                                                   Total Marks _____ / 12

Life Cycle Assessments and Recycling

1. Which of the following factors are analysed as part of a life cycle assessment?

   A colour   D manufacture   G recycling

   B disposal   E popularity   H transport

   C history   F price   I use   [1]

2. Each year thousands of mobile phones are disposed of in landfill sites.
   Many of the materials used in the manufacture of mobile phones could be recycled.

   a) What is meant by the term 'recycled'?   [1]

   b) Explain why it is important to recycle materials.   [4]

   c) Suggest why some of the parts of a mobile phone will not be recycled, even though
      technically it may be possible to do so.   [1]

3. The following factors are considered when making a disposable plastic bag.

   Which factors would **not** be suitable for inclusion in a life cycle assessment?

   A which bags customers prefer to use

   B the environmental impact of disposing of the bags

   C the energy required to dispose of the bags

   D the energy required to make the bags from plant fibres

   E how much should be charged for the bag   [2]

   Total Marks .... / 9

Crude Oil

1. Fractional distillation is used to separate the fractions in crude oil.

   a) Choose words from the list below to label **W**, **X**, **Y** and **Z** on the diagram:

   Bitumen   **Crude oil**   **Diesel**   **Paraffin**   [3]

   b) Where is the **lowest temperature** in the column?   [1]

2. These fractions obtained from fractional distillation can be cracked and used elsewhere to make other chemical products.

   Look at the table:

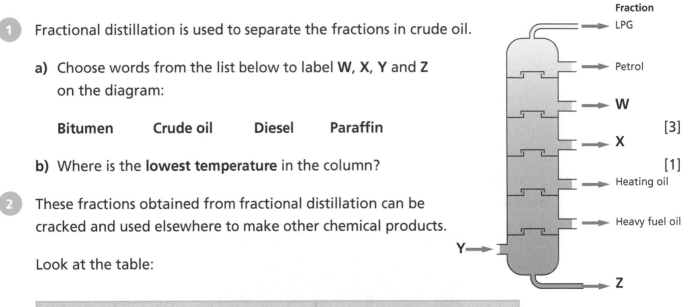

Fraction	Supply (millions barrels per day)	Demand (millions barrels per day)
Petrol	15	40
Paraffin	15	10
Diesel	25	30
Fuel oil	20	10

   Cracking large molecules creates smaller, more useful molecules.

   Which **two** fractions in the table should be cracked to ensure demand is met?   [2]

Total Marks _____ / 6

Interpreting and Interacting with Earth's Systems

1. Look at the graph showing the relationship between temperature and carbon dioxide levels.

   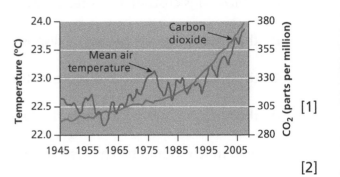

   a) Describe the general trend in air temperature in relation to carbon dioxide levels.   [1]

   b) Explain why carbon dioxide concentrations are linked to temperature increases.   [2]

c) The majority of the world's scientists agree that the causes of climate change are due to the actions of humans.

Which of the following are man-made causes of carbon dioxide emissions?

A   flying aeroplanes   C   planting trees

B   driving cars   D   burning plastics   [3]

2   The table below shows the gases present in the air.

Gas	Percentage in Clean Air
Nitrogen	78
Oxygen	
Carbon dioxide	0.04
Other gases	1

a) Work out the percentage of oxygen in air.   [1]

b) Describe how the process of **photosynthesis** alters the amounts of carbon dioxide and oxygen in the air.   [2]

c) What were the two main gases present in the Earth's early atmosphere?   [2]

Total Marks _____ / 11

Air Pollution and Potable Water

1   a) Explain why carbon monoxide is toxic in mammals.   [2]

    b) What is the main source of carbon monoxide?   [1]

2   Explain why oxides of nitrogen have a negative environmental impact.   [2]

3   Which of the following is the correct explanation for the term **potable water**?

A   Ultra-pure water made from the process of reverse osmosis.

B   Water that is stored in pots and taken from village to village.

C   Water that is safe to drink.

D   Water that needs to be sterilised.   [1]

Total Marks _____ / 6

Matter, Models and Density

You must be able to:

- Explain that matter is made of atoms and describe the structure of an atom
- Understand that nuclei and electrons exert an electric force on each other
- Associate the different behaviours of solids, liquids and gases with differences in forces and distances between atoms and molecules
- Describe and calculate density.

A Simple Model of Atoms

- Matter is made of **atoms**.
- Sometimes an atom can be pictured as a very small ball – the ball is a **model** of an atom.

Inside Atoms – Electrons

- Every atom has even smaller **particles** inside it.
- J. J. Thomson discovered tiny particles with negative electric charge, called **electrons**.
- He suggested that electrons are embedded in atoms, like currants in a plum pudding.
- He also knew that a whole atom is electrically neutral – with no overall **electric charge**.
- So the electrons must be balanced by a positive charge within each atom.
- Thomson suggested that the part of the atom with positive electric charge was like the soft 'sponge' of a plum pudding.
- The plum pudding is Thomson's model of an atom.

Inside Atoms – Nuclei

- Other scientists discovered small positively charged particles that come out from some kinds of radioactive material at very high speed. They called them **alpha particles**.
- Ernest Rutherford asked two of his assistants – Hans Geiger and Ernest Marsden – to fire some of these particles at very thin sheets of gold.
- They expected the alpha particles to shoot straight through the gold atoms, like bullets through cake.
- However, some bounced back. This is called **alpha scattering**.

> ### Key Point
>
> Atoms and small molecules are very small – about 1×10^{-10}m in diameter. That's 0.1nm.

Models of Atoms

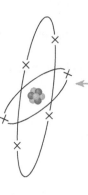

A dot

A ball

A plum pudding: soft positive charge with electrons embedded in it

Nuclear atom: positive nucleus with electrons in orbit

- Atoms must contain something very dense and positively charged to cause the scattering.
- Rutherford concluded that the positive electric charge must be concentrated into a very tiny space deep inside each atom.
- He called this the nucleus of the atom.
- Niels Bohr, from Denmark, improved this nuclear model by describing the pathways of electrons in orbit around the nucleus.

Mass and Charge in Atoms

- The electric charge inside an atom is shared equally between the electrons (negative) and the nucleus (positive). But the nucleus has nearly all of the mass.
- Electric charge can cause forces between atoms.
- The forces hold atoms together to make solids and liquids.
- Some atoms can form groups held together by electric force. These groups are molecules.
- Water is made of molecules – each one has two hydrogen (H) atoms and one oxygen (O) atom, so water is H_2O.
- In gases, the forces between particles are very weak. So gases spread out into the space around them.
- Gases have much lower density than solids and liquids, e.g.
  - The density of air is approximately 1.2kg/m³.
  - The density of solid iron is about 7800kg/m³.
- Density is a ratio of mass to volume, i.e. mass is divided by volume to work out density.
- The unit of density is the kilogram per cubic metre (kg/m³).

LEARN

$$\text{density (kg/m}^3) = \frac{\text{mass (kg)}}{\text{volume (m}^3)}$$

Particles in a Solid, Liquid and Gas

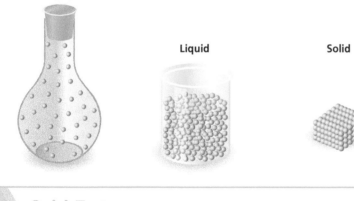

Gas

Liquid

Solid

Quick Test

1. State three facts about electrons.
2. Outline the evidence that led Rutherford to conclude that atoms have nuclei.
3. State the type of electric charge that nuclei have.
4. What holds the atoms of your body together?
5. Suggest why it is easy to walk through air but difficult to walk through walls.

Temperature and State

You must be able to:

- Understand that many materials can change between three states
- Explain that supplying energy to a material can change its internal energy by increasing its temperature or changing its state
- Define and make calculations using specific heat capacity and specific latent heat
- Explain gas pressure and the effects of changing the volume and temperature.

Changes of State

- Some changes of state, e.g. melting (solid to liquid), **sublimating** (solid to gas) and boiling (liquid to gas), need the material to gain energy.
- For other changes of state, e.g. condensing (gas to liquid) and freezing (liquid to solid), the material loses energy.
- There is no change in the structures of atoms during a change of state.
- However, their energy and speed and the distance apart and force between them can change.
- Changes of state are not **chemical changes**. They are **physical changes**. They are **reversible**.
- The mass of material does not change during a change of state. Mass is **conserved**.
- Volume and density usually change during a change of state.
- The **internal energy** of a solid or liquid is linked to the forces between its atoms or molecules, as well as to the motion of the atoms or molecules.
- The internal energy of a gas is simpler. The forces between the atoms or molecules are insignificant, so internal energy is just the total kinetic energy of all of the atoms or molecules in the gas.

Specific Latent Heat

- The amount of energy needed to melt a substance is different for all substances, e.g.
  - It takes about 23 000 joules of energy to melt 1kg of lead.
  - It takes about 335 000 joules of energy to melt 1kg of ice.
- These values are called the **specific latent heat** of melting for lead and water: 23 000 joules per kilogram (J/kg) for lead and 335 000 J/kg for ice.
- A substance has a different value of specific latent heat for the change of state from liquid to gas. This is called the specific latent heat of vaporisation.

> **thermal energy for a change in state =**
> **mass × specific latent heat**

Key Point

Supplying energy to a material can increase its temperature or change its state.

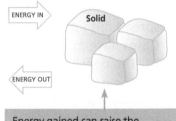

Energy gained can raise the temperature, e.g. from –5°C to 0°C. Extra energy can change the state from solid to liquid or even from solid to gas. If ice loses energy, its temperature goes down.

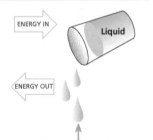

Energy gained can raise the temperature, e.g. from 10°C to 80°C. Further energy gained can change the state, from liquid to gas. Loss of energy can lower the temperature or cause liquid to become solid.

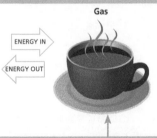

Giving more energy to hot gas raises its temperature. If it loses energy, its temperature can fall or it can change state from gas to liquid.

Specific Heat Capacity

- Energy isn't just needed for a change of state. It is also needed to raise the temperature of a material.
- It takes about 450J to raise the temperature of 1kg of iron by 1°C (or by 1K).
- It takes about 4200J to raise the temperature of 1kg of water by 1°C (or 1K).
- These values are the **specific heat capacities** of iron and water: 450J/kg°C for iron and 4200J/kg°C for water.
- The specific heat capacity of a material is the energy needed to raise the temperature of 1kg by 1°C (or 1K).

> **change in thermal energy =**
> **mass × specific heat capacity × change in temperature**

Gas Pressure

- A gas in a box exerts **pressure** on all of the walls of the box because its molecules **bombard** the walls.
- Most gases have atoms in small groups or molecules. This includes oxygen and hydrogen, where each molecule is a pair of atoms.
- The faster the particles in a gas move, the bigger the pressure. So pressure is related to temperature.
- If the box is made smaller:
  - the particles become closer together and the walls are hit more often by the gas molecules
  - there is more pressure on the walls, so pressure also depends on the volume of the box.

Molecular Bombardment and Gas Pressure

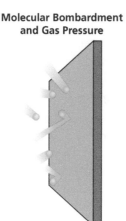

Quick Test

1. Name **one** quantity that changes when a substance changes state and **one** quantity that does not change.
2. Explain whether or not temperature always changes when a substance gains energy.
3. If the specific latent heat of melting of iron is 267 000J/kg, calculate how much energy will be needed to melt 1 tonne (1000kg).
4. If the specific heat capacity of air is 1006J/kg°C, calculate how much energy will be needed to raise the temperature of a room containing 100kg of air by 10°C (if none of the air escapes).

Key Words

sublimate
chemical change
physical change
reversible
conserved
internal energy
specific latent heat
specific heat capacity
pressure
bombard

Journeys

You must be able to:

- Make measurements and calculations of speed, time and distance
- Calculate the kinetic energy of a moving body
- Understand the importance of vector quantities (displacement, velocity and acceleration) when considering motion
- Interpret displacement–time and velocity–time graphs
- Calculate acceleration.

Speed and Velocity

- Miles per hour (mph) can be used to measure the speed of vehicles.
- However, in the standard international **SI system**, metres per second (m/s) is used.
- 1m/s is equal to 3.6 kilometres per hour (km/h or kph) and 2.24mph.
- Distance and speed are **scalar quantities** (without direction).
- They are used when direction is not important.
- When thinking about energy, direction is not important.
- The energy of a moving body is called **kinetic energy**, which is measured in joules (J).

> **kinetic energy (J) = 0.5 × mass (kg) × (speed (m/s))²**

- This equation can be used for predicting journeys:

> **distance travelled (m) = speed (m/s) × time (s)**

- When direction is important, **displacement** and **velocity** are used.
- Quantities with direction are **vector quantities**.
- Arrows of different lengths and directions can be used to compare different displacements and different velocities.
- A negative vector quantity means it is in the reverse direction.

$$\text{velocity} = \frac{\text{displacement}}{\text{time}}$$

Key Point

You must be able to rearrange the distance equation to work out speed and time, i.e.

$$\text{speed} = \frac{\text{distance}}{\text{time}}$$

$$\text{time} = \frac{\text{distance}}{\text{speed}}$$

If the speed on a journey varies, the average speed is given by the total distance divided by the total time.

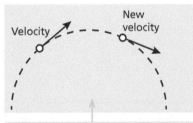

An object moving in a curved path with constant speed has a changing velocity, because the direction of motion is changing.

Graphs of Journeys

- Displacement–time graphs can be used to describe journeys.
- The **slope** or **gradient** of a displacement–time graph is equal to velocity.

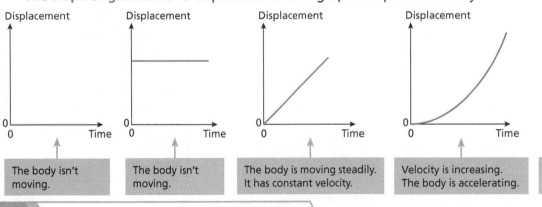

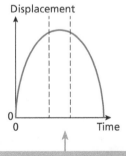

The body isn't moving.	The body isn't moving.
The body is moving steadily. It has constant velocity.	Velocity is increasing. The body is accelerating.
The body turns around and returns to where it started.	

- Velocity–time graphs can be used to describe the same journeys.

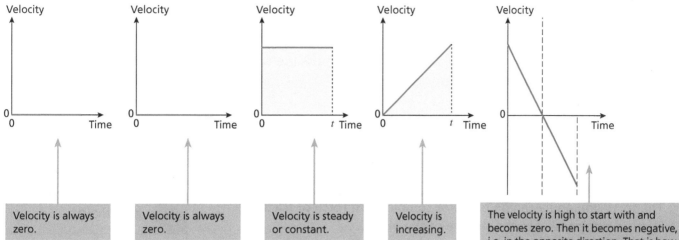

| Velocity is always zero. | Velocity is always zero. | Velocity is steady or constant. | Velocity is increasing. | The velocity is high to start with and becomes zero. Then it becomes negative, i.e. in the opposite direction. That is how the body gets back to where it started. |

- The slope or gradient of a velocity–time graph is equal to acceleration.

HT The enclosed area under a velocity–time graph is equal to displacement. In the third and fourth graphs above, the yellow shaded areas (enclosed areas) are equal to displacement at time t.

Acceleration

- **Acceleration** is rate of change of velocity. It is a vector quantity.

LEARN
$$\text{acceleration (m/s}^2) = \frac{\text{change in velocity (m/s)}}{\text{time (s)}}$$

$$(\text{final velocity (m/s)})^2 - (\text{initial velocity (m/s)})^2 = 2 \times \text{acceleration (m/s}^2) \times \text{distance (m)}$$

- When an object is released close to the Earth's surface, it accelerates downwards.
- Then the acceleration is approximately 10m/s². This is called **acceleration due to gravity**, or acceleration of free fall, (g).

Key Words

SI system
scalar quantity
kinetic energy
displacement
velocity
vector quantity
slope
gradient
acceleration
acceleration due to
    gravity (g)

> ## Quick Test
>
> 1. Calculate the kinetic energy, in joules (J), of a ball of mass 0.5kg moving at 16m/s.
> 2. Calculate how long a 100km journey would take at an average speed of 40kph.
> 3. Calculate the acceleration, in m/s², of a car that goes from standstill (0m/s) to 24m/s in 10s.
> 4. Suggest what the acceleration of a ball that is dropped to the ground will be.

Forces

You must be able to:

- Calculate combinations of forces that act along the same line
- Understand that forces act in pairs of the same size but opposite direction
- Recall that acceleration is not possible without net force, and a net force always produces acceleration.

Contact Force

- You can push and pull objects by making contact with them.
- The force you apply can be partly or completely a force of **friction**.
- Friction acts parallel to the surface of the object.
- You can also push at right-angles to the object. This is called a **normal force**.
- Force is measured in newtons, N.

Non-Contact Force

- Two magnets can attract or **repel** each other without touching.
- Electric force also acts without contact (without touching).
- Gravity provides another force that can act at a distance.

Gravitational and Electric Force

- All objects that have mass experience gravitational force.
- All objects with electric charge experience electric force.
- Gravitational force is always attractive, so there must be only one kind of mass.
- The Earth attracts you. The force, measured in N, is your weight.
- Electric force can be attractive or repulsive, so there must be two kinds of electric charge – positive and negative.

Net Force

- The overall force for a combination of forces is called their **resultant force** or **net force**.
- When there is a net force on an object it *always* accelerates.
- When there is no force at all on an object or the forces are balanced, it *never* accelerates. (It can move, but the motion never changes.)
- A body stays still or keeps moving at constant velocity unless an **external force** acts on it. That idea is called **Newton's first law**.

Direction of Force

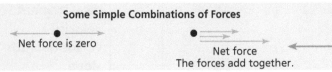

Some Simple Combinations of Forces

Net force is zero

Net force
The forces add together.

- The direction of a force makes a big difference to the effect it has.
  - Two forces of the same size acting in opposite directions do *not* cause acceleration, so the net force is zero.
  - Two forces acting in the same direction add together to produce a bigger net force.

> ### Key Point
> Force is a vector quantity – direction matters. Arrows can be drawn to show forces.

> ### Key Point
> Forces act between pairs of objects. The objects experience force of the same size, but opposite direction.

> ### Key Point
> If net force is not zero, the forces are unbalanced and there is acceleration. If motion is steady and in a straight line, velocity is constant and there is no acceleration.

The forces add together.

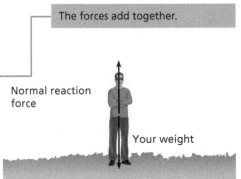

Normal reaction force

Your weight

- When you push an object, you experience a force of the same size and in the opposite direction.
- **Newton's third law** states that for every force there is an equal and opposite force.

Force and Acceleration

- Net force is related to acceleration in a fairly simple way:
  - acceleration is bigger when force is bigger
  - but smaller when mass is bigger.

> **LEARN**
>
> force (N) = mass (kg) × acceleration (m/s²)

- This equation is a form of **Newton's second law**.
- **HT** Resistance to acceleration is called **inertia**.

> **HT** **Key Point**
>
> The bigger the mass of an object, the greater the force needed to produce an acceleration. An object with more mass has more inertia, so it is more difficult to change its velocity.

HT Resistive Force on a Falling Object

- Air resistance creates a **resistive force** opposite to the force of gravity.
- The faster an object falls, the bigger the resistive force.
- Eventually the upwards resistive force becomes as big as the downwards gravitational force.
- The two forces are equal and opposite, so there is no net force.
- When there is no net force there is no acceleration, so a falling object continues to fall at constant velocity.
- That velocity is called the object's **terminal velocity**.

R

W

At first velocity is small so resistive force is small. There is a large net force acting downwards. The skydiver accelerates downwards.

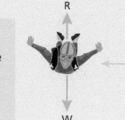

R

W

As velocity increases, resisitive force increases. But there is still a net downwards force so the skydiver continues to gain velocity.

R

W

Eventually, the velocity is so large that resistive force is the same size as the weight. Net force is zero. So acceleration is zero. Velocity stays the same.

> **Key Words**
>
> friction
> normal force
> repel
> resultant force
> net force
> external force
> **Newton's first law**
> **Newton's third law**
> **Newton's second law**
> **HT** **inertia**
> **HT** **resistive force**
> **HT** **terminal velocity**

> **Quick Test**
>
> 1. Name the kind of force that keeps you in your seat.
> 2. Name the kind of force that holds your body together.
> 3. State what is necessary, in terms of forces, for acceleration to happen.
> 4. **HT** When a ball is dropped, the effect of air resistance is ignored because it is so small. Explain why a skydiver, who jumps from a plane, cannot ignore the effect of air resistance.

Force, Energy and Power

You must be able to:

- Select from a range of equations so that you can analyse and predict motion
- HT Understand what momentum is and that it is conserved in collisions
- Understand 'doing work' as mechanical energy transfer.

HT Momentum

- **Momentum** is the product of mass and velocity. It is a vector quantity, so direction is important.

> **LEARN**
>
> momentum (kgm/s) = mass (kg) × velocity (m/s)

- Whenever bodies collide, their total momentum is the same before and after the collision – it is **conserved**.

> **LEARN**
>
> total momentum before = total momentum after

Force, Work and Energy

- When a force acts and a body accelerates as a result of this force, energy is supplied to the body.
- We say that the force does **work** on the body.
- The amount of work done and the energy supplied to the body are the same:

> **LEARN**
>
> work done (J) = energy supplied =
> force × distance (m) (along the line of action of the force)

Key Point

The work done is equal to the energy supplied.

- The unit of energy is the joule (J).

$$1J = 1N \times 1m$$

- Work must be done to overcome **friction** as well as to cause acceleration.
- When there is friction, some or all of the energy causes heating.
- In some situations, some of the work done increases an object's **potential energy**.

> **LEARN**
>
> (in a gravity field) potential energy (J) =
> mass (kg) × height (m) × gravitational field strength, g (N/kg)

Energy Stores and Transfers

- Energy can be stored in different ways, e.g.
  - in the **kinetic energy** of a moving body
  - as **gravitational potential energy**
  - as **elastic potential energy**
  - as **thermal energy**.
- Energy can be taken from these stores and transferred to other systems.
- Sometimes the energy becomes thinly spread out and only heats the surroundings of a process. Then it cannot be usefully transferred.

energy transferred (J) = charge (C) × potential difference (V)

- Wasted energy **dissipates** in the surroundings.

Power

- Energy can be transferred quickly or slowly.
- The rate of transfer of energy is **power**.

power (W) = rate of transfer of energy (J/s)
$$= \frac{\text{energy (J)}}{\text{time (s)}}$$

- Doing work is one way of transferring energy. It is energy transfer involving force and distance.
- So, when work is done:

$$\text{power (W)} = \frac{\text{work done (J)}}{\text{time (s)}}$$

Quick Test

1. HT Two balls of the same mass and with the same speed collide head on. State and explain what their total momentum is before the collision, and what their total momentum is after the collision.
2. An ice skater can move in a straight line at almost constant speed without needing to supply more energy. Explain, in terms of force and distance, how that is possible.
3. When you lean against a wall you are not doing work. Explain why in terms of force and distance.
4. Describe the relationship between power and energy.

Changes of Shape

You must be able to:

- Distinguish between plastic and elastic materials
- Distinguish between linear and non-linear relationships
- Understand that for a linear relationship, the slope or gradient is constant
- Recall that a planet's gravitational field strength depends on its mass.

Extension and Compression

- A pair of forces acting outwards on an object can stretch it, or **extend** it, even if the forces are balanced and there is no acceleration.
- A pair of forces acting inwards on an object can **compress** it.
- Combinations of forces can also bend objects.
- Change of shape can be called **deformation**.

Elastic and Plastic Deformation

- When forces make an object change shape, but it returns to its original shape when the forces are removed, the deformation is said to be **elastic**.
- If the object keeps its new shape when the deforming forces are removed, the deformation is **plastic**.

Extension of Springs

- A spring experiences elastic deformation unless the force applied is large. Then the spring may be permanently stretched.
- For a spring, unless the force is very large, the amount of extension is **proportional to** the size of the deforming force.
- When the force changes, the amount of the extension changes by the same proportion.
- A graph of extension and applied force is a straight line – the relationship between force and extension is **linear**.

Key Point

A spring will behave elastically, unless its elastic limit is exceeded.

Elastic and Plastic Deformation

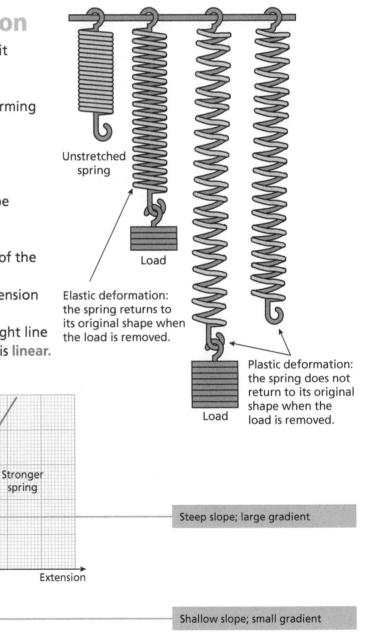

Unstretched spring

Load

Elastic deformation: the spring returns to its original shape when the load is removed.

Plastic deformation: the spring does not return to its original shape when the load is removed.

Load

Force

Weaker spring

0

Extension

Force

Stronger spring

0

Extension

Steep slope; large gradient

Shallow slope; small gradient

- The gradient of the graph is different for individual springs. It is called the **spring constant**.

> **force exerted on a spring (N) =**
> **extension (m) × spring constant (N/m)**

- The graph produced by a rubber band is not a straight line. The relationship is **non-linear**.

Energy Stored by a Stretched Spring

- Stretching a spring involves force and distance (the extension) in the same direction. So work must be done to stretch a spring.

work done = average force × distance = $\left(\dfrac{\text{final force}}{2}\right)$ × distance

work done = energy transferred in stretching (J) =
0.5 × spring constant (N/m) × (extension (m))^2

Mass and Weight

- In Physics, we treat **weight** as a type of force, so we measure it in newtons (N). It is the force on an object due to gravity.
- Different **planets** and **moons** have different **gravitational field strengths** (g), at their surface. This means that objects have different weights on different planets and moons.
- Weight is related to mass:

weight = gravity force (N) =
mass (kg) × gravitational field strength, g (N/kg)

- On Earth, g = 10N/kg; on the Moon, g = 1.6N/kg; near the surface of Jupiter, g = 26N/kg.

Quick Test

1. A pair of forces that act on an object along the same line are balanced (equal size, opposite direction) and cannot accelerate the object. What effect can they have?
2. State which of the following show **elastic** behaviour and which show **plastic** behaviour:
   a) a guitar string
   b) a piece of modelling clay
   c) a saw blade when it is flicked
   d) a saw blade when a large force bends it permanently.
3. Physics distinguishes between mass and weight and uses different units (kg for mass and N for weight). Explain why this distinction is generally not used in everyday life.

Key Words

extend
compress
deformation
elastic
plastic
proportional to
linear
spring constant
non-linear
weight
planet
moon
gravitational field
  strength (g)

Electric Charge

You must be able to:

- Recall that in everyday objects there are many atoms, each with electrons
- Recall that friction can transfer some electrons from one object to another, so that both objects become charged
- Understand that electric current in metal conductors, including wires, is a result of the flow of large numbers of electrons
- Understand that resistance to current produces heat, which transfers energy from the circuit to the surroundings.

Electric Force and Electric Charge

- Electric force can be attractive or repulsive and can act at a distance.
- Electric force acts between bodies that have net charge.
- There are two types of charge – positive and negative.
- When many charged particles, such as electrons, move together they form an **electric current**.
- In a **neutral** atom, the negative charge of the electrons is balanced by the positive charge of the nucleus.
- The unit of charge is the coulomb (C).

> **Key Point**
>
> Electric force holds atoms together. It is the dominant force between individual atoms and between the nuclei of atoms and electrons.

Electrostatic Phenomena

- Forces between objects much larger than atoms can be observed when there is an imbalance of positive and negative charge.
- The observations, or **phenomena**, are called static electricity, or **electrostatics**, because the charged particles are not flowing.
- Friction can **transfer** electrons from one object to another:
  - one object will have an excess of electrons and a negative charge
  - the other object will have a shortage of electrons and a positive charge.
- When objects become charged in this way, then they exert electrical force – attraction or repulsion – on other charged objects.

> **Key Point**
>
> All electrons are the same. They are extremely small and they all have the same negative charge.

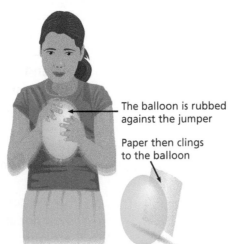

The balloon is rubbed against the jumper

Paper then clings to the balloon

- In the space between objects with different levels of charge, there can be strong forces on the charged particles inside atoms.
- Electrons can then escape from atoms of air, for example.
- Then the air conducts electricity and a spark occurs.

Electric Current

- Metals are **conductors** of electricity.
- When many electrons move in the same direction in a wire, there is an electric current.
- Electric current inside a closed loop of wire can be continuous if the electrons experience continuous force from a source of energy.
- **Batteries** and **cells**, for example, can create continuous force.
- **Ammeters** are used to measure current.
- They are connected into circuits so that the circuit current flows through them.
- Current is measured in amperes or amps (A).
- Current is equal to rate of flow of charge:

$$\text{current (A)} = \frac{\text{charge flow (C)}}{\text{time (s)}}$$

$$\text{charge flow (C)} = \text{current (A)} \times \text{time (s)}$$

- Unless it is isolated (cut off) from other objects, a conductor cannot keep excess electric charge, because electrons flow in or out of it too easily.

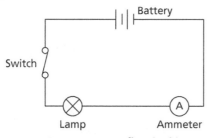

For a current to flow in this circuit there must be an energy supply and a complete loop (or closed circuit).

Key Point

In any electrical conductor, there is always some resistance to the flow of current.

Resistance

- Current can flow through an electrical conductor, such as a metal wire, but there is always some **resistance** to the flow.
- This resistance means that a wire can be heated by an electrical current. Energy passes from the wire to the surroundings.
- Resistance is measured in ohms (Ω).

Quick Test

1. What are the two types of electric force?
2. a) Outline how an object, such as an inflated balloon, becomes electrically charged.
   b) Explain why it is **not** possible to charge a metal spoon in the same way as the balloon.
3. What is the difference between charge and current?
4. Describe how resistance causes transfer of energy from a circuit.

Key Words

electric current
neutral
phenomenon
electrostatics
transfer
conductor
battery
cell
ammeter
resistance

Circuits

You must be able to:

- Understand that resistance to current produces heat, which transfers energy from the circuit to the surroundings
- Understand that a potential difference, or voltage, is needed to keep a current going around a circuit
- Investigate the relationship between current and voltage for different components.

Circuits and Symbols

- A system of symbols is used to represent the different **components** in circuits.
- The components in electrical circuits can be connected in **series** or in **parallel** (see pages 174–175)

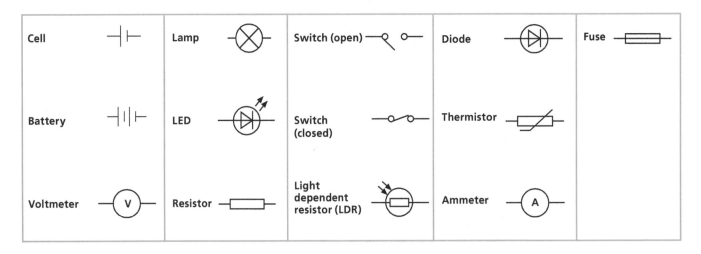

Cell	Lamp	Switch (open)	Diode	Fuse
Battery	LED	Switch (closed)	Thermistor	
Voltmeter	Resistor	Light dependent resistor (LDR)	Ammeter	

Potential Difference or Voltage

- Batteries and other power supplies can replace the energy that is transferred out of the wires and components in a circuit.
- Batteries and other power supplies only work when they are part of a circuit.
- A battery has a **positive terminal** (which attracts negative charge, so it attracts electrons) and a **negative terminal** (which repels negative charge).
- The abilities of batteries or other power supplies to provide energy to moving charge in a circuit can be compared using **potential difference** or **voltage**.
- Potential difference or voltage is measured in volts (V).

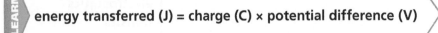

LEARN energy transferred (J) = charge (C) × potential difference (V)

- **Voltmeters** are used to measure potential difference or voltage.
- A voltmeter is connected to two points. The circuit current does not flow through a voltmeter.

> **Key Point**
>
> For continuous current, a continuous potential difference and a continuous loop of conductor (such as wire and other components) are necessary.

> **Key Point**
>
> The potential difference between two points in a circuit is 1V if 1J of energy is transferred when 1C of charge passes between the points.

A voltmeter measures a difference between two points in a circuit.

Current, Potential Difference and Resistance

- An increase in potential difference can increase the current in a circuit.
- An increase in resistance in a circuit can decrease the current.

> **LEARN**
>
> $$\text{current (A)} = \frac{\text{potential difference (V)}}{\text{resistance }(\Omega)}$$
>
> $$\text{potential difference (V)} = \text{current (A)} \times \text{resistance }(\Omega)$$

Current–Voltage Relationships

- If the temperature of a metal wire doesn't change, its resistance doesn't change.
  - Current is proportional to voltage.
  - A current–voltage (I–V) graph is a straight line that passes through the **origin** of the graph.
  - The relationship between current and voltage is linear.
- For a wire that gets hotter as voltage and current get bigger, the resistance increases and the relationship between current and voltage is non-linear.
- This happens in a filament lamp, in which the wire is white hot.
- A **thermistor** behaves in the opposite way to a wire.
- As it becomes hotter, more electrons become free to move, so its resistance goes down.
- For a thermistor the current–voltage relationship is also non-linear.
- Thermistors are sensitive to changes in temperature.
- This means thermistors can be used as electrical temperature **sensors**.
- **Diodes** and **Light Dependent Resistors (LDRs)** also have non-linear current–voltage relationships.
- Diodes only allow current in one direction. When a voltage is in the 'reverse' direction, no current flows.
- For LDRs, when the light intensity increases the resistance goes down.

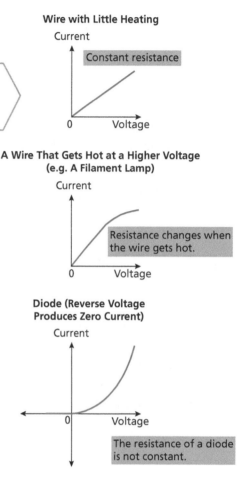

Wire with Little Heating

Current

Constant resistance

0   Voltage

A Wire That Gets Hot at a Higher Voltage (e.g. A Filament Lamp)

Current

Resistance changes when the wire gets hot.

0   Voltage

Diode (Reverse Voltage Produces Zero Current)

Current

0   Voltage

The resistance of a diode is not constant.

Key Words

component
series
parallel
positive terminal
negative terminal
potential difference
voltage
voltmeter
origin
thermistor
sensor
diode
light dependent resistor
(LDR)

> ## Quick Test
>
> 1. Explain why a voltmeter is connected to two points in a circuit that are separated by a component, such as a resistor.
> 2. Outline what can cause the current in a simple circuit to a) increase and b) decrease.
> 3. Explain why a current–voltage graph for a wire becomes curved when the wire becomes hot.
> 4. What happens to the resistance of a thermistor when its temperature increases?

Resistors and Energy Transfers

You must be able to:

- Calculate the resistance of two or more resistors in series or in parallel
- Understand that resistors transfer energy out of circuits by heating and motors are used to transfer energy out of circuits by doing work
- Recall that rate of transfer of energy is power
- Perform calculations on power, energy, voltage, current and time for use of electricity by appliances at home.

Resistors in Series

- Resistors can be connected in **series** – one after the other.
- Since both resistors resist current, their total resistance is greater than their individual resistance.
- Total resistance is the sum of the individual resistances:

> **LEARN**
> **total resistance (R_t) = resistance 1 (R_1) + resistance 2 (R_2)**

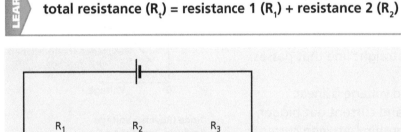

total resistance (R_t) = R_1 + R_2 + R_3 = 2 + 3 + 5 = 10Ω

- The current in each of the resistors must be the same.
- Current is flow and, if there is only one route for it to flow along, it must be the same at all points.

> **LEARN**
> **potential difference (V) = current (A) × resistance (Ω)**

- If the current is the same in two resistors but the resistances are different, the voltages must be different.
- The relative size of the voltages is the same as the relative size of the resistances.

$$\frac{\text{voltage 1 } (V_1)}{\text{voltage 2 } (V_2)} = \frac{\text{resistance 1 } (R_1)}{\text{resistance 2 } (R_2)}$$

> **Key Point**
>
> When two resistors are in series, current must pass through both of them.

Resistors in Parallel

- Resistors can be connected in **parallel** – one alongside the other.
- This gives current two routes to follow, so the total resistance is smaller than either of the resistors:

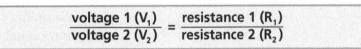

$$\frac{1}{\text{total resistance } (R_t)} = \frac{1}{\text{resistance 1 } (R_1)} + \frac{1}{\text{resistance 2 } (R_2)}$$

- Resistors in parallel have the same voltage.
- If the resistances of two resistors in parallel are different:
  - they do not carry the same current
  - the current will be larger in the smaller resistor.

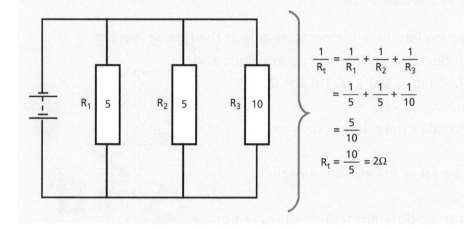

$$\frac{1}{R_t} = \frac{1}{R_1} + \frac{1}{R_2} + \frac{1}{R_3}$$

$$= \frac{1}{5} + \frac{1}{5} + \frac{1}{10}$$

$$= \frac{5}{10}$$

$$R_t = \frac{10}{5} = 2\Omega$$

Transfer of Energy

- Current in resistors heats them. Thermal energy is transferred.
- Motors also transfer energy. They do that by exerting force that can make objects move – they do work.
- Motors are not designed to provide heat, but they do transfer some energy by heating their surroundings.

Electrical Power

- The rate at which a component in a circuit transfers energy is its power, measured in watts (W) or kilowatts (kW).
- Since power is rate of transfer of energy:

$$\text{power} = \frac{\text{energy transferred}}{\text{time}}$$

- Energy can be measured in joules (J), kilojoules (kJ) and kilowatt-hours (kWh).
- In circuits, power is related to current and voltage:

power (W) = potential difference (V) × current (A)
= (current (A))² × resistance (Ω)

Quick Test

1. Calculate the total resistance of a 2Ω resistor and a 4Ω resistor when they are connected a) in series and b) in parallel.
2. Describe how motors transfer energy out of a circuit.
3. A resistor has a current of 1.5A and a potential difference of 12V.
   a) Calculate how much heat energy it transfers to its surroundings in 60 seconds.
   b) What is the power of the resistor?

Review Questions

Controlling Chemical Reactions

1 Jude is carrying out an investigation into how temperature affects the rate of reaction.

He uses the reaction between sodium thiosulfate and hydrochloric acid:

$$Na_2S_2O_3(aq) + 2HCl(aq) \rightarrow 2NaCl(aq) + S(s) + SO_2(g) + H_2O(l)$$

The reactants in this reaction are colourless.

Sulfur produced in the reaction makes the solution turn an opaque yellow colour.

Jude uses a cross drawn under a beaker to help measure the reaction rate.

Jude carries out the reaction at three different temperatures using water baths.

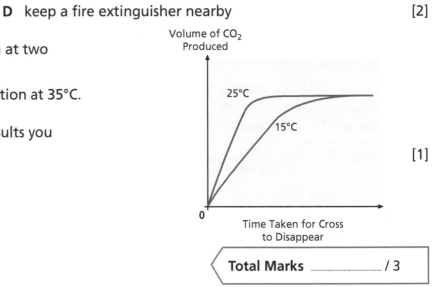

a) Which of the following safety precautions should be taken during this experiment?

    A no naked flames   **B** keep the lab well ventilated

    C wear eye protection   **D** keep a fire extinguisher nearby   **[2]**

b) The graph shows the reaction at two different temperatures.

Jude also carried out the reaction at 35°C.

Draw the line to show the results you would expect for 35°C.   **[1]**

Total Marks _____ / 3

Catalysts and Activation Energy

1 Catalysts are often used in chemical reactions.

Modern car engines use a catalyst in the exhaust system.

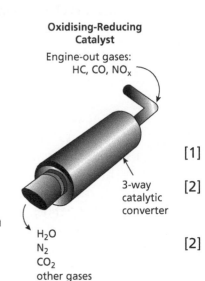

a) What is a catalyst?   **[1]**

b) Suggest why a car engine needs to have a catalytic converter.   **[2]**

c) Draw a graph to show how the **activation energy** for a reaction is affected by the presence of a catalyst.   **[2]**

2 Photosynthesis is represented by the equation:

$$6CO_2(g) + 6H_2O(l) \xrightarrow{\text{chlorophyll}} C_6H_{12}O_6(s) + 6O_2(g)$$

Why is chlorophyll included in this reaction?

A It is the pigment that makes leaves green.

B It is the site of photosynthesis.

C It interacts with catalysts (enzymes) in photosynthesis.

D It makes oxygen. [1]

Total Marks .... / 6

Equilibria

1 HT Which of the following describes Le Chatelier's principle?

A When the conditions of a system are kept the same, the position of the equilibrium changes to try to maintain the conditions.

B When the conditions of a system are altered, the position of the equilibrium changes to try to restore the original conditions.

C When the conditions of a system are altered, the amount of the reactants changes to try to restore the original conditions.

D When the conditions of a system are altered, the amount of the products changes to try to restore the original conditions. [1]

2 HT Look at the following reaction: $N_2(g) + 3H_2(g) \rightleftharpoons 2NH_3(g)$

The reaction is exothermic in the forward direction (reactants → product).

Which of the following will lead to an **increase** in the amount of product produced?

A reducing pressure   B increasing pressure

C reducing temperature   D increasing temperature [2]

3 HT A forward reaction is endothermic. The temperature is increased.

Would this change lead to more reactants or more products being produced? [1]

Total Marks .... / 4

Improving Processes and Products

1 The industrial electrolysis of aluminium ore takes place in large electrolyte cells like the one shown below.

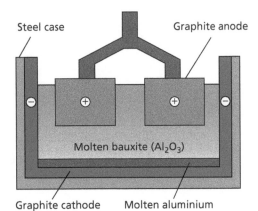

Steel case   Graphite anode

Molten bauxite (Al_2O_3)

Graphite cathode   Molten aluminium

a) Which part of the cell will have to be replaced regularly? [1]

b) **HT** Which of the following shows the correct ionic equations for the reactions that take place at the electrodes?

Cathode:	**Anode:**
A $Al^{3+}(l) + 3e^- \rightarrow Al(l)$	$C(s) + O_2(g) \rightarrow CO_2(g)$
B $C(s) + O_2(g) \rightarrow CO_2(g)$	$Al^{3+}(l) + 3e^- \rightarrow Al(l)$
C $Al^{3+}(l) + 3e^- \rightarrow Al(l)$	$2O^{2-}(aq) \rightarrow O_2(g) + 4e^-$
D $2O^{2-}(aq) \rightarrow O_2(g) + 4e^-$	$Al^{3+}(l) + 3e^- \rightarrow Al(l)$

[1]

Total Marks _____ / 2

Life Cycle Assessments and Recycling

1 A life cycle assessment can be divided into four stages.
The four stages are listed below, but they are in the wrong order:

A Disposal of the product.

B Obtaining raw materials / producing the materials needed for the product.

C Use of the product.

D Making the product.

a) List the stages in the correct order. [1]

The table below shows the results of life cycle assessments for three different types of bottle stops: corks, synthetic (plastic) corks and metal screw caps.

	Totals for 1000 Units		
	Cork	Synthetic Cork	Metal Cap
Energy Use (MJ)	1250	4500	5600
Fossil Fuel Use (kg)	27	54	149
Waste	46	1	90
Greenhouse Gas Emissions (kg CO_2)	40	81	231
Fresh Water Use (litres)	8400	4010	4200

b) Give **two** ways in which the manufacture of synthetic corks is **more** environmentally friendly than the manufacture of corks and metal caps. [2]

c) The metal used to make the caps is aluminium.

Suggest why using aluminium results in the highest CO_2 emissions. [2]

d) Which type of cap uses the least energy? [1]

e) What major advantage do aluminium caps have over the other two types? [1]

Total Marks _____ / 7

Crude Oil

1 The following are fractions of crude oil:

bitumen   petrol   diesel   paraffin

Which fraction has the **lowest** boiling point? [1]

2 Explain the relationship between hydrocarbon chain length and boiling point. [3]

Total Marks _____ / 4

Review Questions

Interpreting and Interacting with Earth's Systems

1 Carbon dioxide levels are increasing globally.

The Kyoto agreement was an international agreement, made in 2012, to reduce CO_2 levels to a certain level.

One way to capture carbon dioxide is to inject it into wells, which are between 800 and 3300m deep, to store the greenhouse gas.

The table below shows:
- the estimated number of wells that are now needed to keep CO_2 at the 2015 levels
- the number of wells predicted by the Kyoto agreement in 2012.

Year	Wells Now Needed to Meet Kyoto Target Levels	Wells Predicted in 2012
2015	100 176	40 332
2020	120 342	60 498
2025	140 508	80 664
2030	160 674	100 830

a) Explain why it is important to reduce the CO_2 levels in the atmosphere. [3]

b) Suggest why the number of wells needed to meet the Kyoto target now is greater than was predicted. [2]

c) Calculate the percentage increase in the wells now needed to meet Kyoto target levels from 2015 to 2030. [1]

d) Scientists believe that human activity has caused climate change.

Give **two** ways in which humans are believed to have increased the amount of CO_2 in the atmosphere. [2]

2 These statements explain how scientists think that our modern-day atmosphere evolved.

1. Nitrogen gas was released from ammonia by bacteria in the soil.
2. The modern atmosphere consists of nitrogen, oxygen and a very small amount of carbon dioxide.
3. Plants evolved and used carbon dioxide for photosynthesis, producing oxygen as a by-product.
4. Volcanoes gave out ammonia, carbon dioxide, methane and water vapour.

5. As the Earth cooled, water fell as rain, which led to the formation of oceans.

   Which is the correct order of this happening?

   A 4, 5, 1, 3, 2   **B** 2, 5, 3, 1, 4

   C 1, 4, 3, 5, 2   **D** 4, 3, 1, 5, 2   [1]

   Total Marks _____ / 9

Air Pollution and Potable Water

1 Explain why carbon monoxide is poisonous to mammals.   [3]

2 Which of the following is **not** a consequence of air pollution from SO_2?

   A acid rain   **B** phytochemical smog

   C fine particulates   **D** lung disease   [1]

3 Give **two** chemical methods for making drinking water from sea water.   [2]

4 The diagram shows a form of reverse osmosis.

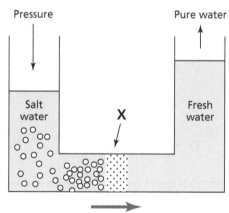

a) Name the part of the diagram labelled **X**.   [1]

b) Suggest where this method of producing water is most likely to be needed.   [2]

Total Marks _____ / 9

Practice Questions

Matter, Models and Density

1. Complete the sentence.

   An atom must be electrically neutral if:

   A it is an ion.

   B it is a molecule.

   C it has equal numbers of protons and neutrons.

   D it has equal numbers of protons and electrons. [1]

2. What is the approximate size of an atom?

   A 1×10^{10}m   B 1×10^{0}m   C 1×10^{-1}m   D 1×10^{-10}m [1]

3. Why are gases less dense than solids and liquids?

   A their particles are further apart

   B their particles are closer together

   C their particles are smaller

   D their particles are bigger [1]

4. An object has a mass of 0.24kg and a volume of 0.0001m³.

   What is the density of the object? [3]

   Total Marks _____ / 6

Temperature and State

1. Which of these is **not** a change of state?

   A evaporation   B expansion   C boiling   D sublimation [1]

2. Which word describes what happens when pressure is applied to a gas to reduce its volume?

   A compression   B expansion   C evaporation   D sublimation [1]

3. Sort the following processes into **two** sets: those that require (or take in) energy and those that release (or give out) energy. [5]

   boiling   condensing   evaporation   freezing   melting

4. Describe what happens to the particles in a gas when it is heated. [2]

5. Why does a scald from steam at 100°C cause more pain and worse damage than a scald from water at 100°C? [2]

6   What additional information, as well as specific latent heat, is needed to calculate
    the amount of energy required to melt an ice cube?   [1]

> **Total Marks** _____ / 12

Journeys

1   Which of these is **not** a unit of speed?

    A mph   **B** km/h   **C** m/s   **D** kg/s   [1]

2   How far can you walk in 3 hours at an average speed of 4km/h?

    A 0.75km   **B** 1.33km   **C** 7km   **D** 12km   [1]

3   Which of these is a unit of energy?

    A joule   **B** newton   **C** pascal   **D** watt   [1]

4   What happens to the speed of the Earth as it orbits the Sun?

    A it decreases   **C** it increases

    B it stays the same   **D** it changes direction   [1]

5   HT What happens to the velocity of the Earth as it orbits the Sun?

    A it decreases   **B** it stays the same

    C it increases   **D** it changes direction   [1]

6   How far can you travel in 1 hour at an average speed of:

    a) 24km/h?   **b)** 40mph?   **c)** 4m/s?   [8]

7   Volume is a scalar quantity but force is a vector quantity.

    a) What is the difference between a vector and a scalar quantity?   [1]

    b) Give another example of a vector quantity.   [1]

    c) Give another example of a scalar quantity.   [1]

8 A plane takes off in Hong Kong and lands in London 12 hours later. The distance of the journey is 6000km.

   a) What is its average speed in km/h? [3]

   b) What is its average speed in m/s? [2]

   c) For most of the flight, speed does not change much, but velocity does.

   In what way does the velocity change? [2]

   Total Marks .... / 23

Forces

1 Which of these forces requires contact?

   A electric or electrostatic force   C gravitational force

   B frictional force   D magnetic force   [1]

2 Which graph shows acceleration?

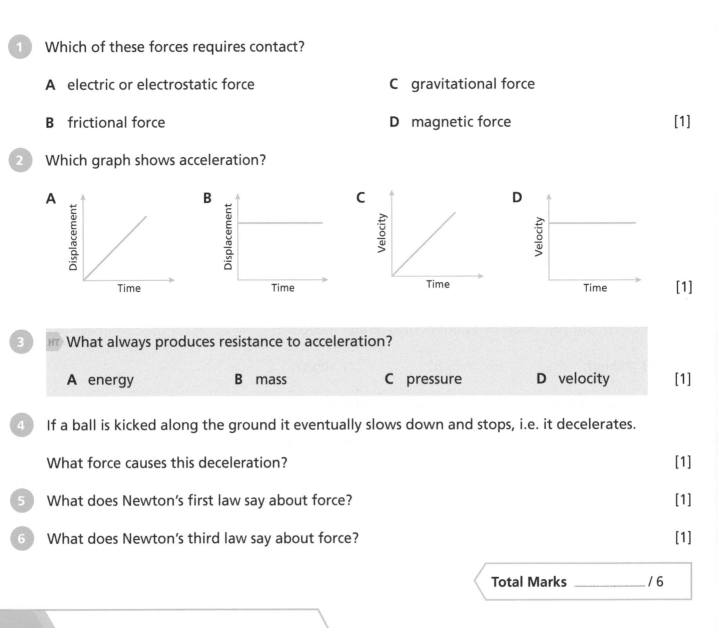

3 HT What always produces resistance to acceleration?

   A energy   B mass   C pressure   D velocity   [1]

4 If a ball is kicked along the ground it eventually slows down and stops, i.e. it decelerates.

   What force causes this deceleration? [1]

5 What does Newton's first law say about force? [1]

6 What does Newton's third law say about force? [1]

   Total Marks .... / 6

Force, Energy and Power

1 Which of these is necessary for doing work?

    A energy   **B** mass   **C** pressure   **D** velocity   [1]

2 If the point of a pin has an area of $1 \times 10^{-7}\,m^2$, how much pressure does it exert on a surface when it is pushed with a force of 10N?

    A $10^{-8}\,Pa$ **B** $10^{-6}\,Pa$ **C** $10^{6}\,Pa$ **D** $10^{8}\,Pa$ [1]

Total Marks _____ / 2

Changes of Shape

1 What word describes the behaviour of a material that keeps its new shape after a deforming force is removed?

    A elastic   **B** plastic   **C** compressed   **D** extended   [1]

2 Here are four force-extension graphs, with the same scales on both axes, for four different springs.

Which graph has the biggest spring constant?

[1]

Total Marks _____ / 2

Electric Charge

1 Why can an object, such as a balloon, become charged when rubbed?

    A friction creates electrons   **C** friction transfers electrons

    B friction destroys electrons   **D** friction gives electrons extra charge   [1]

2 What is produced by the movement of many electrons in the same direction?

    A an electric charge   **B** an electric current   **C** an electric resistance   [1]

3 What is the unit of resistance?

    A amp   **B** coulomb   **C** ohm   **D** volt   [1]

4 Explain why:

    a) metals are good at conducting electricity   [1]

    b) a resistor gets hot when the current is large   [2]

    c) resistance is smaller when two resistors are in parallel than when there is only one of them.   [1]

Total Marks .... / 7

Circuits

1 Which of these statements is correct?

    A The relationship between current and voltage in a wire is always linear.

    B The relationship between current and voltage in a wire is linear provided the wire does not get hot.

    C The relationship between current and voltage in a wire is always non-linear.

    D The relationship between current and voltage in a wire is non-linear provided the wire does not get hot.   [1]

2 Which of these diagrams shows the correct connection of an ammeter and a voltmeter?

A

C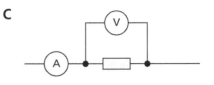

B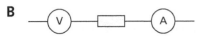

D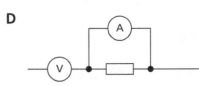

    [1]

3 How does a resistor transfer energy to its surroundings?

    A heating   **B** doing work   **C** storing energy   **D** creating energy   [1]

4 At what rate does a resistor transfer energy if it carries a current of 0.5A and is connected to a voltage of 1.5V?

    A 0.75W   **B** 1.0W   **C** 2.0W   **D** 3W   [1]

5 Explain what each of the following components are used for.

    a) battery or cell   **e)** diode   [2]

    b) ammeter   **f)** thermistor   [3]

    c) voltmeter   **g)** LDR   [3]

    d) resistor   [1]

6 A current of 1.5A passes through a 6Ω resistor. Calculate:

    a) the voltage   [3]

    b) the amount of charge that flows through the resistor in 60 seconds   [3]

    c) the energy transferred by the resistor in 60 seconds   [3]

    d) the rate of energy transfer.   [3]

> **Total Marks** _____ / 25

Resistors and Energy Transfers

1 Which of these circuits has the least resistance? (All of the resistors are identical.)

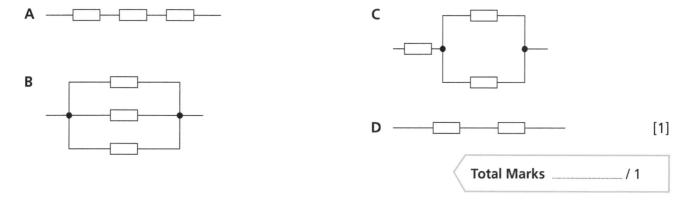

> **Total Marks** _____ / 1

Magnetic Fields and Motors

You must be able to:

- Relate diagrams of magnetic field lines to possible forces
- Recall that strong electromagnets can be made from coils of wire
- Understand that an electric current in a wire creates a magnetic field
- HT Understand that a magnetic field can interact with other magnetic fields to create a force, which can produce a turning effect.

Magnetic Fields

- Magnets can exert force without contact.
- Magnetic force can be attractive or repulsive.
- Every magnet has two poles – a north pole and a south pole.
- The **magnetic field** is the space around a magnet in which its force can act.
- **Magnetic field lines** are used in diagrams to represent direction of force that would act on a small north pole at different places in a magnetic field.
- The distances between magnetic field lines show the strength of the magnetic field at different places.

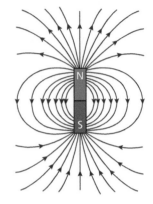

- HT **Magnetic flux density** is a measure of the strength of a magnetic field at a specific point in the field.
- HT The unit of magnetic flux density is the tesla (T).

Permanent and Induced Magnets

- Some iron and steel objects are **permanent magnets**.
- Others become magnetic when they are in a magnetic field – that kind of magnetism is called **induced magnetism**.

The Earth's Magnetic Field

- The Earth has a magnetic field, as if there is a huge magnet inside.
- Compass needles line up along magnetic field lines if they are free to do so.
- A compass needle that can move up and down as well as round and round dips towards or away from the ground.
- These compasses show that in most places the Earth's magnetic field is not parallel to the ground.

Magnetic Field Due to an Electric Current

- Wires with electric current have a magnetic field.
- The strength of the magnetic field at a point around a wire with current depends on the size of the current and also on the distance from the wire.
- A coil of wire with an iron core can be a strong **electromagnet**.
- This can also be called a **solenoid**.

A Solenoid

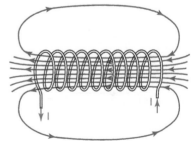

^{HT} Magnetic Force on a Wire

- A conductor (such as a wire) with an electric current experiences a force when it is in a magnetic field.
- The relative directions of the current, the magnetic field lines and the force on the conductor can be predicted using Fleming's left-hand rule.

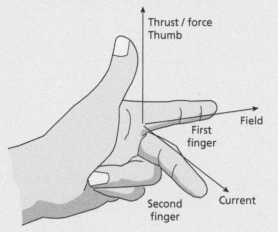

Thrust / force
Thumb

Field

First finger

Current

Second finger

- The force depends on the magnetic flux density, the current and the length of the conductor in the field:

> **force on a conductor (at right-angles to a magnetic field) carrying a current (N) = magnetic flux density (T) × current (A) × length (m)**

- Pairs of forces on a coil of wire can produce rotation – this is how **motors** work.

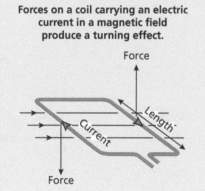

Forces on a coil carrying an electric current in a magnetic field produce a turning effect.

Force

Length

Current

Force

Quick Test

1. Sketch the magnetic field pattern around a bar magnet and use it to show areas where the field is strong and weak.
2. Describe the similarities and differences between the magnetic field around a solenoid and the magnetic field around a permanent bar magnet.
3. ^{HT} Explain what causes the forces on the coil of a motor.

Wave Behaviour

You must be able to:

- Describe wave motion in terms of amplitude, wavelength, frequency and period
- Recall and apply the wave speed formula
- Describe the differences between transverse and longitudinal waves
- **HT** Describe how wave speed may be measured.

Properties of Waves

- All waves have a:
  - **frequency** – the number of waves passing a fixed point per second, measured in hertz (Hz)
  - **amplitude** – the maximum displacement that any particle achieves from its undisturbed position in metres (m)
  - **wavelength** – the distance from one point on a wave to the equivalent point on the next wave in metres (m)
  - **period** – the time taken for one complete oscillation in seconds (s).

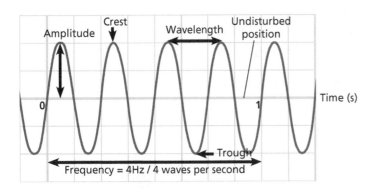

Transverse and Longitudinal Waves

- There are two types of wave: **transverse** and **longitudinal**.
- All waves transfer energy from one place to another.
- For example, if a stone is dropped into a pond, ripples travel outwards carrying the energy. The water does not travel outwards (otherwise it would leave a hole in the middle).
- The particles that make up a wave **oscillate** (vibrate) about a fixed point. In doing so, they pass the energy on to the next particles, which also oscillate, and so on.
- The energy moves along, but the matter remains.
- In a transverse wave, e.g. water wave, the oscillations are perpendicular (at right-angles) to the direction of energy transfer.
- This can be demonstrated by moving a rope or slinky spring up and down vertically – the wave then moves horizontally.
- In a longitudinal wave, e.g. sound wave, the oscillations are parallel to the direction of energy transfer.
- This can be demonstrated by moving a slinky spring moving back and forward horizontally – the wave also moves horizontally.

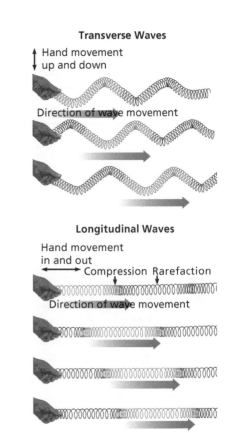

Transverse Waves

Hand movement up and down

Direction of wave movement

Longitudinal Waves

Hand movement in and out

Compression Rarefaction

Direction of wave movement

Sound Waves

- Vibrations from a source pass into a nearby material or **medium** and travel through as **sound waves**.
- Sound waves travel more easily through solids and liquids than through gases.
- However, sound waves can reach our ears by travelling through air.
- Sound waves are longitudinal waves.
- The speed of sound in air is about 330–340m/s.

Wave Speed

- The speed of a wave is the speed at which the energy is transferred (or the wave moves).
- It is a measure of how far the wave moves in one second and can be found with 'the wave equation':

LEARN

wave speed (m/s) = frequency (Hz) × wavelength (m)

- Ripples on the surface of water are slow enough that their speed can be measured by direct observation and timing with a stopwatch.
- As waves are transmitted from one medium to another, their speed and, therefore, their wavelength changes, e.g. water waves travelling from deep to shallow water or sound travelling from air into water.
- As the wave's speed changes, it experiences **refraction**.
- Refraction involves a change in direction of travel of the waves.
- The frequency does not change because the same number of waves is still being produced by the source per second.
- Because all waves obey the wave equation, the speed and wavelength are directly proportional:
  - doubling the speed, doubles the wavelength
  - halving the speed, halves the wavelength.
- A ripple tank can be used to model this behaviour in transverse waves.

Key Point

Sound is emitted by vibrating sources. In a sound wave, particles of the medium only vibrate. They do not travel from the source to our ears.

Refraction

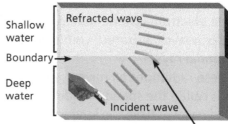

Shallow water · Refracted wave · Boundary → · Deep water · Incident wave

Change in direction and wavelength due to change in wave speed

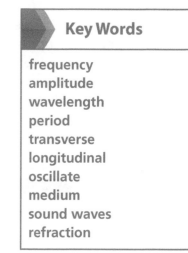

Key Words

frequency
amplitude
wavelength
period
transverse
longitudinal
oscillate
medium
sound waves
refraction

Quick Test

1. What are the units of **wavelength**, **frequency** and **wave speed**?
2. Calculate the frequency of a sound wave with wavelength 2m that travels at 320m/s.
3. Summarise the difference between transverse and longitudinal waves.

Electromagnetic Radiation

You must be able to:

- Understand that vision is based on absorption of light in the eyes
- Use a ripple model of light to illustrate reflection and refraction
- Describe the electromagnetic spectrum, including sources, uses and hazards of different kinds of electromagnetic radiation
- Understand that high frequency radiation can ionise, and that ionisation can cause chemical changes in our bodies that can be harmful.

Light Waves

- Light travels as vibrations of electrical and magnetic fields – it is an **electromagnetic wave**.
- The vibrations are at right-angles to the direction of travel – light waves are **transverse waves**.
- Different frequencies or wavelengths of visible light have different effects on the cells in the **retinas** in our eyes.
- As a result, different electrical changes happen in our brains and we see different colours.
- When light travels from a source (an emitter of light):
  - no material travels
  - it transmits energy from the source to anything that absorbs it.
- When light is completely absorbed, it loses all of its energy and ceases to exist.

> **Key Point**
>
> Electromagnetic radiation transfers energy from the source to anything that absorbs it.

The Electromagnetic Spectrum

- The **electromagnetic spectrum** includes radio waves and **microwaves**, **infrared**, visible light, **ultraviolet (UV)**, **X-rays** and **gamma rays**.
- These radiations all have the same speed in a vacuum.
- It is called the speed of light and is 300 000 000 m/s or 3×10^8 m/s.
- Radio waves have the longest wavelengths and the lowest frequency.
- X-rays and gamma rays have the shortest wavelength and the highest frequency.
- **HT** Electromagnetic waves interact with matter in different ways.
- **HT** How the waves are absorbed or transmitted and refracted or reflected by matter often depends strongly on their wavelength, for example:
  - different surfaces have different colours because they reflect different wavelengths by different amounts
  - the Earth's atmosphere is good at transmitting radio waves and visible light, and good at absorbing high frequency (low wavelength) radiations.

> **Key Point**
>
> Visible light is just a small range of radiations. It is part of a much bigger range, called the electromagnetic spectrum.

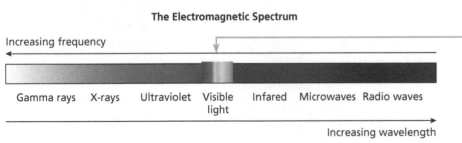

The Electromagnetic Spectrum

Increasing frequency

Gamma rays   X-rays   Ultraviolet   Visible light   Infared   Microwaves   Radio waves

Increasing wavelength

> Different colours of visible light have different wavelengths.

Using Different Kinds of Electromagnetic Radiation

- Radio waves and microwaves are used for communication – for sending signals between mobile phones, between radio stations and radio receivers (or 'radios'), and even to space probes far from Earth.
- Mobile phone networks use microwaves.
- Mobile phones emit and absorb (and so detect) microwaves.

> HT Radio and microwaves are emitted from electrical circuits with a rapidly varying current.
> HT The oscillations or vibrations of the current have the same frequency as the waves.
> HT All radio transmitters, including mobile phones, emit waves in this way.

- Microwaves with a frequency that water molecules are good at absorbing can be used to heat anything that contains water.
- Infrared radiation can cause heating.
- Infrared is emitted by all objects, and the higher their temperature the more energetic the radiation is.
- Infrared cameras can detect this, so they can detect objects that are warmer than their surroundings.
- Paint or ink that reflects UV radiation can be used to make marks on objects. The marks can only be seen using a UV light and UV detector.
- Bones absorb X-rays more strongly than the softer tissue, so shadow images of the human body can be produced, which clearly show bones and any other denser tissue.
- X-rays and gamma rays can be used to kill harmful organisms, such as bacteria.

Ionising Radiation

- X-rays and gamma rays ionise atoms strongly. Some UV radiation can also ionise.
- When materials absorb energy from these radiations, the atoms have enough energy for electrons to escape from them.
- This can cause chemical changes in complicated molecules within our bodies, such as DNA, and lead to cancer. To avoid this, we must limit exposure to these radiations.
- For routine X-ray health checks, the benefits of the very low exposure to radiation outweigh the hazards.

> ### Quick Test
> 1. What do all electromagnetic waves have in common?
> 2. List the parts of the electromagnetic spectrum in order of decreasing wavelength.
> 3. What happens to the frequency of electromagnetic waves as wavelength decreases?
> 4. Explain why some electromagnetic radiations cause ionisation, but others do not.

> ### Key Point
> Radio waves are very useful for transmitting information. TV and radio broadcasting, mobile phones and Wi-Fi systems all use radio waves.

> ### Key Words
> electromagnetic wave
> transverse wave
> retina
> electromagnetic spectrum
> microwave
> infrared
> ultraviolet (UV)
> X-ray
> gamma ray
> HT oscillation
> ionise

Nuclei of Atoms

You must be able to:

- Recall that nuclei have protons and neutrons, and that neutrons have no charge but protons have positive charge
- Use the form $^A_Z X$ to show different nuclear structures and isotopes
- Understand that most nuclei are stable, but some are unstable and can change by emission of particles and energy
- Distinguish between alpha, beta and gamma radiations.

Atoms, Nuclei and Ions

- The nuclei of atoms are very dense and positively charged.
- Electrons, with negative charge, orbit the nuclei in shells.
- The shells of electrons are different distances from nuclei.
- An electron in an atom can gain energy from electromagnetic radiation and move further away from the nucleus.
- When an electron moves closer to the nucleus, the atom emits electromagnetic radiation.
- If an atom loses one of its outer electrons it becomes an ion – it is no longer electrically neutral but has overall positive charge.
- Nuclei contain protons and neutrons.
- The protons have the positive charge.
- Neutrons do not have electric charge – they are neutral.

Absorption of Electromagnetic Radiation by an Atom

Incoming electromagnetic radiation   Electron movement between shells

Nucleus

Electron shells

The electron gains energy and moves to a higher shell.

Emission of Electromagnetic Radiation by an Atom

Outgoing electromagnetic radiation

The electron loses energy and falls to a lower shell.

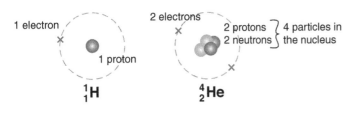

1 electron

1 proton

$^1_1$H

2 electrons

2 protons
2 neutrons } 4 particles in the nucleus

$^4_2$He

Atoms of Different Sizes

- Hydrogen atoms are the smallest atoms and have just one proton and one electron. Some hydrogen atoms can have one or even two neutrons in the nucleus.
- Iron atoms have 26 protons and 26 electrons. They can have 28, 30, 31 or 32 neutrons.
- Forms of the same element, with the same number of protons but different numbers of neutrons, are **isotopes**.
- The number of neutrons affects the mass of an atom, but does not affect the charge.

Isotopes of Iron

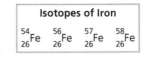

$^{54}_{26}$Fe $^{56}_{26}$Fe $^{57}_{26}$Fe $^{58}_{26}$Fe

Stable and Unstable Nuclei

- Most of the nuclei in the world around us change very rarely – they are stable.
- But some can change, by emitting particles or radiation – they are unstable.
- These changes are called **radioactive emission** or **radioactive decay**.

> ### Key Point
>
> Most of the nuclei of the atoms in your body are stable. They will not change by radioactive decay. A very small proportion of your atoms have unstable nuclei. Your body is a little radioactive. That's normal and natural.

Alpha Emission

- In one kind of emission, two protons and two neutrons leave the nucleus.
- They come out together, and we call the group of four an alpha particle. A continuous flow of alpha particles is **alpha radiation**.
- The alpha particles have enough energy to knock electrons out of atoms as they travel from the material into other materials. This is ionisation.

Beta Emission

- In another kind of emission, a neutron inside the nucleus changes into a proton. An electron is emitted.
- The electron usually has high speed. It is called a beta particle.
- **Beta radiation** is ionising radiation. It is hazardous.

Gamma Emission

- There is a third kind of emission, in which very high energy electromagnetic radiation carries energy away from the nucleus. This is **gamma radiation**.
- Gamma rays also ionise.

Ionisation

- Alpha particles quickly lose energy as they travel through material.
- They don't travel very far, or penetrate, through materials before they slow down and become part of the material.
- Alpha particles pass energy to the material, by ionising its atoms.
- Ionisation in our bodies can cause harmful chemical changes, so alpha radiation is very hazardous if the source is close to our bodies.
- Beta particles and gamma rays also ionise.
- They can travel further than alpha particles through material. They are more penetrating.

Key Point

Some kinds of radioactive material emit alpha particles, some emit beta particles and some emit gamma rays. Some emit alpha and gamma, and some emit beta and gamma.

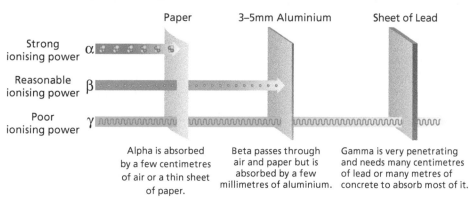

| Paper | 3–5mm Aluminium | Sheet of Lead |

Strong ionising power — α

Reasonable ionising power — β

Poor ionising power — γ

Alpha is absorbed by a few centimetres of air or a thin sheet of paper.

Beta passes through air and paper but is absorbed by a few millimetres of aluminium.

Gamma is very penetrating and needs many centimetres of lead or many metres of concrete to absorb most of it.

Quick Test

1. a) State which **three** of these are the same element?

    $^{1}_{1}Xx$ $^{2}_{1}Xx$ $^{3}_{1}Xx$ $^{3}_{2}Xx$ $^{3}_{3}Xx$

    b) Explain how you can deduce that they are the same element. Mention protons and neutrons.

2. Describe how radioactive emissions cause ionisation of matter.

Key Words

isotope
radioactive emission
radioactive decay
alpha radiation
beta radiation
gamma radiation

Half-Life

You must be able to:

- Distinguish between irradiation and contamination
- Explain the concept of half-life
- **HT** Calculate the proportion of nuclei that are still undecayed after a given number of half-lives
- Write balanced equations for nuclear decay and the emission of alpha, beta and gamma radiation.

Irradiation and Contamination

- Everyday material can be deliberately bombarded with ionising radiations. This is called **irradiation**.
- Irradiation can be used to kill microorganisms, such as bacteria, and it can be used to kill cancerous cells.
- The material that is exposed to the radiation does not become radioactive.
- If radioactive material becomes mixed up with other substances, the substances are **contaminated**. The mixture is radioactive.

Random Change and Half-Life

- A sample of radioactive material contains a very large number of atoms, so it contains a very large number of nuclei.
- The change or decay of a particular nucleus is unpredictable. It can be described as **random**.
- However, in very **unstable** materials the nuclei will change more quickly than the nuclei in more **stable** materials.
- In a sample of material, the number of nuclei that have not yet decayed decreases as, one by one, the nuclei decay.
- This follows a particular pattern – a **decay curve**.
- For a radioactive material, the time taken for the number of undecayed nuclei to reduce by half is always the same, no matter how many nuclei there are to start with.
- This time is called the **half-life** of the material.
- In a material with very unstable nuclei, the half-life is short.

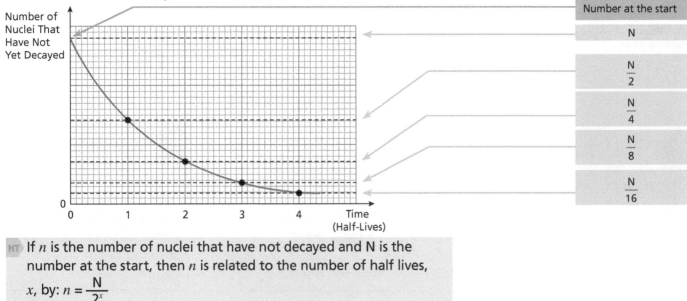

HT If n is the number of nuclei that have not decayed and N is the number at the start, then n is related to the number of half lives, x, by: $n = \dfrac{N}{2^x}$

Nuclear Equations

- Nuclear equations are used to represent radioactive decay:
  - An alpha particle is represented by the symbol $^{4}_{2}He$.
  - A beta particle is represented by the symbol $^{0}_{-1}e$.
- When an alpha particle is emitted:
  - the mass number of the element is reduced by 4
  - the atomic number is reduced by 2.

- This is because 2 protons and 2 neutrons are emitted from the nucleus, e.g.

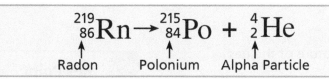

$$^{219}_{86}Rn \rightarrow {}^{215}_{84}Po + {}^{4}_{2}He$$

Radon   Polonium   Alpha Particle

219 = 215 + 4 and 86 = 84 + 2

- With beta decay:
  - the mass number does not change
  - the atomic number is increased by 1.
- This is because a neutron turns into a proton and an electron, and the electron is emitted as the beta particle, e.g.

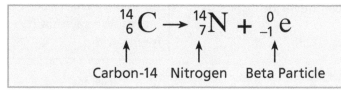

$$^{14}_{6}C \rightarrow {}^{14}_{7}N + {}^{0}_{-1}e$$

Carbon-14   Nitrogen   Beta Particle

14 = 14 + 0 and 6 = 7 – 1

- The emission of a gamma ray does not cause a change in the mass or the charge of the nucleus.
- You need to be able to be able to write balanced decay equations for alpha and beta decay (as shown by the examples above):
  - The mass numbers on the right-hand side must add up to the same number as those on the left.
  - The atomic numbers on the right-hand side must have the same total as those on the left.

Quick Test

1. **HT** If there are N nuclei in a sample of a single radioactive material, how many will there be after three half-lives?
2. In a sample of radioactive material, such as carbon-14, $^{14}_{6}C$, what happens to:
   a) the number of nuclei of the radioactive material?
   b) the total number of nuclei?
3. Fill in the spaces in this nuclear equation:

   $$^{238}_{92}U \longrightarrow {}^{}_{90}Th + {}^{4}_{}He$$

Key Words

irradiation
contaminated
random
unstable
stable
decay curve
half-life

Systems and Transfers

You must be able to:

- Understand that the total energy of a system stays the same unless energy enters or leaves it
- Recall that energy can be stored by systems and can transfer from one system to another
- Calculate the energy stored as thermal energy, as kinetic energy, as elastic potential energy or as gravitational potential energy.

Conservation of Energy

- In a car engine, fuel burns and so the temperature of the gas inside a cylinder rises.
- Pressure also rises and the gas can push a piston.
- If there is no other energy entering or leaving the engine, then the total energy available from the fuel is equal to the energy transferred to the piston.
- In reality, some energy is lost to the surroundings as heat.
- Even so, the energy available from burning fuel can never be destroyed – it just dissipates (spreads out).
- This is an example of **conservation** of energy.

Sources and Transfers of Energy

- Energy is released from a fuel when it is burned (reacting with oxygen in the air). Effectively, the fuel stores energy.
- There are other kinds of **energy store**.
- Water high up in a dam stores energy as **gravitational potential energy**.
- Flowing water is a store of energy until the energy passes from the water, e.g. to a turbine in a hydroelectric system:
  - The flowing water stores energy as **kinetic energy**.
  - The transfer of energy to a turbine is a **mechanical process**. Work is done.
- A stretched spring also stores energy, as does a bow.
- A bow can transfer energy to an arrow, so the arrow gains kinetic energy.
- The stored energy of a stressed object is called **elastic potential energy**.
- Energy flows from a hotter object to cooler ones around it. The hotter object acts as an energy store. The energy transfer processes, such as emission of radiation, are **thermal processes**.
- A battery or electric cell acts as a store of energy.
- Components in a circuit can transfer energy to their surroundings by heating (thermal processes) or by doing work (mechanical processes.)

Energy for Overcoming Resistive Force

- Energy is required to provide force over a distance to produce acceleration of a car.
- Energy is also required to overcome resistive forces – friction and air resistance.

Key Point

The energy of any system stays the same whatever happens inside the system, provided no energy enters or leaves. We say that it is conserved.

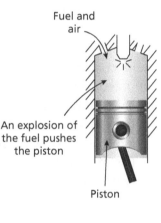

Fuel and air

An explosion of the fuel pushes the piston

Piston

Total energy from fuel and air

Energy that does useful work on the piston

Energy that causes unwanted heating

Key Point

Resistive forces slow down a car and cause transfer of energy to the surroundings. The car will keep slowing until it stops, unless the energy is replaced.

Energy
from
fuel
and air

For a car moving at steady speed on a flat road. Energy from fuel is just used to overcome resistive force. That causes heating.

Energy is dissipated—it's spread thinly into the surroundings.

Energy Calculation Summary

- Stretching a spring transfers energy to it, which is stored as elastic potential energy:

LEARN

energy transferred in stretching (J) = 0.5 × spring constant (N/m) × (extension (M))²

The energy stored in a spring is equal to the work done in stretching it.

- Energy stored by a body raised above the ground is gravitational potential energy:

LEARN

In a gravity field: potential energy (J) =
mass (kg) × height (m) × gravitational field strength, *g* (N/kg)

- The gravitational field strength is about 10N/kg.
- Energy stored by a moving body is kinetic energy:

LEARN

kinetic energy (J) = 0.5 × mass (kg) × (speed (m/s))²

- If one object is hotter than another, the hotter object acts as an energy store.
- Energy can flow from a hotter to a cooler object:

energy available (to pass from hotter to cooler object) =
    mass × specific heat capacity × temperature fall

Use the mass, specific heat capacity and temperature fall of the hotter object.

- Energy transferred by an electrical component:

LEARN

energy transferred (J) =
    charge (C) × potential difference (V)

This equation applies to energy transfer devices, such as a heater or motor. Remember, charge (C) = current (A) × time (s)

Quick Test

1. Energy cannot be destroyed, so describe what happens to the energy of a moving car when it stops.
2. Name **three** different kinds of energy store.
3. Calculate how much energy is effectively stored by:
    a) a kettle containing 0.5kg of water if the water is 70°C hotter than its surroundings
    b) a spring that is stretched by 0.05m by a final force of 15N
    c) 1000kg of water that is 50m above a hydroelectric power station
    d) a 0.03kg bullet moving at 400m/s.
    The specific heat capacity of water = 4200J/kg°C.

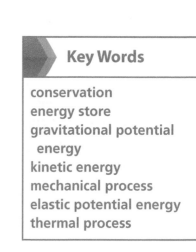

Key Words

conservation
energy store
gravitational potential
 energy
kinetic energy
mechanical process
elastic potential energy
thermal process

Energy, Power and Efficiency

You must be able to:

* Calculate the power of a device from data on energy and time
* Calculate the power of electrical energy transfer from data on current and voltage
* Calculate the efficiency of energy transfer processes.

Power

* Power is the rate of transferring energy:

$$\text{power (W)} = \frac{\text{energy transferred (J)}}{\text{time (s)}}$$

Electrical Energy and Power

* Resistors are used for heating, e.g. in room heaters, cookers and kettles.
* Motors are used for doing work, e.g. in washing machines, vacuum cleaners, hair driers and power drills.
* The power of an electrical device is the rate at which it transfers energy. This is related to potential difference and current:

$$\text{power (W)} = \text{potential difference (V)} \times \text{current (A)}$$

* The unit of power is the watt (W).
* As a watt is quite small, the kilowatt (kW) is often used to measure the power of an electrical component or device.
* Rearranging the power equation gives:

$$\text{energy transferred (J)} = \text{power (W)} \times \text{time (s)}$$

* The standard international unit of energy is the joule (J).
* As a joule is quite small, the **kilowatt-hour (kWh)** is often used as the unit for energy transferred by electrical appliances.
* 1 kilowatt-hour is the energy that a 1 kilowatt appliance transfers in 1 hour.

> **Key Point**
>
> For calculations with answers in joules (J) for energy, the SI system of units, including amp, volt, watt and second, are used. For calculations with answers in kilowatt-hours (kWh) for energy, the kilowatt (kW) is used for power and the hour for time.

Efficiency

> total energy supplied =
>   useful output energy transfer + wasted or dissipated energy

* A ratio can be used to compare the useful energy output to the total energy supplied.

- Multiply the ratio by 100, to create a percentage figure, which represents the **efficiency**.

$$\text{efficiency} = \frac{\text{useful output energy transfer (J)}}{\text{input energy transfer (J)}} \times 100\%$$

- A similar calculation involving power gives the same answer:

$$\text{efficiency} = \frac{\text{useful power output (W)}}{\text{total power input (W)}} \times 100\%$$

Increasing Efficiency, Reducing Dissipation

- We normally want efficiency to be as high as possible. That means that the wasted or dissipated energy must be small.
- Lubrication with oil reduces heating and dissipation in mechanical energy transfers.
- Old **filament lamps** are very inefficient. They produce a lot of unwanted heat.
- Compact fluorescent and **LED** lights produce less heat energy and are more efficient.
- We like our houses to be warm, but thermal energy transfers out from them, heating the world outside. The world outside gets only a very little bit warmer, because it is big. However, the energy lost from inside our homes is important.
- We **insulate** homes as much as possible to reduce energy loss, using materials such as foam boards. These have low **thermal conductivity**.

Key Point

All processes 'waste' some energy. The useful energy output is less than the energy supplied. The 'wasted' energy is spread thinly in the surroundings, often causing its temperature to rise by a small amount. The wasted energy is dissipated.

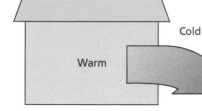

Poor insulation – high rate of energy transfer to the surroundings

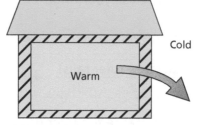

Good insulation – low rate of energy transfer to the surroundings

Quick Test

1. An electrical room heater has a current of 5A and a voltage of 230V.
   a) Calculate its power in kilowatts (kW).
   b) Calculate how much energy the room heater supplies to the room in i) 1 hour and ii) 6 hours. Give your answers in kWh.
   c) Repeat question b), but give your answers in joules (J).
2. Calculate the efficiency of a motor that provides a useful energy output of 1kJ when supplied with 1.25kJ of energy.

Key Words

power
kilowatt-hour (kWh)
efficiency
filament lamp
LED
insulate
thermal conductivity

Physics on the Road

You must be able to:

- Estimate the sizes of some everyday speeds and accelerations
- Describe how to measure human reaction time
- Understand that when a driver makes an emergency stop, the total distance needed to stop is the sum of the thinking distance and the braking distance
- Describe the danger of large deceleration.

Example Speeds

- If you walk quickly you can have a speed of about 1.5 metres per second (m/s). That's about 3.5 miles per hour (mph).
- A steady running speed for a fairly fit person is between 2.0 and 2.5m/s.
- An Olympic sprinter can run at about 10m/s.
- Acceleration at the start of a race is very important. A sprinter who goes from 0 to 1m/s in one second has an acceleration of 10m/s^2. If it takes two seconds, then the acceleration is only 5m/s^2.
- A fast but steady cycling speed is similar, although top cyclists can maintain an average of 14m/s for an hour.
- In a severe hurricane the wind speed can be 35m/s or more, but even a wind speed of 10m/s feels pretty strong.
- A commercial jet travels at about 140m/s, which is 500 kilometres per hour (km/h).

Human Reaction Times

- You can measure your **reaction time** by working with another person:
  1. They hold a ruler with its zero mark level with your open fingers.
  2. They let go at a random time and you close your fingers as quickly as you can.
  3. The distance the ruler falls depends on your reaction time.

$$\text{reaction time} = \sqrt{\frac{(2 \times \text{distance ruler falls})}{g}}$$

- Air resistance doesn't have much effect because the ruler does not reach high speed.
- You can improve the accuracy of your result by repeating the measurement several times and working out a mean (average) value.

Key Point

When we sense something happening, we do not react at that exact instant. There is a short delay – our reaction time.

g is the acceleration of the ruler, which is the acceleration due to gravity and is equal to 10m/s^2.

Stopping Distances of Cars

- When a driver has to stop a car, first he or she must react.
- The distance the car travels during the reaction time is sometimes called **thinking distance**.
- Only after the reaction time does the driver actually apply the brakes. Then the car slows down.
- Thinking distance can be increased by tiredness, drugs, drinking or distractions such as a mobile phone.
- The distance travelled during braking is called **braking distance**.
- Braking distance can be increased by rain, ice, snow, worn tyres or worn brakes.

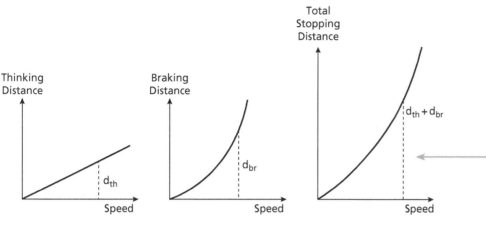

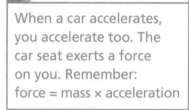

The thinking distance, braking distance and total stopping distance are all bigger at higher speeds.

- The total (or overall) **stopping distance** is the thinking distance plus the braking distance.
- A typical stopping distance at 48km/h (30mph) is 23 metres, but at 96km/h (60mph) it is 73 metres.

Acceleration in a Car

- A car with high **acceleration** can reach more than 30m/s after just 10 seconds.
- The speed increases by an average of 3m/s every second. That's an acceleration of $3m/s^2$.
- The acceleration needs a force.

Deceleration During Accidents

- In a car accident, rapid **deceleration** takes place.
- Deceleration requires force, just as acceleration does, but the force is now in the opposite direction to the motion.
- The decelerating force can hurt and kill.
- Seatbelts and airbags decrease the decelerating force on passengers in an accident.
- A seatbelt is designed to stretch, while ordinary surfaces such as windscreens and steering wheels do not stretch in the same way.
- When a seatbelt stretches, it increases the time during which the deceleration acts, so decreases the deceleration and, therefore, the force.
- Airbags and crumple zones have a similar effect.

> **Key Point**
>
> When a car accelerates, you accelerate too. The car seat exerts a force on you. Remember:
> force = mass × acceleration

> **Key Point**
>
> Safety features such as seatbelts and airbags are designed to reduce the force that decelerates you in a car accident.

> **Key Words**
>
> reaction time
> thinking distance
> braking distance
> stopping distance
> acceleration
> deceleration

> **Quick Test**
>
> 1. Calculate which is faster: a bullet that travels 100m in 0.4s or a plane that travels 1km in 3.3s.
> 2. Name the **two** distances that add up to make the total stopping distance of a car.

Energy for the World

You must be able to:

- Describe the main energy resources that are available
- Sort energy resources into renewable and non-renewable
- Explain why renewable energy resources are better for the future
- Recall that use of renewable resources has increased in the past decade.

Non-Renewable Energy Resources

- Electricity provides a convenient way of supplying energy to our homes, workplaces and leisure places.
- Electricity can be generated by burning **fossil fuels** – oil, coal and gas. Most of the world's electricity is generated in this way.
- However, once we take these fuels from the ground and use them, we can never replace them. They are **non-renewable**.
- **Nuclear fuel** also comes from the ground, from rocks, and is non-renewable.
- Nuclear fuel has the extra problem that it makes radioactive waste, which is very difficult to dispose of safely.

Renewable Energy Resources

- Wood is a **bio-fuel**. It will not run out, provided that we plant new trees to replace the ones we use – it is **renewable**.
- Other bio-fuels are produced from plant crops, such as oil seed rape and oil palm.
- Plants can store energy from sunlight by **photosynthesis**.
- We can use the energy that they store, but so far humans have not been able to create artificial photosynthesis.
- **Wind power** is also renewable.
- Winds in the atmosphere are caused by heating by the Sun, and by more heating in some places than others. As long as there is air and sunshine, there will be wind.
- **Hydroelectric power** stations generate electricity from moving water:
  - The water starts at a high point.
  - It travels down pipes to a lower point where it can turn turbines.
  - The water gets to the higher place by evaporation and rain and, as long as there is water and sunshine, there will always be rain.
  - This energy source is renewable.
- Tides also produce natural movement of water, which can turn turbines.
- Energy can be taken directly from sunlight using **solar panels** that heat water or **solar cells** that generate d.c. electricity.
- Most countries in the world are trying to use more renewable energy and less fossil fuel.

> ### Key Point
>
> Fossil fuels are the remains of organisms from the distant past. The fuels were created by energy storage by the living things, which took millions of years. Humans have used a large proportion of the world's fossil fuels in a much shorter time.

Release of Carbon Dioxide

- The concentration of carbon dioxide in the atmosphere has increased over the last 200 years.
- At the same time, humans have burned more and more fossil fuels.
- The average temperature of the Earth has increased.
- The Earth absorbs energy from the Sun and emits it back into space. There is an energy balance when absorption and emission rates are the same.
- More carbon dioxide in the air makes emission of energy back into space more difficult, so it affects the Earth's energy balance.
- It is difficult to predict what will happen if we continue to add carbon dioxide to the atmosphere. It may produce a continuing **global climate change**.

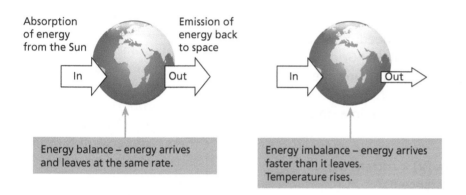

Absorption of energy from the Sun · In · Out · Emission of energy back to space

Energy balance – energy arrives and leaves at the same rate.

Energy imbalance – energy arrives faster than it leaves. Temperature rises.

- Scientists are very worried about climate change.
- Politicians from all over the world have agreed to limit emissions of carbon dioxide from burning fossil fuels.
- There have been some changes, with increased use of renewable energy resources, but politicians sometimes find it hard to agree.
- Nobody is certain that the limits will be enough to prevent possible future climate change.

> **Key Point**
>
> Fossil fuels are not just non-renewable. They also pollute the air. In particular the burning of fossil fuels releases carbon dioxide into the atmosphere, which has a big impact on the atmosphere all around the world.

Quick Test

1. Group the following into **renewable** and **non-renewable energy** resources:
   A wind
   B tide
   C nuclear fuel (uranium)
   D coal
   E bio-fuel oil
   F gas (methane)
   G hydroelectricity.
2. Explain why rapid release of carbon dioxide is a major problem for the future.

> **Key Words**
>
> fossil fuel
> non-renewable
> nuclear fuel
> bio-fuel
> renewable
> photosynthesis
> wind power
> hydroelectric power
> solar panel
> solar cell
> global climate change

Energy at Home

You must be able to:

- Explain that the transmission of energy from power stations uses transformers to reduce heating of the cables that wastes energy
- Explain the roles of live, neutral and earth wires in wiring in homes
- Recall that UK homes use electricity with a.c. frequency of 50Hz and a voltage of about 230V.

Transmission of Energy from Power Stations to Users

- Power stations can be a long way from the places where people live, work and play.
- The energy must be transmitted to where it is needed using cables. The network of cables is called the National Grid.
- If there is a large current in the cables, a large amount of energy will transfer from them by heat, which isn't useful.
- The rate of transmitting the energy, or the power, in a circuit is related to current, voltage and resistance:

> **LEARN**
>
> **power (W) = potential difference (V) × current (A)**
> **= (current (A))² × resistance (Ω)**

- Fortunately, transmission can happen at high power using high voltage but low current. Then there is less heating of the cables and less energy loss.
- At power stations, voltage from the generators is made bigger using **step-up transformers**.
- High voltage in our homes would be dangerous. Near our homes, voltage is made smaller using **step-down transformers**.
- Transformers only work with a.c. (alternating current), so mains electricity is a.c. The current is always changing direction.
- Batteries and solar cells provide d.c. (direct current). This current is always in the same direction.
- In our homes, the voltage is about 230 volts (V). It is still a.c., and one cycle of alternation lasts $\frac{1}{50}$ of a second. The frequency is 50 hertz (Hz).

> **Key Point**
>
> The resistance of the transmission cables is fixed (by their length and other properties), so the power transferred from them is dependent on the square of the current in them.

> **Key Point**
>
> Transformers work using the changing magnetic field due to changing current in their coils. The current changes continuously – it is alternating current (a.c.).

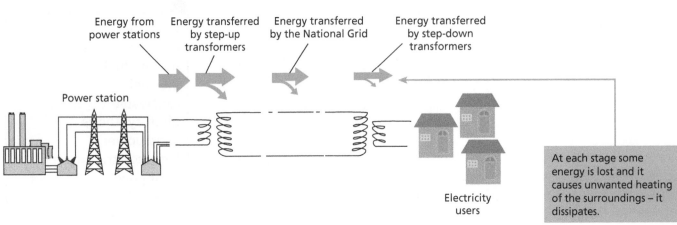

Energy from power stations — Energy transferred by step-up transformers — Energy transferred by the National Grid — Energy transferred by step-down transformers

Power station

Electricity users

At each stage some energy is lost and it causes unwanted heating of the surroundings – it dissipates.

Electrical Wiring at Home

- There is a potential difference or voltage between a **live wire** and a **neutral wire**.
- When a resistor is connected between live and neutral wires it carries a current and it becomes hot – it's a heater.
- When a motor is connected between live and neutral wires the current and the resulting magnetic forces make it spin.
- If you touch a live wire, there is a potential difference between you and anything else that you are touching, such as the ground and there will be a dangerous current through you.
- If a live wire is badly connected inside the metal casing of an appliance then the casing becomes 'live'. Touching it can have the same effect as touching the live wire directly.
- The casing is earthed. It has an extra wire – the **earth wire**.
- There is a potential difference between the live wire and the earth wire. Normally they are not connected together, so no current flows.
- However, if the live wire touches the metal casing, a current can flow between it and the earth wire.
- As there is very little resistance between the live and earth wires, the current will be large.
- The large current can melt a deliberate weak point in the live wire connection – a **fuse**. This cuts off the live wire from the power supply, and makes the appliance safe.
- A fuse in the plug and the wall switch, next to the wall socket, can cut off the appliance from the live wire in the house circuit.
- However, the live wire in the house circuit still provides a potential difference and touching it is dangerous.

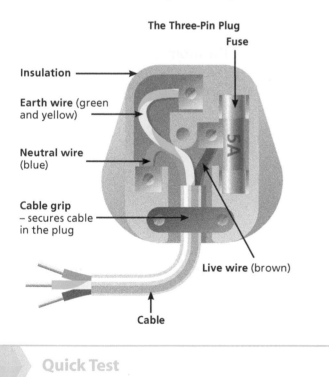

The Three-Pin Plug

Fuse

Insulation

Earth wire (green and yellow)

Neutral wire (blue)

Cable grip – secures cable in the plug

Live wire (brown)

Cable

Quick Test

1. Explain why electrical energy is transmitted using high voltage and low current.
2. Explain why electrical energy is transmitted using a.c. and not d.c.

Key Words

step-up transformer
step-down transformer
live wire
neutral wire
earth wire
fuse

Review Questions

Matter, Models and Density

1 What is the unit of density?

     A kgm   **B** kg/m   **C** kg/m²   **D** kg/m³   [1]

2 Which of these models of an atom is **not** normally used now?

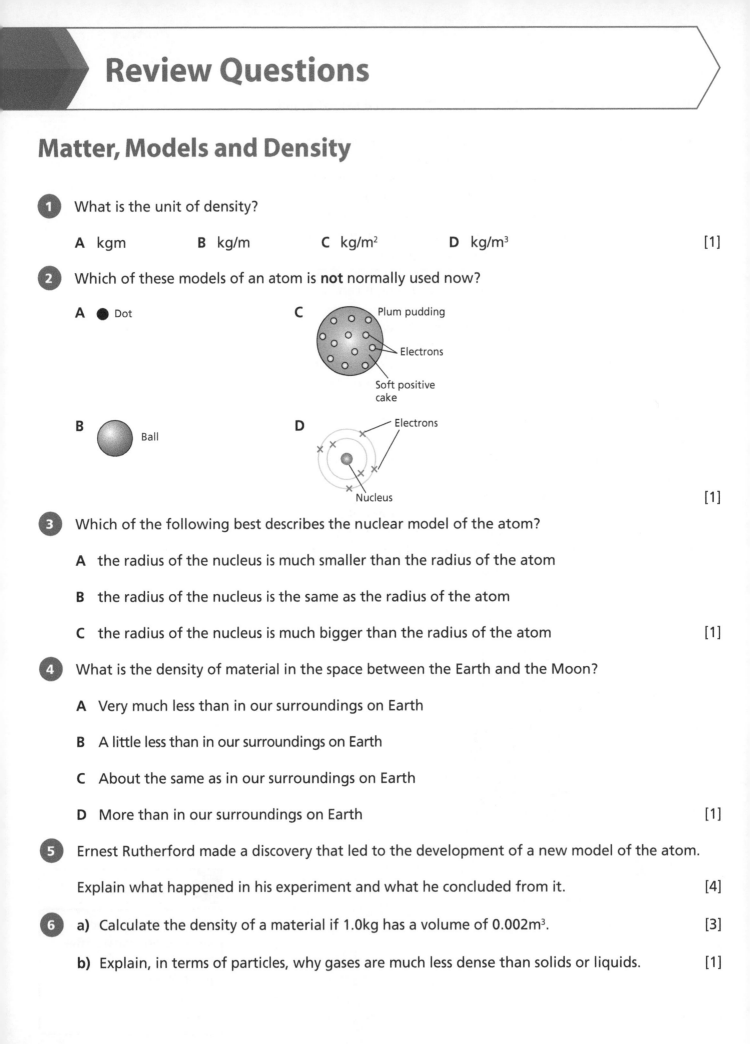

     A Dot

     C Plum pudding / Electrons / Soft positive cake

     B Ball

     D Electrons / Nucleus   [1]

3 Which of the following best describes the nuclear model of the atom?

     A the radius of the nucleus is much smaller than the radius of the atom

     B the radius of the nucleus is the same as the radius of the atom

     C the radius of the nucleus is much bigger than the radius of the atom   [1]

4 What is the density of material in the space between the Earth and the Moon?

     A Very much less than in our surroundings on Earth

     B A little less than in our surroundings on Earth

     C About the same as in our surroundings on Earth

     D More than in our surroundings on Earth   [1]

5 Ernest Rutherford made a discovery that led to the development of a new model of the atom.

     Explain what happened in his experiment and what he concluded from it.   [4]

6 **a)** Calculate the density of a material if 1.0kg has a volume of 0.002m³.   [3]

     b) Explain, in terms of particles, why gases are much less dense than solids or liquids.   [1]

7 Explain the differences between:

a) an atom and a nucleus

c) an electron and a proton [4]

b) an atom and an ion

d) a proton and a neutron. [4]

Total Marks _____ / 20

Temperature and State

1 Which of these does **not** change during a change of state?

    A density   **B** energy   **C** mass   **D** volume   [1]

2 In a gas, what does molecular bombardment cause?

    A melting   **B** freezing   **C** energy   **D** pressure   [1]

3 Which of the following describes a change in state?

    A reversible   **B** irreversible   **C** chemical   [1]

4 Which has most internal energy?

    A an ice cube of mass 0.01kg and a temperature of −1°C

    B a pan of water with mass 0.5kg and a temperature of 80°C

    C a lake of water with mass 1×10^7kg and a temperature of 8°C

    D a cloud of steam with mass 5kg and a temperature of 100°C   [1]

5 Snow can melt slowly, even when the temperature is a few degrees above freezing point.

Why is this?

    A its specific heat capacity is large   **C** energy is needed to melt it

    B its specific heat capacity is small   **D** it gives out energy as it melts   [1]

Total Marks _____ / 5

Review Questions

Journeys

1 What is 100 miles in kilometres?

$$\frac{1 \text{ mile}}{1 \text{ kilometre}} = \frac{8}{5}$$

  A 160km   **B** 62.5km   **C** 60km   **D** 21.6km   [1]

2 How far can you walk in 2.5 hours at an average speed of 1m/s?

  A 1440m   **B** 9000m   **C** 10800m   **D** 12600m   [1]

3 Which of these is the fastest speed?

$$\frac{1 \text{ mile}}{1 \text{ kilometre}} = \frac{8}{5}$$

  A 36km/h   **B** 36000m/h   **C** 10m/s   **D** 25mph   [1]

4 Which graph shows constant velocity?

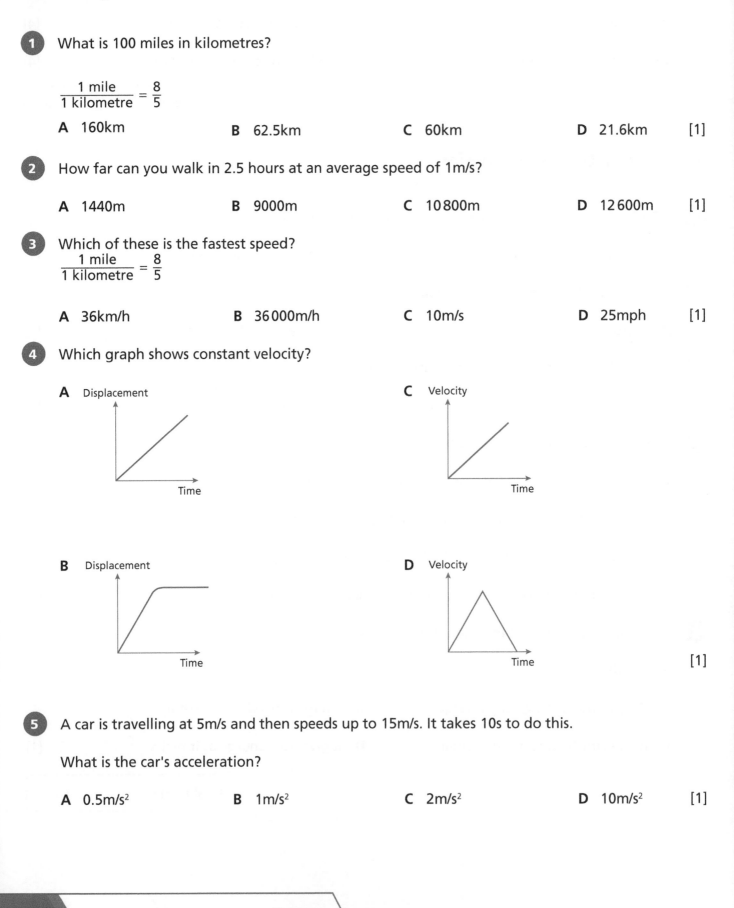

  [1]

5 A car is travelling at 5m/s and then speeds up to 15m/s. It takes 10s to do this.

What is the car's acceleration?

  A 0.5m/s²   **B** 1m/s²   **C** 2m/s²   **D** 10m/s²   [1]

6 The diagram shows the velocity vectors of a car at two times, T_1 and T_2, one minute apart.

T_1 T_2

   a) What has happened to the speed of the car? [1]

   b) What has happened to the velocity of the car? [1]

   c) Has the car accelerated or decelerated? [1]

   d) What can you say about the force acting on the car? [1]

7 a) A sports car can accelerate from standstill (velocity = 0) to 30m/s in 10s.

      What is its acceleration? [3]

   b) If the mass of the car is 1000kg, what is its kinetic energy:

      i) when it is standing still? [1]

      ii) when it is moving at 30m/s? [3]

   c) What average force is needed to accelerate the car? [3]

   d) Use the following equation

      (final velocity (m/s))² − (initial velocity (m/s))² = 2 × acceleration (m/s²) × distance (m)

      to work out the distance the car travels during this acceleration. [2]

   e) Calculate the amount of work that is needed for the car's acceleration. [3]

   f) What is the power at which the car gains energy? [2]

Total Marks .... / 26

Review Questions

Forces

1 What type of force acts on the Earth to keep it in orbit?

    A electric or electrostatic force   **C** magnetic force

    B gravitational force   **D** resistive force   **[1]**

2 Which of these always causes acceleration?

    A unbalanced or net force   **C** constant velocity

    B balanced forces   **D** constant pressure   **[1]**

3 How much force is needed to accelerate a mass of 20kg by 4m/s^2?

    A 5N   **B** 16N   **C** 24N   **D** 80N   **[1]**

4 **a)** When a person steps off a boat, what does their foot do to the boat?   **[1]**

    b) What happens to the boat:

        i) if it is much more massive than the person?   **[1]**

        ii) if it is not much more massive than the person?   **[1]**

    c) Sketch the forces acting on the person and on the boat.   **[3]**

5 Humans cannot jump very high.

Why not?   **[1]**

6 HT Describe the net force or resultant force for each of the following examples.

    a)

                                                         [1]

    b)

                                                         [1]

    c)

                                                         [1]

Total Marks _____ / 13

Force, Energy and Power

1 What is power the same as?

    A rate of change of velocity   **C** rate of energy transfer

    B kinetic energy   **D** strength   [1]

2 A person has to do work to push a cupboard across a room.

    a) If they push it for 2.5m with an average force of 180N, how much work do they do?   [3]

    b) If the cupboard is floating in space, and it experiences a force of 180N over a distance of 2.5m, how much kinetic energy does it gain?   [1]

    c) Why doesn't the cupboard in the room gain kinetic energy?   [1]

    d) On Earth, imagine that the cupboard experiences an **upwards** force equal to its weight, 300N, and rises steadily to a height of 2.5m.

        What can you say about the cupboard's energy?   [3]

Total Marks .... / 9

Changes of Shape

1 Which of the following best describes the relationship between force and extension for an elastic spring?

    A constant   **B** linear   **C** non-linear   **D** varying   [1]

2 **a)** What is the weight on Earth, in newtons (N), of a ball with mass of 1.2kg?   [3]

    b) If g_{moon} and $g_{jupiter}$ are 1.6N/kg and 25N/kg, how much will the ball weigh on **i)** the Moon and **ii)** on Jupiter?   [2]

Total Marks .... / 6

Review Questions

Electric Charge

1 Between which of these pairs of atomic particles will there be a force of attraction?

A $\oplus$ $\oplus$

C Neutral ⚫ $\ominus$

B Neutral ⚫ $\oplus$

D $\oplus$ $\ominus$ [1]

2 Inside atoms, which particles have electric charge?

A neutrons

C neutrons, protons and electrons

B neutrons and protons

D protons and electrons   [1]

3 Why is it difficult to give a static charge to a metal object?

A electrons can easily flow on and off the object

B protons can easily flow on and off the object

C there are no electrons in metals

D electrons in metals can't move   [1]

4 What is the unit of charge?

A amp   **B** coulomb   **C** joule   **D** volt   [1]

5 Which of these is a correct equation?

A current = charge × time

C charge = $\dfrac{\text{current}}{\text{time}}$

B current = $\dfrac{\text{charge}}{\text{time}}$

D time = current × charge   [1]

Total Marks _____ / 5

Circuits

1 What component is this the symbol for?

[1]

2 Which of these will increase the current in a circuit?

    A increasing voltage and keeping resistance the same

    B increasing resistance and keeping voltage the same

    C decreasing voltage and keeping resistance the same

    D decreasing voltage and increasing resistance [1]

3 **a)** Draw a circuit diagram for a circuit you would use to investigate the relationship between current and voltage for a filament lamp. [5]

    b) A filament lamp contains a wire that gets hot.

        Sketch a current–voltage graph for a filament lamp. [4]

    c) Explain why the graph in part **b)** is not perfectly straight. [2]

> **Total Marks** _____ / 13

Resistors and Energy Transfers

1 Which of these pairs are both units of energy?

    A volt and joule   **C** watt and kilowatt

    B watt and joule   **D** joule and kilowatt-hour [1]

2 What happens to:

    a) the resistance of a thermistor when its temperature increases? [1]

    b) the resistance of an LDR when light level decreases? [1]

3 A kettle is rated at 2.0kW.

    a) During 1 week, the kettle is used for a total of 2.5 hours.

        How much energy does it transfer in the week? Give your answer in kWh. [3]

    b) The hour and the kWh are not SI units.

        What are the SI units for the same physical quantities? [2]

> **Total Marks** _____ / 8

Magnetic Fields and Motors

1 Which of these is a correct diagram of a magnetic field pattern?

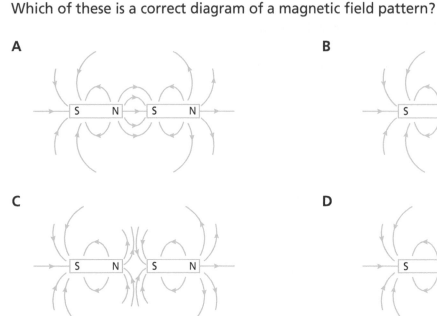

A   B

C   D

[1]

2 A compass is lying close to a coil of wire that is part of a circuit.

When does the compass needle move?

A when the current in the coil is constant

B only when the current in the coil is turned on

C only when the current in the coil is turned off

D when the current in the coil is turned on or off   [1]

3 HT Why does a simple electric motor produce rotation?

   A the forces on opposite sides of a coil are in the same direction

   B the forces on opposite sides of a coil are in opposite directions

   C the current in the coil is constant

   D the current in the coil is changing   [1]

4 What happens to magnetic field strength as distance away from a magnet increases?   [1]

Total Marks _____ / 4

Wave Behaviour

1. What term is given to the number of vibrations per second?

   A amplitude   C speed

   B frequency   D wavelength   [1]

2. If ripples on water have a wavelength of 0.1m and a frequency of 5Hz, what is their speed?

   A 0.5m/s   C 1.5m/s

   B 1.0m/s   D 50m/s   [1]

3. Which of these statements is **true**?

   A Only transverse waves carry energy.

   B Only longitudinal waves carry energy.

   C Transverse and longitudinal waves carry energy.

   D Neither transverse nor longitudinal waves carry energy.   [1]

4. HT What effect do radio waves have on charged particles such as electrons?

   A they directly cause oscillation

   B they directly cause ionisation

   C they directly cause emission of light

   D they directly cause absorption of light   [1]

5. What does a source of waves do?

   A absorbs waves   C reflects waves

   B emits waves   D transmits waves   [1]

Total Marks _____ / 5

Electromagnetic Radiation

1 Which of these have the longest wavelength?

    A radio waves   **C** visible light waves

    B ultraviolet waves   **D** X-rays   [1]

2 Reflection of some frequencies of light but not others by different surfaces gives the surfaces:

    A brightness   **C** texture

    B colour   **D** gloss   [1]

3 Complete the sentence.

The colour that surfaces appear is due to different interactions with different:

    A amplitudes of light.   **C** speeds of light.

    B brightnesses of light.   **D** wavelengths of light.   [1]

4 The frequency of visible light travelling through glass is 4×10^{14} Hz and its wavelength is 5×10^{-7} m. What is its speed?   [3]

5 When waves pass from one medium into another their speed usually changes.

What happens to frequency and wavelength?   [2]

6 How does your voice make sound?

    A by absorbing   **C** by refracting

    B by transmitting   **D** by vibrating   [1]

Total Marks _____ / 9

Nuclei of Atoms

1 How many neutrons are there in a nucleus with this symbol $^{15}_{7}N$?

A 22

C 8

B 15

D 7 [1]

2 What happens to the charge of a nucleus that emits an alpha particle?

A it decreases by two

C it stays the same

B it decreases to zero

D it increases by two [1]

3 Which of these cannot normally change an atom into an ion?

A absorption of energy from electromagnetic radiation

B high-energy collisions

C gravity on Earth [1]

4 What happens when an atom absorbs light?

A electrons can move further from the nucleus

B electrons can move closer to the nucleus

C electrons lose energy

D electrons move more quickly [1]

5 Complete the sentence.

Alpha particles, beta particles and gamma rays are all kinds of:

A electromagnetic radiation.   B light radiation.   C ionising radiation. [1]

6 Sketch an atom and show how:

a) it can gain energy without becoming an ion [2]

b) it can gain energy and become an ion. [1]

Total Marks _____ / 8

Half-Life

1 Different radioactive substances have different half-lives.

What is a half-life? [3]

2 Which graph shows the longest half-life?

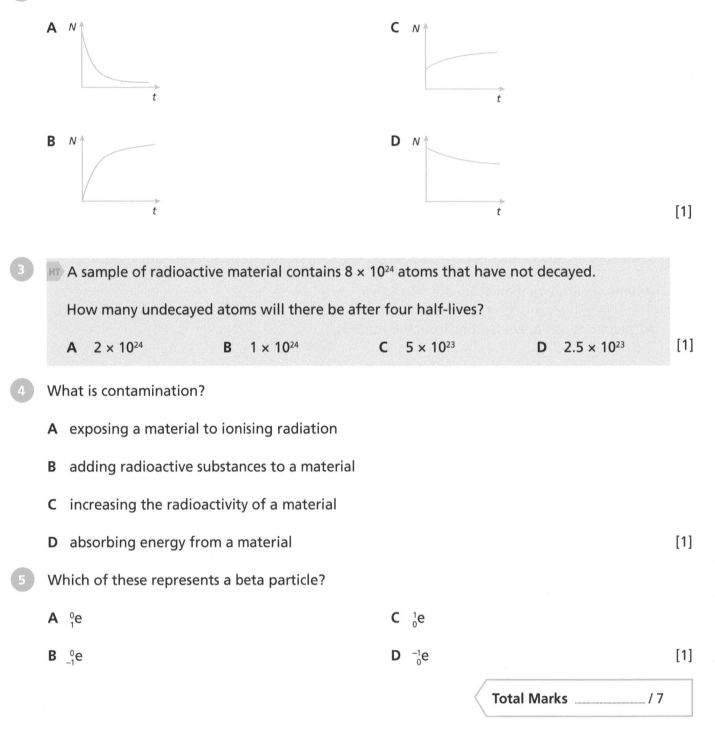

A N

C N

B N

D N

t

[1]

3 HT A sample of radioactive material contains 8×10^{24} atoms that have not decayed.

How many undecayed atoms will there be after four half-lives?

A 2×10^{24} B 1×10^{24} C 5×10^{23} D 2.5×10^{23} [1]

4 What is contamination?

A exposing a material to ionising radiation

B adding radioactive substances to a material

C increasing the radioactivity of a material

D absorbing energy from a material [1]

5 Which of these represents a beta particle?

A $^{0}_{1}e$ C $^{1}_{0}e$

B $^{0}_{-1}e$ D $^{-1}_{0}e$ [1]

Total Marks _____ / 7

Systems and Transfers

1 Which of the following is **not** a store of energy?

    A a wind-up toy   **C** a moving bullet

    B water high in a reservoir   **D** burned ash on the ground   **[1]**

2 How much energy is stored by a spring system, if a total force of 20N produces an extension of 0.5m?

    A 5J   **B** 10J   **C** 20J   **D** 40J   **[1]**

3 Which of these transfers energy?

    A a battery in a circuit that is turned off   **C** a fuel in a car with the engine turned off

    B a resistor in a working circuit   **D** water resting in a high reservoir   **[1]**

4 Which of the following has the most energy?

    A a ball of mass 0.3kg moving at 20m/s   **C** a spring stretched by 0.2m by a force of 300N   **[1]**

    B the same ball at a height of 200m

> **Total Marks** _____ / 4

Energy, Power and Efficiency

1 HT Which of these increases efficiency in a mechanical system?

    A conduction   **B** dissipation   **C** lubrication   **D** convection   **[1]**

2 a) Calculate the efficiency of a motor that does 20J of useful work for each 50J of energy that is supplied to it.   **[2]**

    b) Explain what happens to the 'missing' energy.   **[1]**

3 These diagrams show energy transfers to the same scale.
Which has the highest efficiency? You must explain your answer.

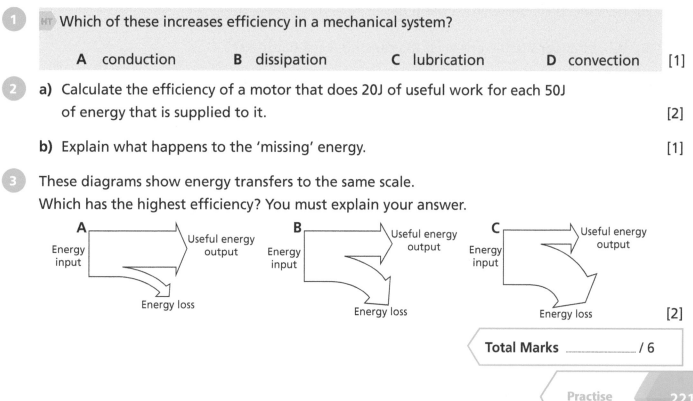

    [2]

> **Total Marks** _____ / 6

Practice Questions

Physics on the Road

1. Which of these is closest to normal walking speed?

   A 0.2m/s   C 2m/s

   B 1.2m/s   D 24m/s   [1]

2. A train is moving at 10m/s and accelerates to 35m/s in 5s.

   Which of these is the train's acceleration?

   A 0m/s²   C 5m/s²

   B 1m/s²   D 10m/s²   [1]

3. In an emergency, what does the total stopping distance of a car **not** depend on?

   A the gradient or slope of the road

   B friction between the tyres and the road

   C the driver's thinking time

   D the speed of other cars   [1]

4. A car of mass 900kg is travelling at 20m/s.

   a) How much kinetic energy does it have?   [3]

   b) What happens to the energy if the car stops on a flat road?   [3]

   c) Why is the stopping distance longer when going downhill?   [1]

5. A cyclist has kinetic energy of 12000J.

   How much force is needed to stop the cyclist in a distance of 20m?   [3]

Total Marks _____ / 13

Energy for the World

1. Which of these is a renewable energy resource?

    A coal   **B** gas   **C** oil   **D** wind   [1]

2. Which graph shows change in the use of wind energy resources in the UK over the last 20 years?

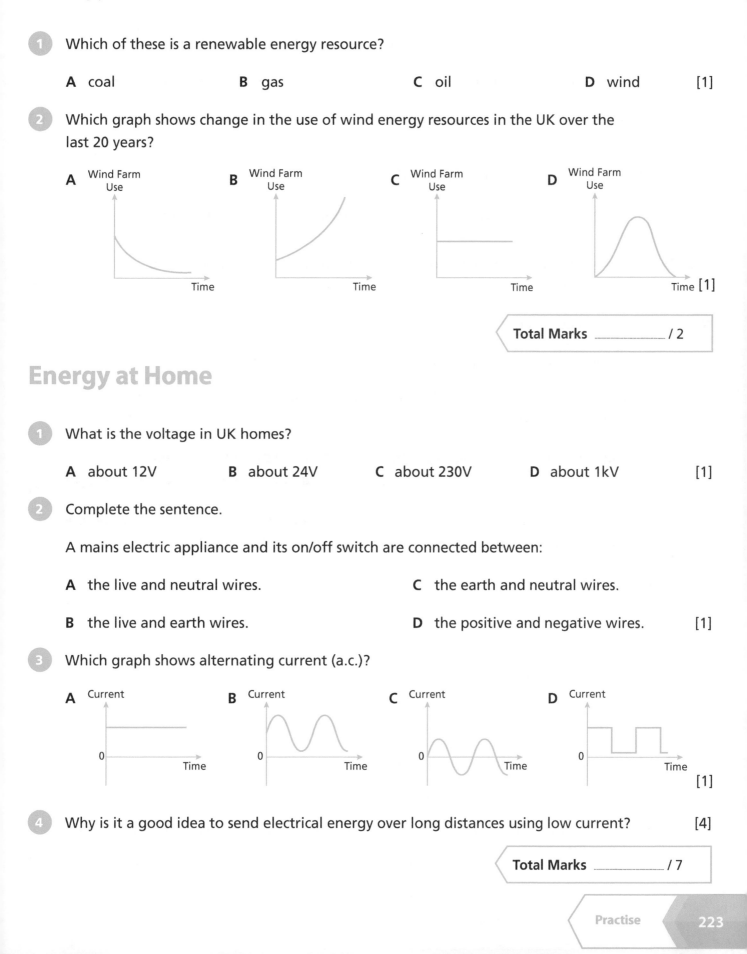

Total Marks _____ / 2

Energy at Home

1. What is the voltage in UK homes?

    A about 12V   **B** about 24V   **C** about 230V   **D** about 1kV   [1]

2. Complete the sentence.

    A mains electric appliance and its on/off switch are connected between:

    A the live and neutral wires.   **C** the earth and neutral wires.

    B the live and earth wires.   **D** the positive and negative wires.   [1]

3. Which graph shows alternating current (a.c.)?

4. Why is it a good idea to send electrical energy over long distances using low current?   [4]

Total Marks _____ / 7

Review Questions

Magnetic Fields and Motors

1 HT What information does magnetic flux density provide?

    A the size of the current in a coil   **C** the strength of magnetic field

    B the size of the current in a magnet   **D** the size of magnetic field   **[1]**

2 HT What is the unit of magnetic flux density?

    A amp   **B** ohm   **C** tesla   **D** volt   **[1]**

3 Describe how you can show that a coil of wire carrying an electric current has a magnetic field around it.   **[2]**

4 A solenoid is a useful electromagnet.

How can you increase the magnetic effect in the space around the coil?   **[3]**

5 HT How much force acts on a wire of length 0.15m that carries a current of 2.5A in a magnetic field of flux density 0.12T?

force on a conductor (at right-angles to a magnetic field) carrying a current =
magnetic flux density × current × length   **[2]**

6 HT In d.c. motors, the force on each side of the coil would not reverse and would not produce continuing rotation without a special connection to the power supply.

Explain this problem.   **[6]**

> **Total Marks** _____ / 15

Wave Behaviour

1 Which of the following best describes a longitudinal wave?

    A direction of travel of the wave and line of vibration of the medium are the same

    B direction of travel of the wave and line of vibration of the medium are opposite

    C direction of travel of the wave and line of vibration of the medium are at right–angles

    D direction of travel of the wave and line of vibration of the medium are perpendicular   **[1]**

2 Which of these statements is **true**?

  A Ripples on water provide a good model of transverse waves.

  B Ripples on water provide a good model of longitudinal waves.

  C Sound waves in air are transverse.

  D Light waves are longitudinal. [1]

3 HT Which of these statements is **true**?

  A Sound waves are always produced by vibration of charged particles in electric circuits.

  B Sound waves and radio waves are always produced by mechanical vibration.

  C Radio waves are always produced by vibration of charged particles in electric circuits.

  D Sound waves and radio waves are always produced by vibration of charged particles in electric circuits. [1]

Total Marks _____ / 3

Electromagnetic Radiation

1 Which of these are particularly hazardous?

  A infrared waves   B radio waves   C visible light waves   D X-rays [1]

2 Which of the following is the same for all electromagnetic waves?

  A amplitude in space   C speed in space

  B frequency in space   D wavelength in space [1]

3 a) Make a large sketch of the electromagnetic spectrum showing the main parts. [7]

  b) Add notes to your sketch to show what different parts of the spectrum can be used for. [7]

  c) Show which end of the spectrum has the longest wavelength and lowest frequency. [1]

  d) Add notes to your sketch to show some hazards of some of the parts. [2]

Total Marks _____ / 19

Review Questions

Nuclei of Atoms

1 Which of these pairs are isotopes of the same element?

A   **C**

B   **D**

⬤ proton
◯ neutron

[1]

2 Which of these are kinds of electromagnetic radiation?

A alpha particles   **C** gamma rays

B beta particles   **D** neutrons   [1]

3 What happens to the charge of a nucleus that emits a gamma ray?

A it decreases   **C** it stays the same

B it decreases to zero **D** it increases   [1]

4 How does an atom become an ion?

A it loses or gains one or more electron

B it loses or gains one or more proton

C it loses or gains one or more neutron

D it loses or gains one or more gamma ray   [1]

5 When do atoms emit light?

A when electrons inside them lose energy

B when electrons inside them gain energy

C when alpha particles inside them lose energy

D when alpha particles inside them gain energy [1]

6 What does 'random' mean?

A in bursts   **C** very fast

B without a pattern   **D** very slowly   [1]

7 Which type of radiation can travel furthest through a material?

    A alpha particles   **B** beta particles   **C** gamma rays   [1]

8 What is irradiation?

    A increasing the radioactivity of a material

    B absorbing energy from a material

    C adding radioactive substances to a material

    D exposing a material to ionising radiation   [1]

Total Marks _____ / 8

Half-Life

1 Which of the following describes the nuclei in radioactive material?

    A stable   **B** unstable   **C** big   **D** small   [1]

2 Replace x and y with the correct numbers in each of the following equations showing radioactive decay.

    a) $^{x}_{6}C \rightarrow \ ^{14}_{y}N + ^{0}_{-1}e$ [2]

    b) $^{238}_{x}U \rightarrow \ ^{y}_{90}Th + ^{4}_{2}He$ [2]

3 Draw a line from each symbol to the correct diagram.

$^{2}_{1}H$

$^{3}_{1}H$

$^{4}_{2}He$

$^{3}_{2}He$

     proton

     neutron

[3]

Total Marks _____ / 8

Review Questions

Systems and Transfers

1 Which of these statements agrees with the principle of conservation of energy?

   A Energy can be created and destroyed.

   B Energy can be created but never destroyed.

   C Energy cannot be created but it can be destroyed.

   D Energy cannot be created or destroyed. [1]

2 Which of the following does **not** supply energy to a system?

   A absorption of light   **C** using fuel to accelerate

   B emission of light   **D** heating [1]

3 What happens when a body accelerates?

   A it must experience a force, work must be done on it and it gains kinetic energy

   B it must experience a force, no work is needed and it gains kinetic energy

   C it must experience a force, work must be done on it and its kinetic energy stays the same

   D it must experience a force, work must be done on it and it loses kinetic energy [1]

4 Which of these is the kinetic energy of a car of mass 800kg travelling at 20m/s?

   A 400J   **B** 820J   **C** 1600J   **D** 160 000J [1]

5 How much energy is stored by 5×10^9kg of water at a height of 500m?

   A 2.5×10^9J   **B** 5×10^9J   **C** 2.5×10^{13}J   **D** 5×10^{13}J [1]

Total Marks _____ / 5

Energy, Power and Efficiency

1 What is dissipation of energy?

   A transfer of energy into surroundings so that it is no longer useful

   B transfer of energy into surroundings so that it can be stored

   C transfer of energy from surroundings so that it can be used [1]

2 A battery-powered motor uses 100J of energy to lift a load. The energy supplied to the load is 60J.

What happens to the other 40J of the supplied energy?

A it does work on the load   **C** it is stored by the motor

B it causes heating and dissipates   **D** it is returned to the battery   [1]

3 Which of these does **not** reduce the rate of unwanted energy transfer?

A lubrication   **C** dissipation

B thermal insulation   **D** streamlining   [1]

4 The table contains two examples of energy storage devices and systems and one example of an energy transfer device and system.

Give **two** more examples of each.

Energy Storage Devices and Systems	Energy Transfer Devices and Systems
battery	resistor
arrow held in a bow	

[4]

5 An electric current carries energy from a supply, such as a battery or the a.c. mains, to energy transfer devices in the circuit.

What are the main types of energy transfer in the following devices?

a) a heater   [1]

b) a motor   [1]

c) a hair drier   [1]

d) a lamp   [1]

6 Which of the following transfers the most energy?

A a 2.0kW kettle used for 5 minutes

B a 0.8kW washing machine used for 30 minutes

C a 400W television used for 2.5 hours   [1]

Total Marks _____ / 12

Review Questions

Physics on the Road

1 Which of the following measurements is closest to 100km/h?

A 100mph

C 30mph

B 60mph

D 10mph   [1]

2 What is reaction time?

A the time between an event happening and a person responding

B the time between a person seeing an event and responding

C the total time taken to stop a car after an event happens

D the braking time of a car after an event happens   [1]

3 Which graph shows how thinking distance changes as speed increases?

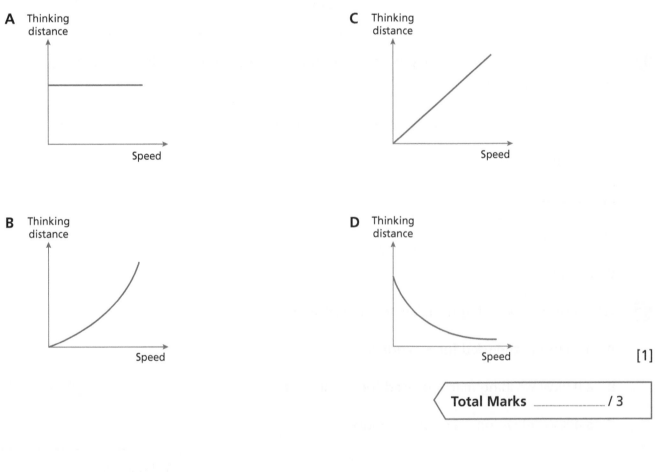

Total Marks _____ / 3

Energy for the World

1 **a)** Copy and complete this table adding at least **one** advantage and **one** disadvantage for each of the five different energy resources.

Energy Resource	Advantages	Disadvantages
coal		
oil		
wind		
solar		
nuclear		

[6]

b) Name **one** other energy resource and list its advantages and disadvantages. [3]

Total Marks _____ / 9

Energy at Home

1 What is the network of cables for transmitting energy over large distances called?

A National Grid

C step-down transformer

B step-up transformer

D power station [1]

2 What immediately happens in an electrical appliance if the live wire becomes connected to the earth wire?

A the voltage becomes very large

C a very large current can flow

B no current can flow at all [1]

3 Which of these is an electrical insulator?

A copper

C PVC

B iron

D steel [1]

Total Marks _____ / 3

Mixed Exam-Style Biology Questions

1 The diagram below shows a heart.

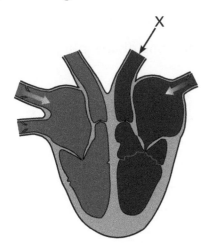

X

What is the name of blood vessel **X**?

A aorta

C pulmonary vein

B pulmonary artery

D vena cava   [1]

2 Which of the bacteria below convert nitrates in the soil to nitrogen in the air?

A putrefying

C nitrifying

B denitrifying

D nitrogen-fixing   [1]

3 The movement of substances from an area of high concentration to an area of low concentration describes which one of the following processes?

A active transport

C transpiration

B diffusion

D osmosis   [1]

4 The following factors affect the distribution of organisms.

Which one is a **biotic** factor?

A light intensity

C number of predators

B pH of soil

D amount of rainfall   [1]

5 Some students wanted to investigate the effect of temperature on respiration of yeast.

They made a simple manometer as shown below.

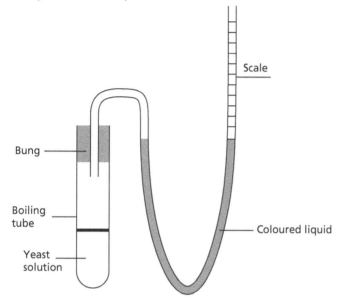

The students placed 0.5g of dried yeast, 0.05g of glucose and 10cm³ of water in the boiling tube. A thin layer of oil was poured onto the top of the water. The boiling tube was then put into a water bath at 20°C.

They left the apparatus for 30 minutes and then recorded how far the coloured liquid had moved on the scale.

They repeated the experiment at 40°C and 60°C.

a) What type of microorganism is yeast?   [1]

b) The yeast is respiring anaerobically in this experiment.

   What will the products be?   [2]

c) Why does the coloured liquid move up the scale?   [1]

d) Suggest **two** variables the students needed to keep the same for each experiment?   [2]

e) What was the dependent variable in this investigation?   [1]

f) The students recorded their results in a table.

A	20	40	60
Number of Divisions on Scale that Coloured Liquid Moved	12	4	0

   i) Suggest a suitable title for row **A**.   [2]

   ii) What can the students conclude from these results?   [1]

iii) Why was there no movement of the coloured water at 60°C? [2]

iv) How could the students improve the accuracy of their results? [1]

g) If the boiling tube at 20°C is left for several weeks, the yeast will eventually die.

Give **two** reasons why the yeast will die. [2]

6 The diagram below shows part of a phylogenetic tree.

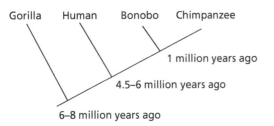

What can be deduced from the tree?

A The gorilla is more closely related to the bonobo than the human.

B The human and chimpanzee share a common ancestor from 1 million years ago.

C The gorilla and human share a common ancestor from 6–8 million years ago.

D Humans are most closely related to the chimpanzee. [1]

7 The diagram below shows the mechanism that plants use to allow gases to enter and exit the leaves and also to control transpiration.

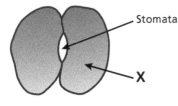

a) What is transpiration? [1]

b) What is the name of cell **X**, which opens and closes the stomata? [1]

c) Which gas will enter through the stomata when the plant is photosynthesising? [1]

8 Some students wanted to investigate if the number of stomata on the upper and lower side of leaves were similar or different.

They collected a leaf from the elephant ear plant, *Saxifrage bergenia*.

They painted the upper and lower surface of the leaf with clear nail varnish and left it to dry.

They then peeled the nail varnish from the leaf and were able to count the number of stomata in several fields of view using a microscope.

a) What is the genus of the plant the students used? [1]

b) The diagram below shows what the students could see in one field of view using a light microscope.

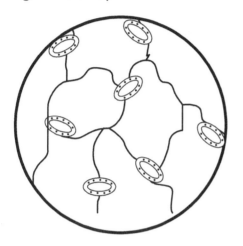

Stomata measure about 30–40µm.

What is likely to be the magnification the students used?

A ×4   **B** × 100   **C** × 10 000 [1]

c) The students recorded their results in a table.

Field of View	1	2	3	4	5
Number of stomata – upper side of leaf	2	3	2	2	1
Number of stomata – lower side of leaf	6	5	7	9	3

   i) Calculate the mean number of stomata per field of view for the upper and lower surface of the leaf.

   Show your working. [4]

   ii) Explain why there are more stomata on the lower surface of leaves compared with the upper surface. [5]

9 Tall corn plants is the dominant trait to short corn plants.
A cross between two plants yielded 72 plants that were tall and 28 that were short.

What are the likely genotypes of the parent plants?

A TT and TT   **B** TT and Tt   **C** Tt and Tt   **D** tt and tt [1]

10 Which one of the following is a communicable disease?

A cancer **B** liver disease **C** AIDS **D** obesity [1]

11 The diagram below shows the components of blood.

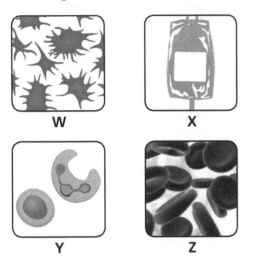

W   X

Y   Z

What is the function of component **W**?

A produce antibodies **C** transport carbon dioxide and glucose

B trap microorganisms **D** help the blood to clot [1]

12 Colour blindness is an example of an X-linked recessive disease.

It is caused by an alteration in one gene which is on the X chromosome.

The altered gene will be expressed unless a 'normal' allele is present on a second X chromosome to 'cancel it out'.

a) Explain why males are much more likely to suffer from colour blindness than females. [2]

b) A female can be a carrier of colour blindness.

What is the chance of a female carrier and a normal male having a son who is colour blind?

A 25% **B** 50% **C** 75% **D** 100% [1]

c) The diagram below shows the incidence of colour blindness in a family.

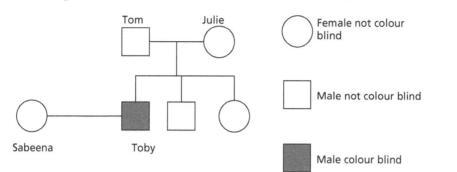

Tom   Julie

Sabeena   Toby

○ Female not colour blind

□ Male not colour blind

■ Male colour blind

Has Toby inherited his colour blindness from his mother or father? [1]

d) Toby marries Sabeena. Sabeena is not a carrier for colour blindness.

Which of the statements below is **true**?

A there is a 50% chance any daughters they have will be carriers

B none of their sons will be colour blind

C there is a 50% chance any daughters will be colour blind [1]

13 Antibiotics are important medicines, which have been used to treat bacterial infections for over 70 years.

However, over recent years, many have become less effective as new antibiotic-resistant superbugs have evolved.

a) Explain how antibiotic-resistant bacteria develop. [3]

b) Some serious illnesses such as tuberculosis are often treated with two different antibiotics simultaneously.

Explain how this regime is likely to reduce the emergence of a resistant strain. [2]

c) Why are antibiotics **not** prescribed for viral infections? [1]

d) The human body has several defence mechanisms to stop bacteria from entering our body.

Give **three** defence mechanisms and explain how each works. [6]

14 Seed banks are an important store of biodiversity.

Which statement is **not** true when considering the use of seed banks to maintain biodiversity compared with storing whole plants?

A only uses a small amount of space

B suitable for storing all types of seeds

C relatively low labour costs

D seeds remain viable for long periods [1]

15 The diagram on the right shows the parts of the nervous system.

Which letter represents the peripheral nervous system?

A W

B X

C Y

D Z [1]

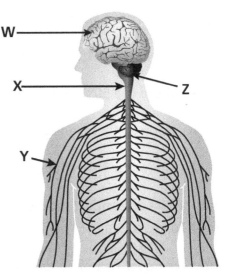

16 Scientists are worried about increased levels of carbon dioxide in the atmosphere.

Until 1950, the level of carbon dioxide in the atmosphere had never exceeded 300 parts per million.

The graph below shows levels of carbon dioxide over the last 10 years.

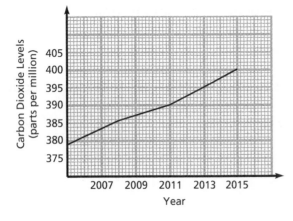

a) What was the level of carbon dioxide in 2008? [1]

b) Use the graph to predict carbon dioxide levels in 2017 if they continue to rise at their current rate. [1]

c) Explain why carbon dioxide levels have increased so much in the last 100 years. [2]

17 A transgenic organism is an organism that contains genes from another organism.

By incorporating a gene for human protein production into bananas, potatoes and tomatoes, researchers have been able to successfully create edible vaccines for hepatitis B, cholera and rotavirus.

a) Which type of human cells are likely to be used as a source of the gene?

A red blood cells

B white blood cells

C brain cells

D platelets [1]

b) HT What enzymes are used to 'cut' the gene from the human cells? [1]

c) HT What enzymes are used to 'stick' the gene into the host cell? [1]

d) Give **two** reasons why people might be reluctant to eat these transgenic plants. [2]

Total Marks _____ / 65

Mixed Exam-Style Chemistry Questions

1 Eliot is carrying out chromatography on inks from different pens.
   He is trying to find out which ink was used to write a note.
   He draws a start line in pencil and then puts a dot of each ink onto the line.
   He then dips the filter paper into ethanol in a beaker.

   a) What is the role of the **paper** in this experiment?

       A mobile phase   **C** solid phase

       B gas phase   **D** stationary phase   **[1]**

   b) Why did Eliot draw the start line in pencil?   **[1]**

   c) The results of Eliot's experiment are shown below.

      Which ink matches sample **X**, which was used on the note?   **[1]**

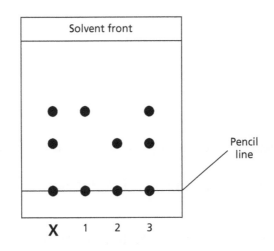

   d) Calculate the R_f value for the ink in **sample 2**.
      Use a ruler to help you.   **[1]**

2 During the electrolysis of molten lithium chloride, what is made at the anode?

   A chlorine

   B hydrogen

   C lithium

   D lithium hydroxide   **[1]**

3 The diagrams below show the structures of two forms of carbon:

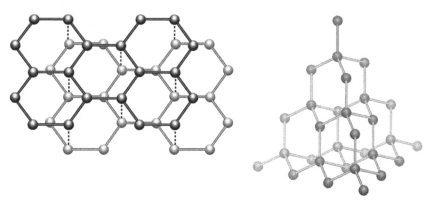

a) Diamond is a very hard material.
   Graphite is not very hard.

   Use ideas about bonding and structure to explain this observation. [4]

b) What is the technical term given to the different forms carbon can take? [1]

c) Describe **one** way of demonstrating that graphite and diamond are different
   physical forms of carbon. [2]

4 Ethane has the formula C_2H_6.
   Look at the three different representations of ethane.

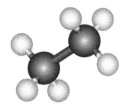

H—C—C—H (Displayed formula)

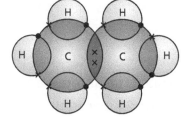

Ball and stick model   Displayed formula   Dot and cross diagram

a) What type of bonding is present in ethane?
   You must explain how the diagrams show this. [2]

b) Describe the limitations of using displayed formulae to represent molecules. [2]

5 The table contains information about some atoms and ions.

Particle	Atomic Number	Atomic Mass	Number of Protons	Number of Neutrons	Number of Electrons	Electronic Structure
A	7	14	7		7	2.5
B		40	20	20	18	2.8.8
C	8		8	8	8	
D	9	19		10	9	2.7

a) Complete the table. [5]

b) Explain why particle **D** is an atom and particle **B** is an ion. [2]

c) Particle A has the electronic structure 2.5.

What does this tell you about the position of particle **A** in the periodic table?
Explain your answer. [4]

d) Another particle has an atomic mass of 40 and contains 21 neutrons.

What is the atomic number of particle **B**? [1]

6 Two isotopes of carbon are:

$^{12}_{6}C$ $^{14}_{6}C$

Explain the difference between isotopes of the same element in terms of
subatomic particles. [4]

7 The table below shows some fractions of crude oil and the range of their boiling points.

Fraction	Boiling Point Range (°C)
Petrol	80–110
Diesel	150–300
Fuel oil	300–380
Bitumen	400+

A hydrocarbon called hexadecane has a boiling point of 289°C.

In which fraction would you expect to find hexadecane? [1]

8 HT Devlin reacts 0.9g of element **X** with 7.1g of chlorine, Cl_2.
There is one product: **X** chloride.

   a) What mass of **X** chloride is produced? [1]

   b) Calculate the number of moles of **X**, chlorine (Cl_2) and **X** chloride involved in the reaction. (The relative atomic mass of **X** = 9. The relative formula mass of chlorine = 71 and **X** chloride = 80.) [1]

   c) Use your answers to construct the **balanced symbol** equation for this reaction. [2]

9 Ras Al-Khair in Saudi Arabia produces around one million cubic litres of drinking water a day from sea water.

Which of the following describes the process of extracting fresh water from sea water?

A chlorination

B filtration

C desalination

D sedimentation [1]

10 Magnesium sulfate is a salt.
It can be made by reacting magnesium oxide, MgO, with sulfuric acid, H_2SO_4.

   a) Construct the **balanced symbol** equation for this reaction. [1]

   b) Naomi suggests another method for preparing magnesium sulfate:

   1. Measure $50cm^3$ of dilute sulfuric acid into a beaker.
   2. Add one spatula of magnesium oxide.
   3. Heat the mixture until only crystals of magnesium sulfate are left.

   Give **one** safety precaution that Naomi should take when following this method. [1]

   c) Naomi's method will **not** make a pure, dry sample of magnesium sulfate.

   How can Naomi make sure that:

   i) The reaction is complete? [1]

   ii) The filtrate contains only magnesium sulfate solution? [1]

11 Look at the diagram of four different atoms:

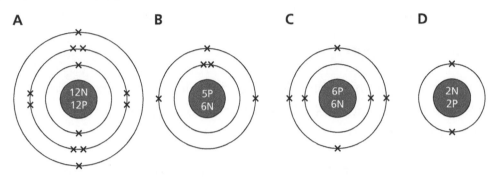

A   B   C   D

a) What is the centre of the atom called? [1]

b) What subatomic particles make up the centre of the atom? [1]

c) Which two elements shown are in the same **period** in the periodic table? [1]

d) What are the names of elements **B** and **C**? [2]

12 Look at the diagrams of compounds below.

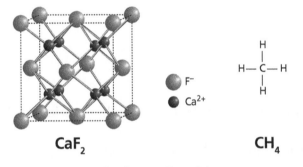

CaF_2 CH_4

a) Explain why the structure of calcium fluoride, CaF_2, means it has a high melting point. [3]

b) Explain why methane, CH_4, has a low melting point and boiling point. [3]

c) Draw dot and cross diagrams to show the ionic bonding in lithium oxide.
You should include the charges on the ions.
The electronic structure of lithium is 2.1.
The electronic structure of oxygen is 2.6. [2]

13 a) What is meant by the term **pure substance** in chemistry? [1]

b) Explain how this is different to the way in which pure is used to describe substances,
e.g. orange juice, in everyday life. [1]

14 Look at the elements from the periodic table and answer the questions that follow.

 $^{9}_{4}Be$ $^{24}_{12}Mg$ $^{27}_{13}Al$ $^{19}_{9}F$ $^{31}_{15}P$

  a) Which element has a relative atomic mass of 9? [1]

  b) Which element has an atomic number of 9? [1]

  c) An isotope of P has two fewer neutrons.

    What is the relative atomic mass of the isotope? [1]

15 **HT** Crude oil is separated into different fractions during fractional distillation.
Each fraction contains hydrocarbon molecules that have similar boiling points.
The boiling point of a hydrocarbon is determined by the number of carbon atoms in it.

	Boiling Point	No. of Carbon Atoms	Size of Intermolecular Forces
A	Low	More than 50	Large
B	High	Less than 20	Large
C	Low	Less than 20	Small
D	High	More than 50	Small

Which letter, **A**, **B**, **C** or **D**, represents the **correct** relationship between boiling point, number of carbon atoms and the size of the intermolecular forces in a molecule? [1]

16 Magnesium and dilute sulfuric acid react to make hydrogen gas:

$Mg(s) + H_2SO_4(aq) \rightarrow MgSO_4(aq) + H_2(g)$

George measures the rate of this reaction by measuring the **loss in mass** of the reaction mixture. He finds that the change in mass is very small and difficult to measure.

  a) Draw a labelled diagram to show a better way of measuring the rate of this reaction. [1]

  b) The reaction between the magnesium and sulfuric acid takes longer than George would like.

    What type of substance can George add to increase the rate of reaction? [1]

  c) Suggest what safety precautions George needs to take when carrying out the experiment. [1]

Mixed Exam-Style Chemistry Questions

17 Iodine is a non-metal.

Which statement is true about iodine **because** it is a non-metal?

A It is a solid at room temperature and pressure.

B It does not conduct electricity.

C It is a halogen.

D It is in Period 5 of the periodic table.   [1]

18 HT Which of the following shows the approximate size of an atom?

A 3×10^3 m

B 3×10^{-3} m

C 3×10^{-13} m

D 3×10^{-23} m   [1]

19 Sea water contains many useful chemicals.
A company in the Middle East, specialising in desalination of sea water, decides that they can also extract particular chemicals and sell them onto the chemical industry.

a) What is desalination?   [1]

b) Sodium chloride and sodium sulfate can both be extracted from sea water.

What is the chemical formula for sodium sulfate?   [1]

c) A technician tests sea water with dilute nitric acid followed by silver nitrate solution.
A white precipitate is formed.

Suggest what conclusion can be drawn from this result.   [1]

d) It is important that people in all parts of the world have access to potable water.
Explain why.   [2]

20 What is the general formula for the reaction between a metal and an acid? [1]

21 Complete the word equations to show the products of each reaction:

a) magnesium + nitric acid [1]

b) zinc + sulfuric acid [1]

c) iron + hydrochloric acid [1]

22 Part of the reactivity series is shown in the diagram below.

Potassium
Sodium
Calcium
Magnesium
Aluminium
Carbon
Zinc
Iron
Tin
Lead
Hydrogen
Copper
Silver
Gold
Platinum

When a metal is reacted with an acid it forms a metal salt, plus hydrogen gas. For example:

lead + sulfuric acid → lead sulfate + hydrogen

calcium + sulfuric acid → calcium sulfate + hydrogen

a) Which of the two reactions has the fastest initial reaction? [1]

b) A metal **X** is reacted with sulfuric acid.
It reacts violently compared with the two reactions above.

Where would **X** be placed on the reactivity series? [1]

Total Marks _____ / 77

Mixed Exam-Style Physics Questions

1 Which one of these statements is **false**?

   A Visible light is electromagnetic radiation.

   B Visible light is ionising radiation.

   C Visible light carries energy.

   D Visible light travels as a transverse wave.   [1]

2 Dissipation of energy is reduced by which of the following?

   A having better insulation of houses

   B reducing the efficiency of an electric motor

   C using hot filament lamps instead of LED lamps

   D using step-up transformers to increase the voltage for long-distance
      transmission of electrical energy   [1]

3 Which one of these statements is **false**?

   A Electrons can orbit the nuclei in atoms.

   B Electrons can flow in wires to make electric current.

   C Electrons are the same as beta particles.

   D Electrons all have positive electric charge.   [1]

4 HT Which of the following has constant speed but changing velocity?

   A a car moving steadily along the road

   B the Earth in orbit around the Sun

   C a sprinter at the start of a race

   D a ball when you drop it to the floor   [1]

5 **a)** Put the particles below in order of size, starting with the largest:

atom   electron   molecule   nucleus   [2]

b) During the change of state from liquid to gas, what happens to:

i)   atoms?   [2]

ii) nuclei?   [1]

c) Describe what happens to the molecules of a gas when it is heated.   [2]

d) The diagram shows a molecule of gas and its velocity before and after colliding with a wall.

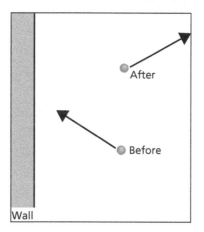

i)   What has happened to the speed of the molecule?   [2]

ii) What has happened to the velocity of the molecule?   [1]

iii) What does the wall do to the molecule?   [1]

iv) What does the molecule do to the wall?   [1]

e) **i)** A molecule of gas has a mass of 4×10^{-26}kg and moves at 100m/s.

Calculate the kinetic energy of the molecule.   [4]

ii) What is the total kinetic energy of a kilogram of the gas if all of
     the molecules move at 100m/s?   [1]

6 This bus has a constant forwards velocity.

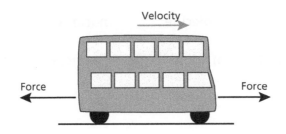

Velocity

Force ←   → Force

a) i)   What is the net force acting on the bus?   [1]

  ii) What provides the forwards force?   [1]

  iii) What provides the backwards force?   [1]

  iv) What is its acceleration?   [1]

b) A girl standing on the bus holds an umbrella so that it is free to swing.

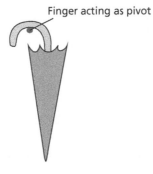

Finger acting as pivot

  i)   The girl notices that when the bus accelerates, the umbrella seems to swing backwards.

     Explain why that happens.   [1]

  ii) What will happen to the umbrella when the bus decelerates to stop at a bus stop?
     You must explain your answer.   [2]

c) i)   A child runs in front of the bus and the driver brakes hard to make the bus decelerate.
     The girl falls forwards.

     Explain why that happens.   [1]

  ii) What is the difference between the reaction time of the driver and the braking time
     of the bus?   [2]

  iii) How do we use reaction time and braking time to work out the stopping time of the bus? [1]

d) The moving bus has kinetic energy.

    What happens to that energy when the bus brakes and stops?   [3]

7 This diagram represents an electric heater.
The supply voltage is 230V.

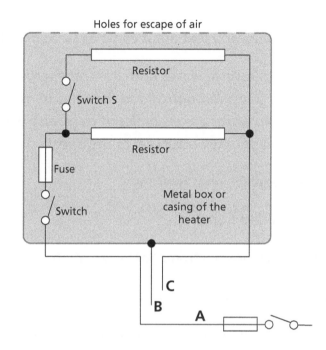

a) What do the resistors do? [2]

b) Are the resistors connected in series or in parallel? [1]

c) What is the purpose of Switch S? [2]

d) What are the names of the three wires: **A**, **B** and **C**? [3]

e) When all the switches are closed, a current of 4.6A passes through each resistor.

   i) What is the value of each resistor? [3]

   ii) The total current through the fuse is 9.2A.

     Is the combined resistance of the two resistors bigger or smaller than your answer to part **i)**?
     Explain your answer. [2]

f) What is the maximum power of the heater? [3]

g) How much energy does the heater provide to a room in 3 hours? [3]

Mixed Exam-Style Physics Questions

8 **a)** Sound waves in water have a frequency of 3000 Hz and a wavelength of 0.5 m.

Work out the speed of sound waves in water. [4]

b) A student has written the following text but it contains same mistakes.

Longitudinal waves and transverse waves both transfer energy. But longitudinal waves need a medium to travel through and so they also transfer substances, such as air. Sound and light waves are examples of transverse waves, which means that their direction of travel is parallel to the direction of vibration.

Identify the mistakes and explain what is wrong. [3]

c) Where are gamma rays emitted from? [1]

d) Radio waves and gamma rays are electromagnetic radiation.

   i) State **two** things that are the same about radio waves and gamma rays. [2]

   ii) State **two** things that are different about radio waves and gamma rays. [2]

e) Visible light is part of the electromagnetic spectrum.

What is the electromagnetic spectrum? [3]

f) How does your body respond to:

   i) visible light? [1]

   ii) infrared radiation? [1]

   iii) ultraviolet radiation? [1]

   iv) X-rays? [1]

   v) radio waves? [1]

9 The diagram shows two houses. The temperature is the same inside them both.
The arrows show rate of energy transfer out from the houses.

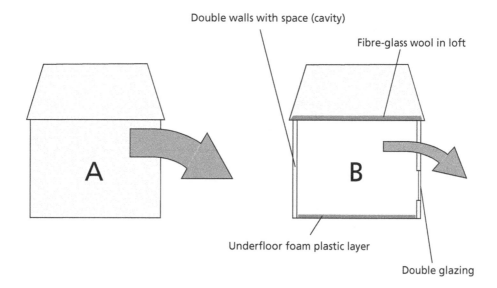

Double walls with space (cavity)

Fibre-glass wool in loft

A

B

Underfloor foam plastic layer

Double glazing

a) Explain why the arrows on the two diagrams are not the same. [2]

b) Why does the person who lives in House **A** pay much more money for heating? [1]

c) What happens to the energy that transfers out from the houses? [2]

d) What is another name for rate of transfer of energy? [1]

e) i) Name **three** energy sources that the people who live in the houses can use for heating. [3]

   ii) State whether the energy sources are based on renewable fuel, non-renewable fuel or a mixture of those. [3]

f) What are the advantages and disadvantages of using wind farms as sources of energy? [4]

g) What are the advantages and disadvantages of using oil as a source of energy? [4]

h) What changes in the use of different energy resources do you expect to happen in the next 10 years?
   Explain your answer. [2]

10 **a)** Explain how atoms can emit different kinds of electromagnetic radiation. [4]

This diagram shows pathways of three kinds of ionising radiation, travelling through air from three different radioactive sources.

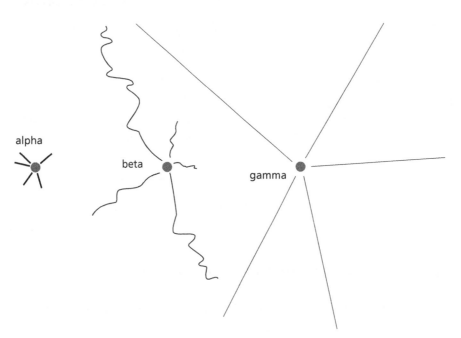

b) What does 'radioactive' mean? [3]

c) What does 'ionising' mean? [2]

d) Which kind of radiation travels furthest through air? [1]

e) **i)**   Which kind of radiation runs out of energy soonest? [1]

    ii) How can you tell this from the diagram? [1]

    iii) Why does it run out of energy? [3]

11 **a)** Describe how you could demonstrate that there is a magnetic field
around a wire that carries a current. [2]

b) Describe how you can use a wire to provide a strong magnetic field. [2]

c) A magnetic effect is produced when there is a current in a wire.

i) What other effect is there? [1]

ii) How does your answer to part **i)** affect transfer of energy from a circuit? [1]

iii) Explain how this transfer of energy from a circuit can be useful. [1]

d) i) The graph of current against potential difference for a wire is like line **A** in the
diagram below.

Explain why it is like line **A** and not line **B**. [1]

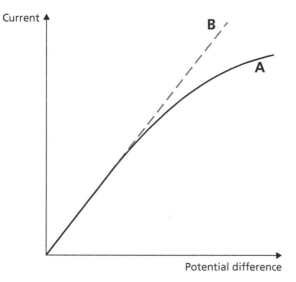

ii) In what way is the graph for a thermistor different to the graph above? [2]

e) Nowadays, LEDs are usually used for lighting rather than wires in light bulbs.
LEDs are much more efficient.

What does 'efficient' mean in this case? [1]

Total Marks .... / 118

ANSWERS

1. **Any three from:** nucleus [1]; cell membrane [1]; cytoplasm [1]; mitochondria [1]; ribosomes [1]
2. For rigidity / support [1]
3. They are the site of photosynthesis [1]
4. A biological catalyst [1]
5. **Any one from:** amylase [1]; carbohydrase [1]
6. They are denatured / lose their shape [1]
7. Genes [1]
8. A [1]; B [1]; D [1]
9. In the nucleus [1]
10. Double helix [1]
11. To release energy [1]
12. oxygen + glucose → [1]; carbon dioxide (+ energy) [1]
13. water + carbon dioxide $\xrightarrow[\text{chlorophyll}]{\text{light}}$ [1]; oxygen + glucose [1]
14. Starch [1]
15. Cellulose [1]; proteins [1] (Accept any other named molecule, e.g. lignin)
16. The net movement of molecules from an area of high concentration to an area of low concentration [1]

> **Net** is a really important word. Molecules will be moving in both directions so it is the overall (the **net**) direction which is important.

17. Between the alveoli [1]; and the blood [1]
18. **Any three from:** large surface area [1]; moist [1]; cells have very thin walls [1]; good supply of blood to carry oxygen away once it has diffused into blood [1]
19. cell, tissue, organ, organ system [2] (1 mark for two words in the correct position)

> The arrow means 'eaten by' and shows the flow of energy through the chain. Do **not** get it round the wrong way.

20. leaf → caterpillar → bird → fox [2] (1 mark if organisms are in the correct order; 1 mark if arrows are pointing in the correct direction)
21. Direction of flow of energy [1] (Accept 'eaten by')
22. Oxygen [1]
23. Carbon dioxide [1]
24. Carbon dioxide [1]
25. No longer exists [1]
26. A diet that contains foods from each food group [1]; in the correct proportions for that individual [1]; includes carbohydrates, fats, proteins, vitamins, minerals, fibre and water [1]

> Stating 'in the correct proportion' is important. For instance, the balance of foods in a bodybuilder's diet would be different from an elderly lady's. You need to show that you understand this.

27. Proteins [1]; fats [1]; carbohydrates [1]; minerals [1]; vitamins [1]
28. **Any three recreational drugs, such as:** alcohol [1]; nicotine [1]; cannabis [1]; heroin [1]
29. Liver [1]
30. Haemoglobin [1]
31. a) C [1]
    b) A [1]
    c) B [1]
    d) Alveoli / air sacs [1]
    e) Diffusion [1]
    f) To trap microorganisms [1]

1. Distilled water = compound [1]; Gold = element [1]; Glucose = compound [1]; Salt water = mixture [1]
2. a) B [1]; C [1]
   b) magnesium + oxygen → magnesium oxide [2] (1 mark for correct reactants; 1 mark for correct product)
3. A correctly drawn diagram of the particles in a solid [1]; liquid [1]; and gas [1]

4. D [1]
5. a) An alkali / base [1]
   b) B [1]
6. a) The elements are very unreactive [1]; and will not react with oxygen or other elements [1]
   b) **Any four from:** Aluminium is very reactive [1]; aluminium is higher than carbon in the reactivity series [1]; aluminium can only be extracted using electricity / electrolysis [1]; electrictiy was not readily available until the beginning of the 19th century [1]; gold and platinum would be found in elemental / unreacted form [1]
7. a) Red [1]
   b) Yellow [1]
   c) Green [1]
   d) Blue [1]
8. a) D [1]
   b) C [1]
   c) B [1]
9. Solid water / ice is less dense than liquid water [1]; this means that under the ice there is liquid water [1]

1. A [1]
2. C [1]
3. B [1]
4. C [1]
5. B [1]
6. D [1]
7. A [1]
8. B [1]
9. B [1]

10. D [1]
11. a) i) 1000kg [1]
       ii) 2400kg [1]
    b) mass = density × volume [1];
       = 1000 × 0.01 [1];
       = 10kg [1]
    c) 10kg [1]
    d) The Moon has weaker gravity / lower gravitational field strength [1]; because it is smaller / less massive (than Earth) [1]
    e) i) reduces it [1]
       ii) reduces it [1]
       iii) no change [1]
12. A2 [1]; B4 [1]; C5 [1]; D8 [1]; E7 [1]; F3 [1]; G1 [1]; H6 [1]

Page 17 Quick Test
1. ×40
2. a) Mitochondria
   b) Chloroplasts
   c) Cell membrane
   d) Cytoplasm
3. **Any two from:** cell wall; vacuole; chloroplasts

Page 19 Quick Test
1. They increase the rate of reaction
2. **Any three from:** temperature; pH; substrate concentration; enzyme concentration
3. Enzymes have an active site in which only a specific substrate can fit

Page 21 Quick Test
1. oxygen + glucose → carbon dioxide + water (+ energy)

> When writing equations, always make sure the reactants are to the left of the arrow and the products are to the right.

2. Mitochondria
3. (Energy), alcohol and carbon dioxide
4. a) Protease
   b) Lipase
   c) Carbohydrase

Page 23 Quick Test
1. a) Water and carbon dioxide
   b) Oxygen and glucose
2. Chloroplasts
3. Enzymes are denatured

Page 25 Quick Test
1. In diffusion, substances move from a high to a low concentration (with a concentration gradient). In active transport, substances move from a low concentration to a high concentration against the concentration gradient. Active transport requires energy.
2. Glucose; oxygen (accept amino acids and carbon dioxide)
3. To replace dead and damaged cells to allow growth and repair
4. a) From human embryos

b) From adult bone marrow

Page 27 Quick Test
1. Large surface area to volume ratio; thin membranes; a good supply of transport medium
2. They are more complex and have a smaller surface area to volume ratio
3. It achieves a higher blood pressure and greater flow of blood to the tissues

Page 29 Quick Test
1. They have a small lumen and their walls are thick and muscular with elastic fibres
2. Substances are exchanged between blood and the tissues
3. Biconcave, which increases the surface area for absorbing oxygen; contains haemoglobin, which binds to oxygen; it does not have a nucleus – therefore, more space for oxygen

Page 31 Quick Test
1. Nitrates; phosphates; potassium
2. **a)** Water (and minerals)
   b) Glucose
3. Wind velocity; temperature; humidity

Pages 32–37 Practice Questions

Page 32 Cell Structures
1. Plant: all boxes ticked **[1]**; Animal: crosses against cell wall and chloroplasts, all other boxes ticked **[1]**; Bacteria: crosses against nuclear membrane and chloroplasts, all other boxes ticked **[1]**
2. **a) i)** Black / blue **[1]**
      ii) Pink **[1]**
   b) 10 × 40 = 400 **[1]**
3. Cell B **[1]**

Page 33 What Happens in Cells
1. **a)** Double helix **[1]**
   b) In the nucleus **[1]**
2. **a)**

Substrate **[1]**

Enzyme **[1]** Active site **[1]**

   b) Denatured / lose its shape **[1]**; will not work **[1]**
   c) Any one from: pH **[1]**; substrate concentration **[1]**; enzyme concentration **[1]**

Page 33 Respiration
1.

	Aerobic	Anaerobic
Where it Occurs	mitochondria **[1]**	cytoplasm **[1]**
Energy Release	high **[1]**	low **[1]**
Breakdown of Glucose	complete **[1]**	incomplete **[1]**

2. glucose **[1]** + oxygen **[1]** → water **[1]** + carbon dioxide **[1]** (+ energy)

Page 34 Photosynthesis
1. carbon dioxide **[1]** + water **[1]** —light energy→ glucose **[1]** + oxygen **[1]**
2. **a)** Palisade cell **[1]**
   b) Chloroplasts **[1]**
   c) Chlorophyll **[1]**

Page 34 Supplying the Cell
1. **a) Any two from:** glucose **[1]**; oxygen **[1]**; water **[1]**
   b) Diffusion **[1]**
   c) Unicellular **[1]**
2. net **[1]**; high **[1]**; low **[1]**; against **[1]**; energy **[1]**
3. C **[1]**
4. Growth **[1]**; repair **[1]**
5. Bone marrow **[1]**
6. Embryos are 'killed' / embryos are disposed of (Accept any other sensible answer) **[1]**

Page 36 The Challenges of Size
1. **a)** The lungs **[1]**
   b) Red blood cell **[1]**
   c) Alveolus **[1]**
   d) Carbon dioxide **[1]**
   e) Oxygen **[1]**

Page 36 The Heart and Blood Cells
1. **a)** A = pulmonary artery **[1]**; B = aorta **[1]**; C = atrium **[1]**; D = ventricle **[1]**
   b) Atrioventricular valve **[1]**
   c) To stop blood flowing backwards from ventricle to atria **[1]**
   d) The lungs **[1]**
2. Capillary – smallest blood vessels **[1]**; and thin permeable walls **[1]**
   Artery – thick muscular walls **[1]**; and small lumen **[1]**
   Vein – large lumen **[1]**; and has valves **[1]**
3. **Any three from:** carbon dioxide **[1]**; glucose **[1]**; nutrients **[1]**; hormones **[1]**; antibodies **[1]**; water **[1]**; minerals **[1]**; vitamins **[1]**

Page 37 Plants, Water and Minerals
1. **a)** active transport **[1]**
   b) osmosis **[1]**
   c) diffusion **[1]**
   d) active transport **[1]**
2. To open / close stomata **[1]**; to control water loss in plants **[1]**
3. **a)** Increase **[1]**
   b) Increase **[1]**
   c) Decrease **[1]**

Pages 38–49 Revise Questions

Page 39 Quick Test
1. Central and peripheral
2. **a)** Carries impulses from sense organs to the CNS
   b) Carries impulses from the CNS to an effector
   c) Carries impulses from the sensory neurone to the motor neurone

Always talk about nerve **impulses**, not messages.

3. **a)** Light
   b) Sound
4. Impulse arrives and triggers the release of neurotransmitters into the gap. Neurotransmitters cross the gap and bind to receptors on the next neurone, which triggers a new electrical impulse. The message goes from electrical to chemical and back to electrical.

Remember, at the synapse the impulse changes from electrical to chemical then **back** to electrical.

Page 41 Quick Test
1. **a)** Testosterone
   b) Thyroxine
2. Hormonal messages give a slower response; they target a more general area; their effects are longer lasting
3. FSH encourages ovaries to produce oestrogen

Page 43 Quick Test
1. Oestrogen; progesterone
2. They are very effective; they can reduce the risk of getting some types of cancer
3. They do not protect against sexually transmitted diseases; they can have side effects
4. FSH can be given to women to stimulate the release of eggs, either for natural fertilisation or IVF
5. Social: involves people
   Economic: involves cost
   Ethical: involves morals

When talking about **auxins** always mention that they are found in / the response is in the **tip** of the root or shoot.

Page 45 Quick Test
1. Enzymes have an optimum temperature and their action slows / stops if the temperature is above or below the optimum temperature

Enzymes are not denatured by low temperatures, they just have a slower rate of action.

2. Type 2
3. It causes glucose to be converted to glycogen for storage and increases the permeability of cell membranes to glucose

Make sure you understand the difference between the terms glucose, glycogen and glucagon. Do not get them mixed up.

Page 47 Quick Test
1. Water is needed to stay alive; the recycling of water influences climate

and maintains habitats; recycling water brings fresh supplies of nutrients to habitats

2. Nitrogen-fixing; nitrifying; denitrifying; putrefying

3. Photosynthesis; dissolved in seas

Page 49 Quick Test

1. A community is all the organisms (plants and animals) within an ecosystem; a population is all the individuals of the same species within a community

2. **Any two biotic factors:** food; predators; pollinators; disease; human activity, e.g. deforestation

   Any two abiotic factors: temperature; light; moisture or pH of soil; salinity of water

3. **Any three from:** food; space; water; mates

Pages 50–53 Review Questions

Page 50 Cell Structures

1. Four correctly drawn lines **[3]** (2 marks for two lines; 1 mark for one)
   chloroplast – contains chlorophyll
   cell wall – gives support
   cell membrane – controls movement of substances in and out of the cell
   mitochondria – contains the enzymes for respiration

Page 50 What Happens in Cells

1. a) The temperature of the apple–pectinase solution **[1]**

   b) The volume of juice produced **[1]**

   > The independent variable is usually the one you change / control. The dependent variable changes as the independent variable is changed.

   c) **Any two from:** size of pieces of apple **[1]**; mass of apple **[1]**; volume and strength of pectinase added **[1]**

   > You will not get a mark for just saying 'apple'. You need to describe what you are changing about the apple and you also need to state what you are changing about the pectinase solution.

   d) A graph with a correctly labelled *x*-axis **[1]**; correctly labelled *y*-axis **[1]**; and all points plotted accurately **[1]**

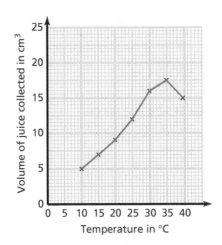

   e) The optimum temperature for the enzyme was 35°C **[1]**

   f) **Any one of:** Smaller intervals between temperatures **[1]**; use of two or more different fruits **[1]** **(Accept any other sensible answer)**

Page 51 Respiration

1. a) Lactic acid **[1]**

   b) His muscles are demanding more energy so respiration must go faster **[1]**; so his breathing rate increases to provide more oxygen for respiration **[1]**; if the demand is great he may go into anaerobic respiration **[1]**; in which case the heavy breathing will be to pay back the oxygen debt **[1]**

   c) i) Aerobic **[1]**; mitochondria **[1]**
      ii) oxygen + glucose **[1]**; → carbon dioxide + water (+ energy) **[1]**

2. a) Fungus **[1]**

   b) Glucose **[1]**

   c) Carbon dioxide gas is being produced **[1]**

Page 51 Photosynthesis

1. a) A = beaker **[1]**; B = funnel **[1]**

   b) Use water of different temperatures **[1]**; count the number of bubbles in a given time **[1]**

   c) **Any two from:** stopwatch **[1]**; kettle / Bunsen burner **[1]**; thermometer **[1]**

   d) There is another limiting factor / the amount of carbon dioxide / the light is a limiting factor **[1]**

Page 52 Supplying the Cell

1. They have a large surface area **[1]**

2. a) Glucose **[1]**

   b) Red blood cell **[1]**

   c) A capillary **[1]**

3. Stem cells are found in meristematic tissues in plants **[1]**; Stem cells are found in the tips of shoots and roots **[1]**

Page 52 The Challenges of Size

1. a) Diffusion **[1]**

   b) A specialised transport system is needed to transport materials over long distances in the monkey **[1]**;

Paramecium has larger surface area / volume ratio **[1]**; so can rely on diffusion **[1]**

> When asked about transport of materials in large and small organisms, always refer to surface area to volume ratio.

Page 53 The Heart and Blood Cells

1. a) A **[1]**

   b) A **[1]**

   c) B **[1]**

   d) B **[1]**

   e) A correctly drawn arrow labelled L **[1]**

Page 53 Plants, Water and Minerals

1. Transpiration **[1]**

2. a) Xylem **[1]**

   b) Water / minerals **[1]**

   c) Made of dead cells with end walls removed **[1]**

3. Active transport **[1]**

Pages 54–57 Practice Questions

Page 54 Coordination and Control

1. a) Light **[1]**

   b) Ears **[1]**

   c) Smell **[1]**

   d) Tongue **[1]**

   e) and f) **Any two from:** touch **[1]**; pressure **[1]**; pain **[1]**; temperature change **[1]**

   > The skin cannot sense temperature, it can only sense a **temperature change**.

2. electrical **[1]**; synapse **[1]**; neurotransmitters **[1]**; receptors **[1]**

3. **A** = Receptor **[1]**
   B = Relay neurone **[1]**
   C = Response **[1]**

Page 55 The Endocrine System

1. Four correctly drawn lines **[3]** (2 marks for two correct lines; 1 mark for one)
   Pancreas – Insulin
   Thyroid – Thyroxine
   Testes – Testosterone
   Pituitary – Anti-diuretic Hormone

2. thyroid **[1]**; metabolism **[1]**; pituitary **[1]**; thyroid stimulating hormone **[1]**; negative **[1]**

Page 55 Hormones and Their Uses

1. a) Less effective than the combined pill **[1]**

   b) Causes a thick sticky mucus which stops sperm reaching egg **[1]**

   c) The ovaries **[1]**

   d) Do not protect against STDs / can have side effects **[1]**

Page 56 Maintaining Internal Environments
1. a) Diffusion [1]
   b) Respiration [1]
2. A [1]; D [1]
3. Sandwich C [1]; because the glucose level does not go as high as for the others [1]; and the rise in the glucose level is slower [1]; therefore, the person will need to take less insulin / there is less chance of this person going hyperglycaemic [1]
4. B [1]

Page 57 Recycling
1. a) i) E [1]
      ii) D [1]
      iii) G [1]
      iv) A [1]
      v) F [1]
   b) Death / excretion [1]
   c) **Any one from:** combustion / burning [1]; evaporation of seas [1]

Page 57 Interdependence
1. Biotic: animals [1]; bacteria [1]; trees [1]; detritivores [1]
   Abiotic: rivers [1]; soil [1]; sea [1]
2. a) Hen harriers and owls [1]
   b) Prey–predator [1]
   c) Numbers will decrease [1]; because stoats are eating their food supply [1]
   d) **Any one from:** nesting space [1]; water [1]

Pages 58–75 Revise Questions

Page 59 Quick Test
1. An organism's complete set of DNA, including all of its genes
2. A capital letter shows a dominant allele; a lower case letter shows a recessive allele
3. **Any two from:** height; weight; skin colour; intelligence (Accept any other sensible answer)
4. a) **Any one from:** height; weight (Accept any other sensible answer)
   b) **Any one from:** blood group; right or left handed; shoe size (Accept any other sensible answer)

Page 61 Quick Test
1. Haploid cells have half the number of chromosomes found in a diploid cell, as they have just one copy of each chromosome, e.g. sperm or egg cell; diploid cells have two copies of each chromosome, e.g. normal body cell.
2. Meiosis produces gametes with half the number of chromosomes as body cells; during fertilisation, two gametes from different individuals fuse together; this produces a new combination of genes creating variation
3. Eggs will always carry an X chromosome; sperm cells may carry an X or a Y chromosome; there is a 50% chance that an egg will fuse with a sperm carrying an X chromosome = XX

(female); and there is a 50% chance that an egg will fuse with a sperm carrying a Y chromosome = XY (male)

Page 63 Quick Test
1. There is natural variation within any population; organisms with characteristics best suited to the environment are likely to survive, breed and pass on their successful genes to the next generation; animals with poor characteristics less well suited to the environment are less likely to survive and breed

> Don't forget 'breed and pass on genes'. It is an easy way to get two marks.

2. A bacterium mutates to become resistant to the antibiotic that is being used; the antibiotic kills all the sensitive bacteria; the resistant bacteria multiply creating a population of antibiotic-resistant bacteria
3. Organisms that are found to have very similar DNA will share common ancestors

Page 65 Quick Test
1. Increased biodiversity offers greater opportunity for medical discoveries; it boosts the economy; it ensures sustainability (ecosystems more likely to recover after a 'natural disaster')
2. a) **Any two from:** hunting; overfishing; deforestation; farming single crops
   b) **Any two from:** creating nature reserves; reforestation; sustainable fishing
3. Tourism that aims to reduce the negative impact of tourists on the environment

Page 67 Quick Test
1. Measurements should be repeated
2. A result that does not fit the pattern of the rest of the results
3. Errors that are made every time that may be due to faulty equipment
4. It checks the design of the experiment and the validity of the data

Page 69 Quick Test
1. The genes may get into wild flowers or crops; they may harm the consumers; we don't know the long-term effects
2. To cut the required gene from a strand of DNA
3. A loop of DNA

Page 71 Quick Test
1. **Any four from:** bacteria; viruses; fungi; protoctista; parasites
2. **Any three from:** air; water; food; contact; animals
3. **Any three from:** the skin; platelets; mucous membranes; stomach acid; eyelashes / tears

Page 73 Quick Test
1. White blood cells recognise the antigens on the dead or weakened pathogen that is in the vaccination. They make antibodies but also form memory cells. In a subsequent infection, memory cells produce antibodies quickly and in large numbers.

> Antibodies are produced **rapidly** and in **large numbers** in subsequent infections. These two words may gain you two marks.

2. Antibiotics act on bacteria; antivirals act on viruses; antiseptics are used on skin and surfaces to kill microorganisms
3. Measles, mumps and rubella

Page 75 Quick Test
1. Too much carbohydrate / fat can lead to obesity, which leads to high blood pressure and cardiovascular disease; a high fat diet can lead to increased levels of cholesterol, which coats the smooth lining of the arteries, causing them to narrow, restricting blood flow and leading to heart attacks or strokes and high blood pressure; a diet high in salt can lead to high blood pressure.

> Remember, cholesterol coats **arteries** not veins.

2. Cancer is when cells grow and divide uncontrollably
3. Change in lifestyle; medication; surgery
4. Stem cells may develop into tumours / may be rejected by the patient

Pages 76–79 Review Questions

Page 76 Coordination and Control
1. a) Eyes [1]
   b) Voluntary response [1]
   c) Muscles in feet / legs [1]
2. a) Motor neurone [1]
   b) Arrow drawn pointing from left to right, i.e. ⟶ [1]

Page 76 The Endocrine System
1. a) Accept any answer between: 35–45 [1]
   b) Accept any answer between: 48–53% [1]
   c) i) Follicle–stimulating hormone [1]; and luteinising hormone (LH) [1]
      ii) The likelihood of conceiving would still be very low [1]

Page 77 Hormones and Their Uses
1. progesterone [1]; oestrogen [1]; more [1]; side effects [1]

Page 77 Maintaining Internal Environments
1. a) Active transport [1]
   b) Glucose enters cells [1]; Glucose is converted to glycogen [1]

c) i) Glucose is required for respiration [1]; more energy is required in exercise therefore more respiration [1]

> Energy is **released** in respiration; it is not used or made.

ii) The pancreas [1]; produces glucagon [1]; converts glycogen to glucose [1]; blood glucose levels are restored [1]

2. a) Glands [1]
   b) In the blood [1]

Page 78 Recycling
1. a) Nitrogen-fixing [1]
   b) They turn nitrogen in the air into nitrates [1]; which plants need for growth [1]
2. a) The soil will be rich in nitrates [1]
   b) Bacteria [1]; fungi [1]
   c) They break down matter into smaller pieces [1]; which increases the surface area [1]
3. a) Denitrifying bacteria [1]
   b) Nitrogen-fixing [1]

Page 79 Interdependence
1. **Any two from:** amount of space available [1]; amount of food / prey [1]; amount of water [1]; numbers of predators [1]
2. a) Seaweed [1]
   b) Mussel / crab / gull [1]
   c) natural [1]; community [1]; population [1]
   d) i) Biotic: accessibility of the beach to humans [1]; the amount of fishing in the area [1]
      Abiotic: the aspect of the beach [1]; temperature [1]; slope of beach [1]
      ii) The oil has killed the mussels, which feed on the seaweed [1]

Page 80 Genes
1. cell, nucleus, chromosome, gene [1] (1 mark for three in the correct place)
2. a) Dominant [1]
   b) Heterozygous [1]
   c) ee [1]

Page 80 Genetics and Reproduction
1. 23 + [1]; 23 [1]; = 46 [1]
2. 4 [1]
3. a) 1 = FF [1]; 2 = Ff [1]; 3 = Ff [1]; 4 = ff [1]
   b) 2 [1]; 3 [1]
   c) 4 [1]

Page 81 Natural Selection and Evolution
1. A bacterium mutates to become resistant [1]; sensitive bacteria are killed by the antibiotic [1]; resistant bacteria multiply rapidly to form large population [1]
2. a) C and D [1]
   b) E and F [1]
   c) C and G [1]

Page 82 Monitoring and Maintaining the Environment
1. Four correctly drawn lines [3] (2 marks for two correct lines; 1 mark for one)
   The variety of water invertebrates in the canal – Pond net
   The variety of plants growing by the side of the canal – Quadrat
   The variety of flying insects in the long grass by the side of the canal – Sweep net
   The variety of invertebrates found under the hedges along the side of the canal – Pitfall trap
2. $\frac{12 \times 10}{4}$ [1]; = 30 [1]

Page 82 Investigations
1. **Answer must have:**
   Suitable number / range of temperatures (minimum of 3 temperatures at between 20 and 50°C) [1]
   Suitable number of peas used at each temperature (minimum 5) [1]
   Suitable time period to leave peas (2–3 days) [1]
   Simple method (must include adding water) [1]
   At least **two** other controlled variables (amount of water / size of peas / type of peas / amount of light / same growth medium, e.g. soil or cotton wool) [1]

Page 83 Feeding the Human Race
1. a) C, B, A [1]
   b) **A** = ligase enzyme [1]; **B** = restriction enzyme [1]
2. B, E, A, C, D [1]
3. a) Loop / ring of DNA [1]
   b) Plasmids can be inserted into bacteria and will replicate each time the bacteria divides [1]

Page 83 Monitoring and Maintaining Health
1. Four correctly drawn lines [3] (2 marks for two correct lines; 1 mark for one)
   Malaria – Animal vector
   Cholera – By water
   Tuberculosis – By air
   Athlete's foot – Contact
2. A [1]; C [1]

> Do not confuse AIDS and HIV. HIV is the name of the virus. AIDS is the disease it causes.

3. a) A = microorganism, B = antibody, C = antigen [2] (1 mark for one correct)
   b) lymphocyte [1]

Page 84 Prevention and Treatment of Disease
1. a) Bacteria [1]
   b) Because there will still be microorganisms present even when he is feeling better [1]; without antibiotics these could start to multiply again, or even mutate into resistant strains [1]
2. D, A, B, C [3] (2 marks for two in correct places; 1 mark for one)
3. To test for efficacy / efficiency [1]; toxicity / safety [1]; dosage [1]
4. a) Bacteria / microorganisms [1]
   b) i) Antiseptic [1]
      ii) **Any one from:** gloves [1]; masks [1]; gowns [1]; all equipment sterilised [1]

Page 85 Non-Communicable Diseases
1. a) Statins [1]
   b) A stent [1]
2. High fat diet can lead to high cholesterol [1]; which narrows arteries [1]; causing high blood pressure, which puts a strain on heart [1] **OR** high salt intake [1]; causes high blood pressure [1]; which puts a strain on heart [1] **OR** eating more kJ energy / calories than used [1] leads to obesity [1]; which causes cardiovascular disease [1]
3. **Any four from:** drink less alcohol [1]; eat less [1]; lower salt intake [1]; switch to low fat products [1]; exercise more [1]
4. **Any two from:** Type 2 diabetes [1]; joint problems [1]; heart disease [1]; high blood pressure [1]; high cholesterol [1]
5. C [1]

Page 87 Quick Test
1. Different forms of an element with the same number of protons but a different number of neutrons
2. 8 (14 – 6 = 8)
3. New experimental evidence has led to changes

Page 89 Quick Test
1. 58.3 g
2. Mixtures of substances in solution
3. CH_3

Page 91 Quick Test
1. By atomic number; by number of electrons in the outermost shell
2. Magnesium, 2.8.2

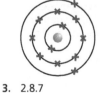

3. 2.8.7

Page 93 Quick Test

1. The electrons in the outermost shells of the metal atoms are free to move, so there are a large number of electrons moving between the metal ions.
2. **Any two from:** Distances between electrons and the nucleus are not realistic; bonds appear to be physical structures; bond lengths are not in proportion to the size of the atom; they do not give a good idea of the 3D shape of the atoms.
3. A 3D arrangement of a large number of repeating units (molecules / atoms) joined together by covalent bonds.

Page 95 Quick Test

1. A compound that contains carbon
2. It is a giant covalent molecule in which every carbon atom forms bonds with four other carbon atoms (the maximum number of bonds possible).
3. Allotropes are different physical forms of the same element, for example, graphite, graphene and diamond are all allotropes of carbon (they only contain carbon atoms but have different structures and, therefore, different properties)

Pages 96–101 Review Questions

Page 96 Genes

1. a) Combination [1]
   b) Genetics [1]
   c) Environment [1]
   d) Combination [1]
   e) Environment [1]
2. a) i) No sickle cell anaemia [1]
      ii) No sickle cell anaemia but a carrier [1]
      iii) Sickle cell anaemia sufferer [1]
   b) Aa [1]
3. a) Continuous: height [1]; weight [1] Discontinuous: eye colour [1]; shoe size [1]; freckles [1]
   b) i) Bar correctly plotted to show 7 people [1]
      ii) **Accept:** 0, 1 or 2 [1]

Page 97 Genetics and Reproduction

1. Meiosis [1]
2. a) Correct gametes [1]; correct offspring [1]

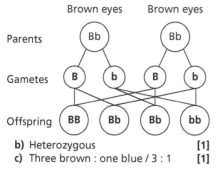

   b) Heterozygous [1]
   c) Three brown : one blue / 3 : 1 [1]

> Ratios can be expressed in a number of ways: 3:1, 75%:25%, 0.75:0.25

   d) 100% [1]

Page 97 Natural Selection and Evolution

1. a) Genus [1]
   b) New DNA evidence shows it to be very closely related to grey wolf [1]
   c) New fossils discovered [1]
   d) A group of similar animals that can breed to produce fertile offspring [1]

Page 98 Monitoring and Maintaining the Environment

1. a) Advantage: can be applied to soil and will be taken in by plants [1] Disadvantage: can be washed away into water courses, rivers, etc. and can get into animals that drink the water [1]
   b) They do not affect the nervous system of mammals [1]
   c) The birds are consuming the pesticide when they drink water and this is harming them [1]; the birds are starving because the pesticide is killing their food source [1]
   d) Decrease = 12000 − 11600 = 400 [1]; % decrease = $\frac{400}{12000}$ × 100 = 3.33% [1]

Page 99 Investigations

1. a) Use the tape measure to mark a transect line from the river outwards across the field [1]; at suitable intervals, place a quadrat [1]; count the number of meadow buttercups in the quadrant [1]; measure the moisture content of the soil using the moisture meter [1]; repeat the transect in different areas / repeat and calculate and average [1]
   b) i) Mean = 39 ÷ 10 = 3.9 [1]; mode = 4 [1]; median = 4 [1]
      ii) 1 [1]; **plus, any two from:** chance [1]; error made in counting [1]; net swept at different level [1]; net swept in different direction / angle [1]

Page 99 Feeding the Human Race

1. a) Plants which have had their DNA changed, usually by addition of useful gene from another organism [1]
   b) 2002 and 2010 [1]
   c) $\frac{180}{30}$ × 100 [1]; = 600% [1]
   d) **Any two from:** insecticide resistance [1]; drought resistance [1]; able to grow in salty water [1]; increased nutritional content; increased shelf life [1]
2. Restriction enzyme [1]; to cut the gene from the maize plant DNA [1]; Ligase

---

enzyme [1]; to paste the gene into rice plant DNA (or accept plasmid) [1]

Page 100 Monitoring and Maintaining Health

1. a) Athlete's foot [1]
   b) Flu [1]
   c) Tuberculosis [1]
2. a) Phagocytosis [1]

> Remember, the word **phagocytosis** means to devour the cell.

   b) A phagocyte [1]
3. a) Form a clot [1]; which seals the wound [1]
   b) Trap microorganisms and dirt [1]

Page 101 Prevention and Treatment of Disease

1. a) Dead / weakened measles virus [1]
   b) Tanya already had antibodies against the measles virus [1]; she also has memory cells [1]; which can quickly produce large numbers of antibodies [1]
   c) In the air / moisture droplets [1]
   d) Measles is caused by a virus [1]; antibiotics cannot be used for viral infections / can only be used for bacterial infections [1]
2. **Any two from:** tested on cells [1]; animals [1]; computer models [1]; healthy volunteers [1]
3. To compare the results of people taking the drug to those who have not taken the drug [1]

Page 101 Non-Communicable Diseases

1. a) Correctly labelled axes [1]; correctly plotted points [1]; joined by a smooth curve [1]

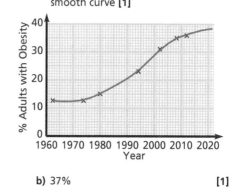

   b) 37% [1]
   c) **Any two from:** heart disease [1]; type 2 diabetes [1]; high blood pressure [1]; liver disease [1]

Answers

Page 102 Particle Model and Atomic Structure

1. a) D [1]
    b) The temperature is increasing [1]; so the particles in the solid ice start vibrating more quickly [1]
2. There is nothing in the gaps / empty space between the water molecules [1]
3. a) Ne [1]
    b) Ca [1]
    c) 28 + 2 = 30 [1]

> It is usual for mass numbers to be rounded to the nearest whole number.

4. a) A heating curve drawn showing a straight line at 100°C [1]; and a straight line at 0°C [1]; sloping down in direction left to right [1]; with correctly labelled axes [1]

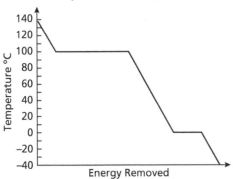

    b) The water molecules slow down [1]; as they lose kinetic energy [1]
5. a) charge = – / negative [1]; relative mass = 0.0005 / zero / negligible [1]
    b) charge = + / positive [1]; relative mass = 1 [1]
    c) charge = 0 / neutral [1]; relative mass = 1 [1]

Page 103 Purity and Separating Mixtures

1. a) Containing one type of atom or molecule only [1]
    b) Every substance has a specific melting point at room temperature and pressure [1]; if the substance melts at a different temperature, it indicates that there are impurities [1]
2. a) Distance moved by the solvent = 28 [1]; R_f = $\dfrac{\text{distance moved by the compound}}{\text{distance moved by the solvent}}$,

    R_f (pink) = $\dfrac{7.5}{28}$ = 0.27 [1];

    R_f (purple) = $\dfrac{17.5}{28}$ = 0.63 [1]
    b) D [1]
3. C_4H_4S [1]
4. C = $\dfrac{84}{12}$ = 7 [1]; H = $\dfrac{16}{1}$ = 16, O = $\dfrac{64}{16}$ = 4 [1];

    $C_7H_{16}O_4$ [1]

> Look for common factors to see whether an empirical formula can be simplified further.

5. A [1]

Page 103 Bonding

1. a) H = 1, Na = 11, Mg = 12, C = 6, O = 8, K = 19, Ca = 20, Al = 13 [2] (1 mark for 6–7 correct; 0 marks for 5 or fewer correct)
    b) (2 × 39.1) + 16 = 94.1 [1] (Accept 94)
    c) 16 × 2 = 32 [1]
2. a) They all have the same number of electrons in their outer shell [1]
    b) They all have the same number of shells / they all have two shells [1]
3. a) C [1]
    b) B = 2.8.1 [1]; E = 2.8.7 [1]
    c) Electrons [1]

Page 104 Models of Bonding

1. a) An atom or molecule that has gained or lost electrons [1]
    b) A correctly drawn sodium ion (2.8) [1]; and chloride ion (2.8.8) [1]

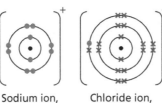

Sodium ion, Na^+ Chloride ion, Cl^-

2. a) A bond formed by the sharing [1]; of two outer electrons [1]
    b) Two correctly drawn chlorine atoms (each with 7 electrons) [1]; overlapping and sharing two electrons [1]

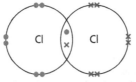

3. a) A large structure that is made up of repeating units (atoms / molecules) [1]; that are covalently bonded together [1]
    b) The atoms in the water / H_2O molecules are covalently bonded [1]; but each water molecule is separate – there is no covalent bond between the water molecules [1]

Page 105 Properties of Materials

1. a) **Any three from:** Carbon has four electrons in its outer shell [1]; it can form covalent bonds [1]; with up to four other atoms [1]; and can form chains [1]
    b) An allotrope is a different form of an element [1]
    c) **Any three from:** graphite [1]; diamond [1]; fullerene / buckminsterfullerene [1]; graphene [1]; lonsdaleite [1]; amorphous carbon [1]

2. a) It conducts electricity because it has free electrons [1]; and it is stronger than steel [1]
    b) Diamond does not conduct electricity [1]
    c) A diagram showing each carbon joined to four other carbon atoms [1]; with a minimum of five atoms shown in tetrahedral arrangement [1]

Diamond

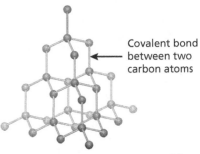

Covalent bond between two carbon atoms

3. a) The ions must be either molten [1]; or dissolved in aqueous solution [1]
    b) In crystalline form, the distance between the ions is at its smallest / the ions are close together [1]; so the electrostatic forces are very high and have to be overcome for the crystal to melt [1]
4. C [1]

Page 107 Quick Test

1. $Ca(OH)_2$
2. $2Na(s) + Cl_2(g) \rightarrow 2NaCl(s)$
3. 2Mg, 2S, 8O (2 × O_4)

Page 109 Quick Test

1. BaO, CuF_2, $AlCl_3$
2. Al_2O_3
3. $Ba^{2+}(aq) + CO_3^{2-}(aq) \rightarrow BaCO_3(s)$

> The $BaCO_3$ formed is insoluble.

Page 111 Quick Test

1. mass = number of moles × relative molecular mass
2. 18g
3. 2.2×10^{-22}g

Page 113 Quick Test

1. The minimum amount of energy needed to start a reaction
2.

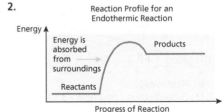

3. Exothermic

Page 115 Quick Test

1. Carbon dioxide
2. Zinc nitrate
3. copper oxide + sulfuric acid →
   copper sulfate + water

Page 117 Quick Test

1. Universal indicator can show a range of pHs from 1 to 14. Litmus paper only shows if something is an acid or alkali.
2. The volume of acid needed to neutralise the alkali; the pH when a certain amount of acid has been added.
3. pH is a measure of the number of H^+ ions in solution.

Page 119 Quick Test

1. In solution (aq) / molten
2. The anode
3. Set up an electrolytic cell using a silver anode. Place the key at the cathode. Add a solution containing silver ions as the electrolyte.

Page 121 Quick Test

1. Iodine
2. The outermost electron in caesium is much further away from the nucleus than in lithium, so the force of attraction is weaker (it can lose this electron more easily)
3. Lithium

Pages 122–125 Review Questions

Page 122 Particle Model and Atomic Structure

1. a) It has evaporated **[1]**
   b) Increase the temperature of the room / switch on a fan **[1]**
   c) i) A diagram showing more than five particles, with at least 50% of the particles touching **[1]**
   ii) A diagram showing more than five particles, with none of the particles touching **[1]**

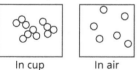

In cup   In air

2. D **[1]**
3.

Subatomic Particle	Relative Mass	Relative Charge
Proton **[1]**	1	+1 (positive) **[1]**
Neutron	1 **[1]**	0 (neutral)
Electron	Negligible **[1]**	–1 (negative) **[1]**

4. a) Protons **[1]**
   b) The positively charged protons are at the centre of the atom **[1]**; so, only alpha particles travelling close to the centre of the atom will be repelled / deflected by them **[1]**

> Alpha particles are positively charged. Like charges repel each other.

5. **He:** 2 (4 – 2) **[1]**; **Na:** 12 (23 – 11) **[1]**; **V:** 28 (51 – 23) **[1]**

Page 123 Purity and Separating Mixtures

1. a) Measure its boiling point **[1]**; compare the boiling point with data from a data book / known values **[1]**
   b) The water is not pure **[1]**; it contains other substances **[1]**
2. a) CH_2O **[1]**

> In $C_6H_{12}O_6$ the common factor of all the numbers is 6, so divide by six to simplify the formula.

   b) CH_2O **[1]**

> Collect all the atoms of the same element together first, and then simplify: $CH_3COOH \rightarrow C_2H_4O_2 \rightarrow CH_2O$

   c) C_2H_4O **[1]**
3. a) (6 × 12) + (12 × 1) + (6 × 16) = 180 **[1]**
   b) (2 × 12) + (4 × 1) + (2 × 16) = 60 **[1]**
   c) (1 × 12) + (2 × 16) = 44 **[1]**
   d) (2 × 1) + (32.1) + (4 × 16) = 98.1 **[1]**
   (Accept 98)
4. Alloy **[1]**
5. C **[1]**

Page 124 Bonding

1. Mendeleev **[1]**
2. a) A **[1]**; E **[1]**
   b) A **[1]**; B **[1]**
   c) B **[1]**
   d) A = 3 **[1]**; B = 11 **[1]**; C = 13 **[1]**; D = 18 **[1]**; E = 4 **[1]**
3. a) **Any three from:** malleable **[1]**; sonorous **[1]**; ductile **[1]**; form a positive cation **[1]**; shiny **[1]**
   b) Metal oxide **[1]**
   c) An ionic compound **[1]**

Page 124 Models of Bonding

1. C **[1]**
2. One carbon atom drawn **[1]**; with covalent bonds with four hydrogen atoms **[1]**

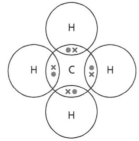

3. Ball and stick models give a better picture of the 3D shape of the molecule **[1]**; and the bond angles / directions **[1]**

4. A correctly drawn magnesium atom, 2.8.2 **[1]**; and magnesium ion (2.8)$^{2+}$ **[1]**

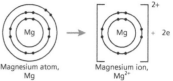

Magnesium atom, Mg   Magnesium ion, Mg^{2+}

Page 125 Properties of Materials

1. a) A different physical structure to other forms of the element **[1]**
   b) Graphite is made of layers of atoms **[1]**; these are held together by weak forces **[1]**; so the layers can separate / slide over each other easily **[1]**; preventing the surfaces from rubbing together **[1]**
   c) In diamond all possible covalent bonds have been used / each carbon atom is bonded to four other carbon atoms **[1]**; so it is extremely hard **[1]**; and has a high melting point **[1]**

> You need to mention high melting point to get the third mark. Drill bits get hot due to frictional forces, so it is an important property.

   d) **Any one from:** electrical components **[1]**; solar panels **[1]**
2. Four correctly drawn lines **[3]** (2 marks for two correct lines; 1 mark for one) sodium chloride – conducts electricity in solution, graphite – conducts electricity in solid, hydrogen – low melting point, fullerene – superconductor

Pages 126–131 Practice Questions

Page 126 Introducing Chemical Reactions

1. In a chemical reaction, the mass of the reactants will always equal the mass of the products **[1]** (Accept: No atoms are made or destroyed)
2. a) The number of atoms of that element present (the element before the number) **[1]**
   b) i) C = 6, H = 12, O = 6 **[1]**
   ii) C = 3, H = 6, O = 2 **[1]**
   iii) H = 2, O = 2 **[1]**
   iv) Ca = 1, N = 2, O = 6 **[1]**
3. solid = (s), liquid = (l), gas = (g), aqueous = (aq) **[1]**
4. a) $2Mg(s) + O_2(g) \rightarrow 2MgO(s)$ **[2]**
   (1 mark for correct balancing; 1 mark for correct state symbols)
   b) $4Li(s) + O_2(g) \rightarrow 2Li_2O(s)$ **[2]**
   (1 mark for correct balancing; 1 mark for correct state symbols)
   c) $CaCO_3(s) + 2HCl(aq) \rightarrow CaCl_2(aq) + CO_2(g) + H_2O(l)$ **[2]** (1 mark for correct balancing; 1 mark for correct state symbols)
   d) $4Al(s) + 3O_2(g) \rightarrow 2Al_2O_3(s)$ **[2]** (1 mark for correct balancing; 1 mark for correct state symbols)

Answers

Page 126 Chemical Equations

1. a) 2+ **[1]**
   b) 2– **[1]**
   c) 3+ **[1]**
   d) 2– **[1]**
2. a) $2H^+(aq) + 2e^- \rightarrow H_2(g)$ **[1]**
   b) $Fe^{2+}(aq) + 2e^- \rightarrow Fe(s)$ **[1]**
   c) $Cu^{2+}(aq) + 2e^- \rightarrow Cu(s)$ **[1]**
   d) $Zn(s) \rightarrow Zn^{2+}(aq) + 2e^-$ **[1]**
3. $Ag^+(aq) + Cl^-(aq) \rightarrow AgCl(s)$ **[2]** (1 mark for the correct ions and product; 1 mark for the correct charges)

Page 127 Moles and Mass

1. One mole of a substance contains the same number of particles as the number of atoms in 12g of the element carbon-12 **[1]**
2. B **[1]**
3. g/mol **[1]**
4. a) number of moles of Mo =
   $$\frac{mass}{relative\ molecular\ mass} =$$
   $$\frac{287.7g}{95.9g/mol}\ \textbf{[1]};\ = 3mol\ \textbf{[1]}$$
   b) mass = relative molecular mass × number of moles = 50.9g/mol × 5 mol **[1]**; = 254.5g **[1]** (Accept 255g)

> Rearrange the equation
> $$moles = \frac{mass}{relative\ molecular\ mass}$$
> to work out the mass.

5. $(6 \times 12) + (12 \times 1) + (6 \times 16) = 180$ **[1]**
6. a) mass of one atom (V) =
   $$\frac{atomic\ mass}{Avogadro's\ constant} = \frac{50.9g}{6.022 \times 10^{23}}$$
   [1]; $= 8.5 \times 10^{-23}$g **[1]**
   b) mass of one atom (Mo) =
   $$\frac{95.9g}{6.022 \times 10^{23}}\ \textbf{[1]};\ = 1.6 \times 10^{-22}g\ \textbf{[1]}$$
   c) mass of one atom (Cs) =
   $$\frac{132.9g}{6.022 \times 10^{23}}\ \textbf{[1]};\ = 2.2 \times 10^{-22}g\ \textbf{[1]}$$
   d) mass of one atom (Bi) =
   $$\frac{209g}{6.022 \times 10^{23}}\ \textbf{[1]};\ = 3.5 \times 10^{-22}g\ \textbf{[1]}$$
7. $2H_2(g) + O_2(g) \rightarrow 2H_2O(l)$,
   1mol of H_2 → 1mol of H_2O, so
   5mol of H_2 → 5mol of H_2O **[1]**;
   1mol of H_2O = $(2 \times 1) + 16 = 18$g, so
   5mol of H_2O = $5 \times 18 = 90$g **[1]**

Page 128 Energetics

1. a) Exothermic **[1]**
   b) Endothermic **[1]**
2. **Any two from:** heating (water / central heating) **[1]**; produce electricity **[1]**; make sound **[1]**; make light **[1]**
3. The minimum energy required to start a reaction **[1]**
4. A correctly drawn reaction profile for an exothermic reaction **[1]**

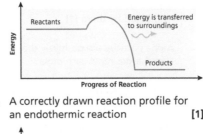

5. A correctly drawn reaction profile for an endothermic reaction **[1]**

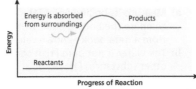

6. making chemical bonds **[1]**
7. Bond breaking: 432 + 155 = 587kJ/mol **[1]**; bond making: 2 × 565 = 1130kJ/mol **[1]**; bond breaking – bond making (ΔH) = 587 – 1130 = –543kJ/mol **[1]**; the reaction is exothermic **[1]**
8. a) Endothermic **[1]**
   b) Exothermic **[1]**
   c) Endothermic **[1]**
   d) Endothermic **[1]**
9. C **[1]**

Page 129 Types of Chemical Reactions

1. A **[1]**; C **[1]**

> Remember, oxidisation is the addition of oxygen.

2. Oxidation is loss of electrons **[1]**; and reduction is gain of electrons **[1]**
3. a) $2Na(s) + Cl_2(g) \rightarrow 2NaCl(s)$ **[2]** (1 mark for correct balancing; 1 mark for correct state symbols); sodium is oxidised and chlorine is reduced **[1]**
   b) $2Mg(s) + O_2(g) \rightarrow 2MgO(s)$ **[2]** (1 mark for correct balancing; 1 mark for correct state symbols); magnesium is oxidised and oxygen is reduced **[1]**
   c) $2Li(s) + Br_2(g) \rightarrow 2LiBr(s)$ **[2]** (1 mark for correct balancing; 1 mark for correct state symbols); lithium is oxidised and bromine is reduced **[1]**
   d) $CuO(s) + H_2(g) \rightarrow Cu(s) + H_2O(l)$ **[2]** (1 mark for correct balancing; 1 mark for correct state symbols); copper is reduced and hydrogen is oxidised **[1]**
4. a) $H^+(aq)$ **[1]**
   b) $OH^-(aq)$ **[1]**
5. acid + base → salt + water **[1]**
6. a) $H_2SO_4(aq) + 2NaOH(aq) \rightarrow$
   $\qquad Na_2SO_4(aq) + 2H_2O(l)$ **[2]** (1 mark for correct reactants; 1 mark for correct products)
   b) Na^+ **[1]**; SO_4^{2-} **[1]**
   c) $H^+(aq) + OH^-(aq) \rightarrow H_2O(l)$ **[2]** (1 mark for correct ions; 1 mark for correct product)

Page 130 pH, Acids and Neutralisation

1. An acid that does not fully dissociate when dissolved in water **[1]**

2. a) $2mol/dm^3\ H_2SO_4$ **[1]**
   b) $3mol/dm^3\ HNO_3$ **[1]**
3. 1000 times greater ($10 \times 10 \times 10$ or 10^3) **[1]**

Page 130 Electrolysis

1. a) Cations **[1]**
   b) Anions **[1]**
2. Table salt is a solid at room temperature and pressure and electrolysis only works if the ion is in solution or molten **[1]**
3. Set up an electrolytic cell using a nail as the cathode **[1]**; and copper for the anode **[1]**; fill with copper(II) sulfate solution and apply an electric current **[1]**
4. Because they do not react with the products of electrolysis or the electrolyte **[1]**

Page 131 Predicting Chemical Reactions

1. C **[1]**
2. 1 **[1]**
3. fluorine, chlorine, bromine, iodine **[1]**
4. a) A correctly drawn sodium atom (2.8.1) **[1]**; and potassium atom (2.8.8.1) **[1]**

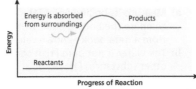

Na   K

   b) Potassium moves around more violently **[1]**; and burns with a lilac flame **[1]**
   c) D **[1]**
5. Beryllium, Be **[1]**

> Group 2 elements are more reactive going down the group.

6. $Zn^{2+}(aq)$, $H^+(aq)$, $SO_4^{2-}(aq)$ **[1]**
7. Collect the gas in a test tube and bring a lit splint to the edge of the tube **[1]**; if it burns with a squeaky pop, hydrogen is present **[1]**

Pages 132–147 Revise Questions

Page 133 Quick Test

1. Rate of reaction increases
2. Increase the pressure
3. A large block of calcium carbonate has a low surface area to volume ratio compared to powder, which consists of lots of small particles with a high surface area to volume ratio.

Page 135 Quick Test

1. The steeper the line / the greater the gradient, the faster the rate of reaction
2. Catalysts reduce the activation energy (making a reaction more likely)
3. Enzymes

Page 137 Quick Test

1. Temperature; pressure; concentration
2. False

3. There will be a decrease in the yield of products and an increase in amount of reactants, restoring the original conditions according to Le Chatelier's principle.

Page 139 Quick Test

1. Heat (in a blast furnace) with carbon
2. The oxygen that forms at the anode reacts with the carbon producing $CO_2(g)$, which then escapes, so the anode gradually wears away.
3. **Any two from:** can be bred to tolerate the metal ion; they reproduce quickly; they are very efficient / only target a specific ion

Page 141 Quick Test

1. Recycling is when the materials in an object are used for a new purpose – the refilled bottles are being reused, not recycled.
2. With some plastics, the materials may be contaminated, so it would be impractical / too expensive to recycle them.
3. To determine the environmental impact of a product throughout its life cycle, from sourcing materials through manufacture, use and disposal

Page 143 Quick Test

1. The breaking down of long-chain hydrocarbons into smaller molecules
2. The longer the chain, the greater the total amount of intermolecular forces that have to be overcome
3. **Any three from:** refinery gases / LPG – bottled gas; petrol – fuel for cars; naphtha – making other chemicals; kerosene / paraffin – aircraft fuel; diesel – fuel for cars / lorries / buses; fuel oil – fuel for power stations / ships; bitumen – tar for road surfaces / roofs (Accept any other sensible use for each fraction)

Page 145 Quick Test

1. Volcanic activity
2. Photosynthesis
3. Methane

Page 147 Quick Test

1. The burning / incomplete combustion of fossil fuels
2. Chlorine kills any bacteria in the water
3. Water that is safe to drink

Pages 148–151 Review Questions

Page 148 Introducing Chemical Reactions

1. a) H = 2, S = 1, O = 4 **[1]**
   b) Cu = 1, N = 2, O = 6 **[1]**
   c) C = 3, H = 6, O = 2 **[1]**
   d) C = 2, H = 6 **[1]**
2. a) $CuO(s) + H_2SO_4(aq) \rightarrow$
   $CuSO_4(aq) + H_2O(l)$ **[2]**
   (1 mark for correct reactants; 1 mark for correct products)
   b) $2Mg(s) + O_2(g) \rightarrow 2MgO(s)$ **[2]**

(1 mark for correct reactants; 1 mark for correct products)
   c) $Mg(OH)_2(aq) + 2HCl(aq) \rightarrow$
   $MgCl_2(aq) + 2H_2O(l)$ **[2]**
   (1 mark for correct reactants; 1 mark for correct products)
   d) $CH_4(g) + 2O_2(g) \rightarrow CO_2(g) + 2H_2O(l)$ **[2]**
   (1 mark for correct reactants; 1 mark for correct products)
3. a) $Pb(s) \rightarrow Pb^{2+}(aq / l) + 2e^-$ **[1]**
   b) $Al^{3+}(aq / l) + 3e^- \rightarrow Al(s)$ **[1]**
   c) $Br_2(l) + 2e^- \rightarrow 2Br^-(aq)$ **[1]**
   d) $Ag^+(aq) + e^- \rightarrow Ag(s)$ **[1]**

Page 148 Chemical Equations

1. The reacting species **[1]**

> Spectator ions are not included in ionic equations.

2. a) $Ag^+(aq) + Cl^-(aq) \rightarrow AgCl(s)$ **[1]**
   b) $Mg^{2+}(aq) + CO_3^{2-}(aq) \rightarrow MgCO_3(s)$ **[1]**

Page 149 Moles and Mass

1. a) $\frac{6.9g}{6.9g/mol} = 1mol$ **[1]**
   b) $\frac{62g}{31g/mol} = 2mol$ **[1]**
2. $(1 \times 14) + (4 \times 1) + (1 \times 35.5) =$
   $53.5g/mol$ **[1]**
3. $BaCl_2 + MgSO_4 \rightarrow BaSO_4 + MgCl_2,$
   5mol of $BaCl_2$ makes 5mol of $BaSO_4$,
   5mol of $BaSO_4 =$
   $5 \times (137.3 + 32 + (4 \times 16))g/mol$ **[1]**; =
   1166.5g **[1]**

> The stoichiometry of the reaction is 1 : 1 ratio reactant to product.

4. B **[1]**

Page 149 Energetics

1. Energy is taken in from the environment / surroundings **[1]**
2. a) bond breaking: 436 + 243 = 679kJ/mol, bond making: 2 × 431 = 862kJ/mol **[1]**; bond breaking – bond making (ΔH) = 679 – 862 = –183kJ/mol **[1]**; the reaction is exothermic **[1]**
   b) A correctly drawn reaction profile for an exothermic reaction **[1]**

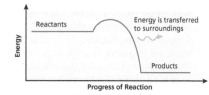

Page 150 Types of Chemical Reactions

1. a) $2AgNO_3(aq) + Cu(s) \rightarrow$
   $Cu(NO_3)_2(aq) + 2Ag(s)$ **[2]**
   (1 mark for correct reactants; 1 mark for correct products)
   b) Silver **[1]**
   c) Copper nitrate **[1]**

Page 150 pH, Acids and Neutralisation

1. a) $HNO_3(aq) + NaOH(aq) \rightarrow$
   $NaNO_3(aq) + H_2O(l)$ **[2]**
   (1 mark for correct reactants; 1 mark for correct products)
   b) $Na^+(aq)$ **[1]**; $NO_3^-(aq)$ **[1]**
   c) $H^+(aq) + OH^-(aq) \rightarrow H_2O(l)$ **[2]** (1 mark for correct ions; 1 mark for correct product)
2. A strong acid dissociates completely **[1]**

Page 150 Electrolysis

1. Molten or in solution **[1]**
2. a) C **[1]**
   b) anion = oxygen **[1]**; cation = hydrogen **[1]**
   c) i) $2O^{2-}(aq) \rightarrow O_2(g) + 4e^-$ **[1]**
   ii) $2H^+(aq) + 2e^- \rightarrow H_2(g)$ **[1]**

Page 151 Predicting Chemical Reactions

1. C **[1]**
2. B **[1]**
3. caesium, rubidium, potassium, sodium, lithium **[1]**
4. a) i) 2.1 **[1]**
   ii) 2.8.2 **[1]**
   iii) 2.8.3 **[1]**
   iv) 2.6 **[1]**
   b) More bubbles should be drawn than in the Mg tube (at least 5 more) to show a more vigorous reaction **[1]**
   c) Hydrogen **[1]**
5. Collect the gas in a test tube and place a glowing splint into the tube **[1]**; if the splint relights, the gas is oxygen **[1]**

Pages 152–157 Practice Questions

Page 152 Controlling Chemical Reactions

1. a) C **[1]**
   b) The higher the temperature, the faster the rate of reaction **[1]**

Page 152 Catalysts and Activation Energy

1. Catalysts lower the activation energy for a reaction **[1]**; this means that the reaction will happen more quickly when the catalyst is present **[1]**
2. a) Catalase is the catalyst, so it is written over the arrow and not as a reactant **[1]**

> Remember, a catalyst is not used up in a reaction – it is not a reactant.

   b) Catalase is made of amino acids / is a protein / is an organic molecule **[1]**; manganese(IV) oxide is an inorganic molecule / catalyst **[1]**
3. a) i) A correct diagram showing particles at a low temperature **[1]**
   ii) A correct diagram showing particles at a high temperature **[1]**

Answers

Low Temperature

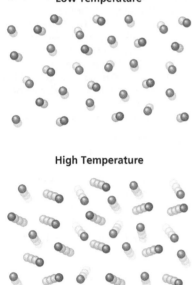

High Temperature

b) The faster reactant particles move, the greater the chance of successful collisions with other reactant particles **[1]**; successful collisions lead to product particles being formed **[1]**; also, there will be more reactant particles with energy greater than the activation energy **[1]**

Page 153 Equilibria
1. Nothing can enter or leave a closed system **[1]**; and the temperature and pressure remain the same **[1]**
2. C   **[1]**
3. a) The amount of product would increase **[1]**
   b) Nothing **[1]**
   c) More reactant ($CH_3COOH(aq)$) will be produced **[1]**

Page 154 Improving Processes and Products
1. B **[1]**
2. a) copper(II) oxide + carbon → copper + carbon dioxide **[1]**
   b) For safety reasons / to prevent the hot contents from being ejected from the tube **[1]**
3. D **[1]**
4. B **[1]**
5. $Cu^{2+} + 2e^- \rightarrow Cu(s)$ **[1]**
6. a) $Al^{3+}(l)$ and $O^{2-}(l)$ **[1]**
   b) i) $Al^{3+}(l) + 3e^- \rightarrow Al(l)$ **[1]**
      ii) $2O^{2-}(l) \rightarrow O_2(g) + 4e^-$ **[1]**
7. **Any three from:** bacteria reproduce rapidly **[1]**; the bacteria accumulate a specific ion **[1]**; the process is very efficient **[1]**; the process is much cheaper than extracting metals in a furnace (displacement) **[1]**

Page 155 Life Cycle Assessments and Recycling
1. A, C, E and F **[1]**
2. a) Recycled means to put the materials to another use / purpose **[1]**

b) Extracting new materials uses a lot of energy, costs money and contributes to environmental pollution **[1]**; throwing away products means the materials / resources are no longer available **[1]**; recycling means that the materials that would have been lost are now used in a different way **[1]**; this often involves less energy, lower costs and less waste than obtaining new materials **[1]**
c) The cost (in money and pollution) is greater than the benefit gained from recycling the parts **[1]**
3. A **[1]**; E **[1]**

Page 156 Crude Oil
1. a) W = paraffin, X = diesel, Y = crude oil, Z = bitumen **[3]** (2 marks for two correct; 1 mark for one correct)
   b) At the top of the column **[1]**
2. Paraffin **[1]**; fuel oil **[1]**

Page 156 Interpreting and Interacting with Earth's Systems
1. a) As carbon dioxide levels increase, so does air temperature **[1]**
   b) Carbon dioxide is a greenhouse gas **[1]**; it traps heat energy in the atmosphere, leading to an increase in air temperature **[1]**
   c) A **[1]**; B **[1]**; D **[1]**
2. a) 20.96% **[1]** (Accept 21%)
   b) Photosynthesis leads to the reduction of carbon dioxide **[1]**; and the production of oxygen **[1]**
   c) Ammonia **[1]**; carbon dioxide **[1]**

Page 157 Air Pollution and Potable Water
1. a) Carbon monoxide binds to the red blood cells instead of oxygen **[1]**; starving body cells of oxygen **[1]**
   b) Burning / incomplete combustion of fossil fuels **[1]**
2. Oxides of nitrogen can form acid rain **[1]**; and photochemical smog **[1]**
3. C **[1]**

Pages 158–175 Revise Questions

Page 159 Quick Test
1. **Any three from:** Electrons are very small; have negative charge; in atoms they orbit the nucleus; they give an atom its negative charge to balance the positive charge of the nucleus; in metals some electrons are free to move between atoms but in electrical insulators there are no free electrons
2. Alpha particles were fired at atoms and some bounced back. That is only possible if atoms contain something very dense / positively charged.
3. Positive
4. Electric force, between atoms
5. The electric forces between gases are almost zero, but the forces are strong in

solids so it's very hard to push particles of a solid apart.

Page 161 Quick Test
1. Quantity that changes: internal energy (also, volume and density may change) Quantity that does not change: **(any one from)** mass; temperature; chemical composition; atomic structure

   > Remember that in changes of state, atoms or molecules do not change. The forces between them and their internal energy change.

2. No. Changes in state involve changes of internal energy without the need for temperature change.
3. thermal energy for a change in state
   = mass × specific latent heat
   = 267 000 × 1000
   = 2.67×10^8 J
   (Accept 267 000 000 J or 267 000 kJ)
4. change in thermal energy
   = mass × specific heat capacity × change in temperature
   = 100 × 1006 × 10
   = 1.01×10^6 J to 3 significant figures
   (Accept 1 006 000 J or 1006 kJ)

Page 163 Quick Test
1. kinetic energy = 0.5 × mass and speed²
   = 0.5 × 0.5 × 16²
   = 64 J

   > Do calculations like these step-by-step. Always start by writing the equation. Then put in the numbers. Do the arithmetic and don't forget to write the final unit. If you try to do everything all at once you'll often get confused and make mistakes.

2. time $= \dfrac{\text{distance}}{\text{speed}}$
   $= \dfrac{100}{40}$
   = 2.5 hours
3. acceleration $= \dfrac{\text{change in velocity}}{\text{time}}$
   $= \dfrac{(24-0)}{10}$
   = 2.4 m/s²
4. 10 m/s², which is the acceleration of free fall

Page 165 Quick Test
1. Gravity, or gravitational force (i.e. weight)
2. Electric or electrical force
3. Unbalanced (or net or resultant) force

   > Remember that unbalanced force **always** causes acceleration of a body.

4. The skydiver falls a long distance and accelerates to high velocity. Resistive

force increases as velocity increases. (Also, the surface area of the skydiver is large enough that the air resistance is substantial.)

Page 167 Quick Test
1. Zero and zero.

> The two balls have the same mass and the same speed. Because they are traveling in opposite directions, their momentums cancel each other out – the total is zero.

2. The ice skater experiences low resistive force (friction) so loses little energy, and needs to do little work to replace lost energy.
3. A person leaning on a wall exerts a force on it, but the wall doesn't move and there is no distance involved. Work = force × distance = force × 0 = 0.
4. Power is rate of transferring energy / power measures how quickly energy is transferred.

Page 169 Quick Test
1. They can change its shape / cause extension or compression.

> If the forces acting on a body are balanced then there can be no acceleration, but shape can change and this can be extension or compression.

2. a) elastic
   b) plastic
   c) elastic
   d) plastic
3. Physics concerns the whole universe, and weight for an equivalent mass is very different in different places. On the Earth's surface, the weight of a particular mass is much the same everywhere so the distinction is not relevant.

Page 171 Quick Test
1. Electric force can be attractive or repulsive.
2. a) Electrons are transferred to or from the surface by friction.
   b) Electrons can flow within the metal, and even our skin can conduct some electricity. So when we hold the spoon, charge can flow to or from it and it doesn't stay charged.
3. Current is rate of flow of charge.
4. Current through resistance causes heating, so the surroundings are heated.

Page 173 Quick Test
1. A voltmeter measures the potential difference between two points.
2. a) An increase in voltage, or a decrease in resistance
   b) A decrease in voltage, or an increase in resistance

3. resistance = $\frac{voltage}{current}$. A metal wire's resistance increases when it is hot, so the ratio of voltage to current increases.
4. It decreases.

Page 175 Quick Test
1. a) total resistance $= 2 + 4$
   $= 6\,\Omega$
   b) $\frac{1}{total\ resistance} = \frac{1}{2} + \frac{1}{4}$
   $= \frac{3}{4}$
   total resistance $= \frac{4}{3}$
   $= 1.33\,\Omega$
2. They transfer energy by doing work (= force × distance) on objects outside the circuit, such as by lifting loads. (They may also become warm, and transfer some energy to the surroundings by heating.)
3. a) energy transfer =
   current × voltage × time
   $= 1.5 × 12 × 60$
   $= 1080\,J$
   b) power = current × voltage
   $= 1.5 × 12$
   $= 18\,W$
   OR
   power $= \frac{energy}{time}$
   $= \frac{1080}{60}$
   $= 18\,W$

Pages 176–181 Review Questions

Page 176 Controlling Chemical Reactions
1. a) B [1]; C [1]
   b) A graph drawn with a line steeper than the 25°C line [1]

Volume of CO$_2$ Produced

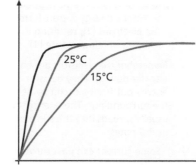

0 Time Taken for Cross to Disappear

Page 176 Catalysts and Activation Energy
1. a) A chemical that reduces the activation energy, making it more likely that the reaction will take place [1]
   b) Gases burned in a car engine are harmful for the environment [1]; the catalytic converter enables the reactions to take place leading to

non-harmful gases and acceptable levels of gases [1]
   c) One graph line drawn showing the reaction without a catalyst [1]; a second correctly drawn graph line showing the reaction with a catalyst [1]

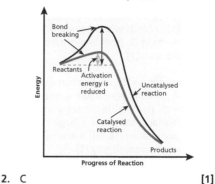

2. C [1]

Page 177 Equilibria
1. B [1]
2. B [1]; C [1]
3. More products [1]

Page 178 Improving Processes and Products
1. a) The carbon / graphite anode [1]
   b) C [1]

Page 178 Life Cycle Assessments and Recycling
1. a) B, D, C, A [1]
   b) Less waste [1]; less fresh water used in its manufacture [1]
   c) **Any two from:** more energy is required to melt the ore (bauxite / aluminium oxide) [1]; anode reacts with oxygen to form carbon dioxide [1]; more fossil fuel used in production [1]
   d) Cork [1]
   e) Aluminium is easily recycled [1]

Page 179 Crude Oil
1. petrol [1]
2. The longer the hydrocarbon chain length, the higher the boiling point [1]; this is because there are more intermolecular forces between the chains [1]; which make it more difficult to separate the chains [1] (Accept reverse argument, i.e. shorter chains have lower boiling points)

Page 180 Interpreting and Interacting with Earth's Systems
1. a) Carbon dioxide is a greenhouse gas [1]; increased carbon dioxide levels leads to greater levels of heat energy being trapped in the atmosphere [1]; this results in increased air temperature and climate change [1]
   b) Levels of CO$_2$ have risen faster than was expected [1]; so reducing the levels will require more wells [1]
   c) 60.4% increase [1]
   d) Carbon dioxide is released through the burning of fossil fuels

in factories and cars **[1]**; large-scale deforestation has removed trees that would normally take in carbon dioxide for photosynthesis **[1]**

2. A **[1]**

Page 181 Air Pollution and Potable Water

1. Carbon monoxide is similar in shape to O_2 **[1]**; it binds with haemoglobin **[1]**; preventing O_2 from being transported to the cells **[1]**
2. B **[1]**
3. Desalination **[1]**; reverse osmosis **[1]**
4. a) A partially permeable membrane / semi-permeable membrane **[1]**
   b) In places that have a large population **[1]**; but a low rainfall / poor water supply **[1]**

Page 182 Matter, Models and Density

1. D **[1]**
2. D **[1]**
3. A **[1]**
4. density $= \dfrac{mass}{volume}$ **[1]**;

   $= \dfrac{0.24}{0.0001}$ **[1]**;

   $= 2400 kg/m^3$ **[1]**

Page 182 Temperature and State

1. B **[1]**
2. A **[1]**
3. Requiring energy: boiling **[1]**; evaporation **[1]**; melting **[1]** Releasing energy: condensing **[1]**; freezing **[1]**
4. They gain energy / they gain kinetic energy **[1]**; they move faster **[1]**
5. Steam has extra energy due to its state as well its temperature **[1]**; it transfers more energy to the skin **[1]**

> Remember that energy transfers to or from materials during changes of state **and** during changes of temperature.

6. mass (of the ice cube) **[1]**

Page 183 Journeys

1. D **[1]**
2. D **[1]**
3. A **[1]**
4. B **[1]**
5. D **[1]**
6. a) distance = speed × time **[1]**;
      = 24 **[1]**; kilometres **[1]**
   b) distance = speed × time
      = 40 **[1]**; miles **[1]**
   c) distance = speed × time
      = 4 **[1]**; × 3600 **[1]**;
      = 14400m **[1]**

> There are 60 × 60 = 3600 seconds in an hour.

7. a) A scalar has size, a vector has size and direction **[1]**
   b) **Any one from:** displacement **[1]**; force **[1]**; velocity **[1]**; acceleration **[1]**; momentum **[1]**
   c) **Any one from:** energy **[1]**; distance **[1]**; area **[1]**; speed **[1]**
8. a) speed $= \dfrac{distance}{time}$ **[1]**;

   $= \dfrac{6000}{12}$ **[1]**;

   $= 500 km/h$ **[1]**
   b) $500 \times \dfrac{1000}{3600}$ **[1]**; $= 139 m/s$ **[1]**
   c) Direction changes **[1]**; including change due to curvature of the Earth **[1]**

Page 184 Forces

1. B **[1]**
2. C **[1]**
3. B **[1]**
4. Friction / resistive force **[1]**
5. A body can't accelerate without force / the acceleration of a body is proportional to the resultant force acting on it / force = mass × acceleration **[1]**
6. Any body that exerts a force on another itself experiences an equal force in the opposite direction **[1]**

Page 185 Force, Energy and Power

1. A **[1]**
2. D **[1]**

Page 185 Changes of Shape

1. B **[1]**
2. A **[1]**

Page 185 Electric Charge

1. C **[1]**
2. B **[1]**
3. C **[1]**
4. a) Metals have electrons that are free to move, they are not all attached to individual atoms **[1]**
   b) Resistors resist the flow of electrons, so kinetic energy is transferred from the electrons **[1]**; resulting in an increase in temperature **[1]**

> Remember that resistors are energy transfer devices. They transfer energy out from a circuit by heating the surroundings. The energy then usually spreads into the surroundings, or dissipates.

   c) Some current can pass through each of the resistors **[1]**

Page 186 Circuits

1. B **[1]**
2. C **[1]**
3. A **[1]**
4. A **[1]**
5. a) To supply energy / create a potential difference or voltage **[1]**
   b) To measure current **[1]**
   c) To measure voltage / potential difference **[1]**
   d) To oppose current / to control current (or voltage) / or to provide heating **[1]**
   e) To allow current in only one direction **[1]**
   f) To change current (or voltage) **[1]**; depending on temperature **[1]**
   g) To change current (or voltage) **[1]**; depending on light brightness **[1]**
6. a) voltage = current × resistance **[1]**;
      = 1.5 × 6 **[1]**;
      = 9V **[1]**
   b) charge moved = current × time **[1]**;
      = 1.5 × 60 **[1]**;
      = 90C **[1]**
   c) energy transferred =
      current × voltage × time **[1]**;
      = 1.5 × 9 × 60 **[1]**;
      = 810J **[1]**
   d) rate of transfer of energy
      = power
      $= \dfrac{energy}{time}$ **[1]**;

      $= \dfrac{810}{60}$ **[1]**;

      = 13.5W **[1]**
   OR
      = current × voltage **[1]**;
      = 1.5 × 9 **[1]**;
      = 13.5W **[1]**

> Remember, units matter – never forget to include the unit with your answer.

Page 187 Resistors and Energy Transfers

1. B **[1]**

Page 189 Quick Test

1. Sketch should show **looped** lines (magnetic field lines) that are closest together near the poles of the magnet. Arrows on the lines should point **away** from the north pole of the magnet and towards the south pole.

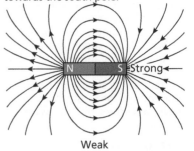

2. The shape of the magnetic field lines around a solenoid is similar to the shape around a permanent bar magnet but the field around the solenoid can be turned on and off by turning the current on and off.

3. Magnetic force due to interaction of the magnetic field around motor coil resulting from electric current and an external / separate magnetic field.

Page 191 – Quick Test

1. Wavelength = m; frequency = Hz; wave speed = m/s
2. $\frac{320}{2}$ = 160Hz
3. In longitudinal waves the oscillations are parallel to the direction of energy transfer; in transverse waves the oscillations are at right-angles to the direction of energy transfer

Page 193 – Quick Test

1. All electromagnetic waves have the same speed in a vacuum.
2. radio waves, microwaves, infrared, visible (red to violet), ultraviolet, X-rays, gamma rays
3. Frequency increases as wavelength decreases
4. Radiations with concentrated (or high) energy can pass this energy to electrons in atoms, so that the electrons become free from the atoms. Radio, microwave, infrared, visible and some UV radiations do not have sufficiently concentrated energy / Some UV, and X-rays and gamma rays do.

Electromagnetic waves with higher frequency also have more concentrated energy.

Page 195 Quick Test

1. a) ^1_1Xx, ^2_1Xx, ^3_1Xx
   b) Isotopes of the same element have the same number of protons, shown by the lower number, even if they have different total numbers of protons and neutrons, shown by the upper number.
2. Radiations carry a high concentration of energy, and when this energy is absorbed by atoms one or more of their orbital electrons can escape, leaving the atom as a positive ion.

Page 197 Quick Test

1. $\frac{N}{8}$
2. a) It decreases (through decay)
   b) It does not change
3. 234, 2

Page 199 Quick Test

1. The energy is transferred to the environment, such as in moving air, and eventually spreads thinly causing heating of the environment (air, road). The energy dissipates / is dissipated.
2. **Any three from:** gravitational potential energy of a raised mass; kinetic energy of a moving mass; elastic potential energy of a stretched or compressed object; thermal energy of a hot object; electric potential energy of a battery

3. a) energy available to transfer to surroundings = mass × specific heat capacity × temperature difference
   = 0.5 × 4200 × 70
   = 147 000J **OR**
   = 147kJ
   b) energy stored = (average) force × distance (extension)
   = $\left(\frac{15}{2}\right) \times 0.5$
   = 0.375J
   c) potential energy = mass × height × g
   = 1000 × 50 × 10
   = 500 000J **OR**
   = 500kJ
   d) kinetic energy = 0.5 × mass × speed²
   = 0.5 × 0.03 × 400²
   = 2400J **OR**
   = 2.4kJ

Page 201 Quick Test

1. a) power = current × potential difference
   = 5 × 230
   = 1150W = 1.15kW
   b) i) energy = power × time
   = 1.15 × 1
   = 1.15kWh
   ii) energy = power × time
   = 1.15 × 6
   = 6.9kWh
   c) i) energy = power × time
   = 1150 × 3600
   = 4 140 000J **OR**
   = 4.14 × 10⁶J
   ii) 4 140 000 × 6 = 24 840 000J **OR** 2.48(4) × 10⁷J

2. efficiency = $\frac{\text{useful output energy transfer}}{\text{input energy transfer}} \times 100\%$
   = $\frac{1}{1.25} \times 100\%$
   = 80%

Page 203 Quick Test

1. speed = $\frac{\text{distance}}{\text{time}}$
   For the bullet: speed = $\frac{100}{0.4}$
   = 250m/s
   For the plane: speed = $\frac{1000}{3.3}$
   = 303m/s
   The plane is faster.
2. Thinking distance and braking distance

During thinking time speed is constant, but during braking time speed is decreasing.

Page 205 Quick Test

1. Renewable: A, B, E and G
   Non-renewable: C, D and F

2. Carbon dioxide makes it harder for the atmosphere / Earth to emit energy back into space. This can cause global climate change.

Page 207 Quick Test

1. High current in cables causes heating. That means that the system loses a lot of energy. By transmitting at low current this energy loss is reduced. The power output from the power station is voltage × current, so if the current is to be low then the voltage needs to be high.
2. Transformers can transfer energy from one circuit to another. They allow energy to be transmitted at high voltage (stepped up) and then received by users at lower voltage (stepped down). But they don't work if the current in the transformers themselves is steady – it has to change. a.c. is continuously changing.

Pages 208–215 Review Questions

Page 208 Matter, Models and Density

1. D [1]
2. C [1]
3. A [1]
4. C [1]
5. Alpha particles (or small and fast positive particles) [1]; bounced back / deflected [1]; from gold atoms [1]; so it was concluded that atoms must contain a very dense region of positive charge [1]

6. a) density = $\frac{\text{mass}}{\text{volume}}$ [1];
   = $\frac{1.0}{0.002}$ [1];
   = 500kg/m³ [1]
   b) Yes, because its density is less than that of water (1000kg/m³) [1]
7. a) **Any two from:** an atom has electrons / a nucleus does not [1]; an atom is neutral / a nucleus has positive charge [1]; an atom has low density / a nucleus has high density [1]; a nucleus has a very much smaller radius than an atom [1]
   b) An atom is neutral / an ion is charged or has charge [1]; an atom has equal numbers of protons and electrons, but an ion has different numbers [1]

Remember that in a neutral atom the numbers of protons and electrons are equal, but in an ion they are not.

   c) **Any two from:** an electron is much smaller / a proton is much bigger [1]; an electron has negative charge / a proton has positive charge [1]; electrons orbit the nucleus / protons are in the nucleus [1]
   d) Protons have positive charge [1]; neutrons are neutral [1]

Answers

Page 209 Temperature and State
1. C [1]
2. D [1]
3. A [1]
4. C [1]
5. C [1]

Page 210 Journeys
1. A [1]
2. B [1]
3. D [1]
4. A [1]
5. B [1]
6. a) It has decreased [1]
   b) It has decreased and changed direction [1]
   c) Decelerated [1]
   d) It is a braking / resistive force [1]

7. a) acceleration = $\dfrac{\text{change in speed}}{\text{time}}$ [1];

   $= \dfrac{30}{10}$ [1];

   $= 3\,\text{m/s}^2$ [1]

   b) i) 0 [1]
   ii) kinetic energy =
   $0.5 \times \text{mass} \times \text{speed}^2$ [1];
   $= 0.5 \times 1000 \times 30^2$ [1];
   $= 450\,000\,\text{J}$ or $450\,\text{kJ}$ [1]

   c) force = mass × acceleration [1];
   $= 1000 \times 3$ [1];
   $= 3000\,\text{N}$ [1]

   d) distance $= \dfrac{(30^2 - 0)}{(2 \times 3)}$ [1];

   $= 150\,\text{m}$ [1]

   e) work done = force × distance [1];
   $= (3000 \times 150)$ [1];
   $= 450\,000\,\text{J}$ or $450\,\text{kJ}$ [1]

   f) power $= \dfrac{\text{energy or work}}{\text{time}}$ [1];

   $= 45\,000\,\text{W}$ or $45\,\text{kW}$ [1]

Page 212 Forces
1. B [1]
2. A [1]
3. D [1]
4. a) Pushes / exerts a force [1]
   b) i) little effect [1]
   ii) it accelerates / moves [1]
   c) **Diagram to show:** forces of the same size [1]; opposite directions [1]; acting on person and boat [1]

5. Gravity is too strong / weight is too big [1]
6. a) Large force to the right [1]
   b) Smaller force than answer a to the right [1]
   c) No net force [1]

Page 213 Force, Energy and Power
1. C [1]
2. a) work = force × distance [1];
   $= 180 \times 2.5$ [1];
   $= 450\,\text{J}$ [1]
   b) 450 J [1]
   c) Because of resistive forces / friction [1]
   d) It gains gravitational potential energy [1]; of 300×2.5 [1] $= 750\,\text{J}$ [1]

Page 213 Changes of Shape
1. B [1]
2. a) weight = mass × g [1];
   $= 1.2 \times 10$ [1];
   $= 12\,\text{N}$ [1]
   b) i) 1.9(2) N on the Moon [1]
   ii) 30 N on Jupiter [1]

Page 214 Electric charge
1. D [1]
2. D [1]
3. A [1]
4. B [1]
5. B [1]

Page 214 Circuits
1. A diode [1]
2. A [1]
3. a) Circuit drawn with: cell or battery and switch [1]; ammeter [1]; filament lamp [1]; the components all in a series [1]; voltmeter in parallel with the lamp [1]

   b) Graph drawn with: current labelled on y-axis [1]; voltage labelled on x-axis [1]; line is straight at low current [1]; line becomes curved, towards the voltage axis [1]

   c) Resistance is higher when the wire is hot [1]; due to increased difficulty of electron flow [1]

Page 215 Resistors and Energy Transfers
1. D [1]
2. a) It decreases [1]
   b) It increases [1]
3. a) energy = power × time [1];
   $= 2 \times 2.5$ [1];
   $= 5\,\text{kWh}$ [1]
   b) Second [1]; joule [1]

> Remember that the kilowatt-hour, kWh, is a unit of energy but the kilowatt, kW, is a unit of power.

Pages 216–223 Practice Questions

Page 216 Magnetic Fields and Motors
1. A [1]
2. D [1]
3. B [1]
4. It decreases [1]

Page 217 Wave Behaviour
1. B [1]
2. A [1]
3. C [1]
4. A [1]
5. B [1]

Page 218 Electromagnetic Radiation
1. A [1]
2. B [1]
3. D [1]
4. speed = frequency × wavelength [1];
   $= (4 \times 10^{14}) \times (5 \times 10^{-7})$ [1];
   $= 2 \times 10^8\,\text{m/s}$ [1] (Accept 200 000 000 m/s)

> Using powers of ten for very large and very small numbers is easier than writing numbers in full and counting the zeros.

5. Frequency: stays the same [1]; wavelength: changes [1]
6. D [1]

Page 219 Nuclei of Atoms
1. C [1]
2. A [1]
3. C [1]
4. A [1]

> To move further away from a nucleus, an electron must gain energy. If it escapes completely the atom is ionised.

5. C [1]
6. a) Atom with electron orbits [1]; representation of movement of electron to higher orbit [1]
   b) Representation of removal of electron from atom [1]

Page 220 Half-Life
1. The time taken [1]; for half [1]; of the radioactive nuclei to decay [1]
2. D [1]
3. C [1]
4. B [1]
5. B [1]

Page 221 Systems and Transfers
1. D [1]
2. A [1]
3. B [1]
4. B [1]

Page 221 Energy, Power and Efficiency
1. C [1]
2. a) efficiency = $\dfrac{\text{useful output energy transfer}}{\text{input energy transfer}} \times 100$

   $= \dfrac{20}{50} \times 100$ [1];

   $= 40\%$ [1]
   b) It causes heating / temperature rise / it dissipates [1]
3. **A** [1]; the 'lost energy' is a smaller proportion [1]

Page 222 Physics on the Road
1. B [1]
2. C [1]
3. D [1]
4. a) kinetic energy = 0.5 × mass × speed² [1];

   $= 0.5 \times 900 \times 20^2$ [1];

   $= 180\,000\text{J}$ or 180kJ [1]
   b) It transfers to the surroundings / dissipates [1]; (mostly) through the brakes [1]; (mostly) causing heating [1]
   c) The car has gravitational potential energy to dissipate as well as its kinetic energy. [1]
5. work = force × distance,

   force = $\dfrac{\text{work}}{\text{distance}}$ [1];

   $= \dfrac{12\,000}{20}$ [1]; $= 600\text{N}$ [1]

Page 223 Energy for the World
1. D [1]
2. B [1]

Page 223 Energy at Home
1. C [1]
2. A [1]
3. C [1]
4. Wires get hot and the bigger the current the hotter they get [1]; the heating transfers energy to the surroundings [1]; that energy is wasted [1]; so using low current reduces energy loss to the surroundings. [1]

Pages 224–231 Review Pages

Page 224 Magnetic Fields and Motors
1. C [1]
2. C [1]
3. Bring a compass close to the coil [1]; the compass will point along the magnetic field lines of the coil [1]
4. More turns in the coil [1]; more current in the coil [1]; have an iron core inside the coil [1]

5. force =

   magnetic flux density × current × length

   $= 0.12 \times 2.5 \times 0.15$ [1];

   $= 0.045\text{N}$ [1]
6. The force on one side of the coil [1]; must be upwards to match its upwards movement [1]; and downwards to match its downwards movement [1]; the field is always in the same direction [1]; so the current must reverse in direction [1]; every half turn [1]

Page 224 Wave Behaviour
1. A [1]
2. A [1]
3. C [1]

Page 225 Electromagnetic Radiation
1. D [1]

Spectrum means 'range'. Electromagnetic waves have a range of wavelengths and range of frequencies, but they all have the same speed in space.

2. C [1]
3. a) **The sketch should show:** radio [1]; microwaves [1]; infrared [1]; visible [1]; ultraviolet [1]; X-rays [1]; gamma rays [1] **(this order required, or its reverse)**

   Radio

   Microwave

   Infrared

   Visible

   Ultraviolet

   X-Ray

   Gamma Ray
   b) Radio – communications [1]; microwaves – communications / cooking [1]; infrared – heating [1]; visible – seeing [1]; ultraviolet – security marking [1]; X-rays – medical imaging [1]; gamma rays – killing bacteria / treating tumours / medical imaging [1]
   c) The radio end of the spectrum [1]
   d) **Any two from:** ultraviolet – skin damage / cancer [1]; X-rays – tissue damage / cancer [1]; gamma rays – tissue damage / cancer [1]

Page 226 Nuclei of Atoms
1. A [1]
2. C [1]
3. C [1]
4. A [1]
5. A [1]

Less penetrating radiations lose energy more quickly as they travel through a medium or substance. They pass the energy to the medium.

6. B [1]
7. C [1]
8. D [1]

Page 227 Half-Life
1. B [1]
2. a) $x = 14$ [1]; $y = 7$ [1]
   b) $x = 92$ [1]; $y = 234$ [1]
3. Four correctly drawn lines [3] (2 marks for two correct lines; 1 mark for one)

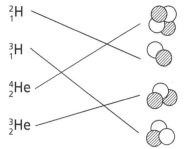

$^{2}_{1}\text{H}$

$^{3}_{1}\text{H}$

$^{4}_{2}\text{He}$

$^{3}_{2}\text{He}$

Page 228 Systems and Transfers
1. D [1]
2. B [1]
3. A [1]
4. D [1]
5. C [1]

Page 228 Energy, Power and Efficiency
1. A [1]
2. B [1]
3. C [1]
4. Energy storage devices and systems – **any two from:** water in a reservoir [1]; fuel [1]; a hot kettle [1]; any moving object [1]; Energy transfer devices and systems – **any two from:** motor [1]; lamp [1]; engine [1]; any electrical appliance [1]

A resistor with a steady current receives energy just as quickly as it passes energy to the surroundings by heating. It doesn't store energy.

5. a) heating [1]
   b) working / kinetic energy [1]
   c) heating and working / kinetic energy [1]
   d) emitting light [1]
6. C [1]

Page 230 Physics on the Road
1. B [1]
2. A [1]
3. C [1]

Answers

Page 231 Energy for the World

1. a) 1 mark for each advantage / disadvantage (maximum of 6 points; for full marks there must be a balance of advantages and disadvantages)

Energy resource	Advantages	Disadvantages
coal	• cheap • still a lot in the ground	• causes pollution • releases carbon dioxide
oil	• cheap • convenient, widely usable • still a lot available	• causes pollution • releases carbon dioxide
wind	• renewable • clean	• requires large wind farms
solar	• renewable • good for local supply	• only suitable for mass use if large solar farms are built • expensive
nuclear	• uses little fuel • no polluting emissions	• produces radioactive waste material • expensive

   b) **Any one resource, plus one advantage and one disadvantage, from:** gas [1]; advantage – widely available [1]; disadvantage – burning releases carbon dioxide [1]; bio-fuel [1]; advantage – renewable [1]; disadvantage – large land use needed [1]; hydroelectric [1]; advantage – renewable [1]; disadvantage – needs high dammed reservoirs [1]; tides [1]; advantage – renewable / always available [1]; disadvantage – barrage needed across estuary [1]

Page 231 Energy at Home

1. A [1]
2. C [1]
3. C [1]

1. A [1]
2. B [1]
3. B [1]
4. C [1]
5. a) A fungus [1]
    b) Ethanol [1]; carbon dioxide [1]
    c) Pressure caused by carbon dioxide gas [1]
    d) **Any two from:** amount of yeast [1]; amount of glucose [1]; volume of water [1]; incubation time [1]
    e) Units that liquid moved up the scale [1]
    f) i) Temperature [1]; °C [1]
       ii) The yeast respires at a faster rate at a temperature of 20°C [1]
       iii) The enzymes [1]; in the yeast had been denatured / were destroyed by the high temperature [1]
       iv) Use a smaller range of temperatures / smaller divisions on the scale [1]
    g) The yeast will run out of glucose [1]; the yeast is killed by increasing concentrations of ethanol [1]
6. C [1]
7. a) Movement of water through a plant and its evaporation from the leaves [1]
    b) Guard cell [1]
    c) Carbon dioxide [1]
8. a) *Saxifrage* [1]
    b) B [1]
    c) i) Upper: $\frac{10}{5}$ [1]; = 2 [1]; lower: $\frac{30}{5}$ [1]; = 6 [1]
       ii) On a hot day plants will open their stomata [1]; to allow exchange of gases [1]; for photosynthesis [1]; plants will lose less water through evaporation [1]; if the stomata are not in direct sunlight / are shaded [1]
9. C [1]
10. C [1]
11. D [1]
12. a) Males only have one X chromosome [1]; so if the allele for colour blindness is present, there is no chance of a second allele to cancel it out [1]
    b) B [1]
    c) Mother [1]
    d) B [1]
13. a) A mutation occurs that makes the bacterium resistant [1]; the sensitive bacteria are killed by the antibiotic [1]; the resistant bacteria survive and multiply to form a resistant population [1]
    b) If a bacterium mutates to develop resistance to one antibiotic [1]; it is still likely to be killed by the second antibiotic [1]
    c) Antibiotics are not effective against viruses / only kill bacteria [1]
    d) **Any three defences, plus a correct explanation for a total of 6 marks:** skin [1]; physical barrier [1] OR platelets [1]; form clot to seal wounds [1] OR mucous membranes [1]; trap microorganisms [1] OR acid in stomach [1]; kills microorganisms [1] OR tears / sweat [1]; contain antimicrobial substances [1]
14. B [1]
15. C [1]
16. a) **Accept: 385.5 OR 386ppm** [1]
    b) **Accept any value between:** 404–406ppm [1]
    c) Increase in the burning of fossil fuels [1]; deforestation [1]
17. a) B [1]
    b) Restriction enzymes [1]
    c) Ligase enzymes [1]
    d) Worried about safety for consumption / long-term effects not known [1]; dislike the idea of eating something that contains human genes [1]

> If asked for concerns about genetically engineered foods, your answer must include concerns about safety when **eaten / consumed**.

1. a) D [1]
    b) The line should not move – if it had been drawn in pen, the ink would have moved with the solvent front. [1]
    c) Ink 3 [1]
    d) $\frac{13}{40}$ = 0.33 (Accept answers between 0.30 and 0.35) [1]
2. A [1]
3. a) Graphite has a giant covalent structure in which each carbon atom is bonded to three other carbon atoms [1]; the graphite structure is in layers, held together by weak forces, so the layers separate easily [1]; diamond has a giant covalent structure in which each carbon atom is covalently bonded to four other carbon atoms / the maximum number of carbon atoms [1]; this makes it very strong [1]
    b) Allotrope [1]
    c) Burn / react diamond and graphite with oxygen [1]; to show that they produce the same product: carbon dioxide [1]
4. a) Covalent bonds [1]; the dot and cross diagram shows that the electrons are shared / the ball and stick model and the displayed formula use a single lines or sticks to represent a covalent bond [1]
    b) The bond angles are incorrect [1]; the model does not show the 3D shape of a molecule [1]
5. a) Particle A: 7 neutrons [1]; Particle B: atomic number of 20 [1]; Particle C: atomic mass of 16 [1]; and electronic structure 2.6 [1]; Particle D: 9 protons [1]
    b) D is an atom because the number of electrons equals the number of protons (the overall charge is neutral) [1]; B is an ion because it has two electrons less than the number of protons (so it has an overall charge) [1]
    c) It is in Group 5 [1]; and Period 2 [1]; the electronic configuration shows the numbers of electrons in each shell / energy level – the number of shells (two) gives the period number [1]; the number of electrons in the outermost shell / energy level (five) gives the group number [1]
    d) 20 [1]

6. The number of neutrons varies **[1]**; as neutrons have mass this means that the atomic mass changes **[1]**; whilst the number of electrons stays the same **[1]**; as the number of protons is the same in each isotope **[1]**
7. Diesel **[1]**
8. a) 0.9g + 7.1g = 8.0g **[1]**
   b) 0.1mol of X, 0.1mol of chlorine / Cl_2 and 0.1mol of X chloride **[1]**

> **X** has a relative atomic mass of 9g/mol and Devlin reacts 0.9g with chlorine. Start by working out the number of moles actually reacted.

   c) $X + Cl_2 \rightarrow XCl_2$ **[2]** (1 mark for the correct reactants; 1 mark for the correct product)
9. C **[1]**
10. a) $MgO(s) + H_2SO_4(aq) \rightarrow MgSO_4(aq) + H_2O(l)$ **[1]**
    b) **Any one from:** wear eye protection / safety goggles **[1]**; beware of spitting acid when heating the mixture **[1]**; do not boil the mixture **[1]**
    c) i) Keep adding magnesium oxide until no more dissolves (all acid has reacted) **[1]**
       ii) Pass the mixture through a filter funnel to remove unreacted magnesium oxide **[1]**
11. a) The nucleus **[1]**
    b) Protons and neutrons **[1]**
    c) B and C **[1]**
    d) B = boron **[1]**; C = carbon **[1]**
12. a) The compound is ionic **[1]**; the ions of calcium and fluoride are arranged in a giant lattice structure in the solid form of the ion **[1]**; this means that it is extremely difficult to break the strong electrostatic forces between ions to melt it **[1]**
    b) Methane has a simple molecular structure **[1]**; it only has very weak intermolecular forces between molecules, so is easy to separate from other methane molecules **[1]**; so the melting point and boiling points are low **[1]**
    c) Correctly drawn lithium ion **[1]**; and oxide ion **[1]**

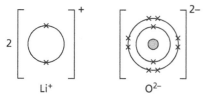

Li⁺   O²⁻

13. a) A substance that contains only one type of atom or molecule **[1]**
    b) In everyday life, pure means a substance comes from one source with no other contaminants **[1]**
14. a) Be **[1]**
    b) F **[1]**

c) 29 **[1]**
15. C **[1]**
16. a) Correctly drawn diagram **[1]**

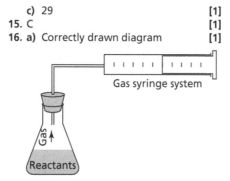

Gas syringe system

Gas

Reactants

    b) Add a catalyst **[1]**
    c) Wear eye protection **[1]**
17. B **[1]**
18. C **[1]**
19. a) The removal of salt from seawater to produce pure water **[1]**
    b) Na_2SO_4 **[1]**
    c) Chloride ions must be present in the water **[1]**
    d) Fresh water is needed for people to survive **[1]**; if the water is contaminated with disease-causing organisms or high in salts then the person drinking it is likely to become ill and die **[1]**
20. metal + acid → salt + hydrogen **[1]**
21. a) magnesium nitrate + hydrogen **[1]**
    b) zinc sulfate + hydrogen **[1]**
    c) iron(II) chloride + hydrogen **[1]**
22. a) calcium + sulfuric acid → calcium sulfate + hydrogen **[1]**
    b) Above calcium **[1]**

Pages 248–255 Mixed Exam-Style Physics Questions

1. B **[1]**
2. A **[1]**
3. D **[1]**
4. B **[1]**
5. a) molecule, atom, nucleus, electron **[2]** (1 mark if there is one error)
   b) i) The forces between them become weaker **[1]**; they move further apart **[1]**
      ii) There is no change **[1]**
   c) They gain energy **[1]**; and move faster **[1]**
   d) i) No change **[1]**
      ii) It changes (because direction changes) **[1]**
      iii) Exerts / applies a force **[1]**
      iv) Exerts / applies a force **[1]**
   e) i) energy = 0.5 × mass × (speed)² **[1]** = 0.5 × (4 × 10⁻²⁶) × 100² **[1]**; = 2 × 10⁻²² **[1]**; J **[1]**
      ii) 5000J **[1]**
6. a) i) Zero **[1]**
      ii) Driving force / engine **[1]**
      iii) Resistance (air, friction) **[1]**
      iv) Zero **[1]**
   b) i) Inertia / resistance to acceleration / need for a force for acceleration **[1]**

ii) Swing forward **[1]**; inertia / resistance to deceleration / need for a force for deceleration **[1]**
   c) i) Inertia / resistance to deceleration / need for a force for deceleration **[1]**
      ii) Reaction time – time between event and driver acting **[1]**; braking time – time between driver acting and vehicle stopping **[1]**
      iii) Add them together **[1]**
   d) Heating **[1]**; in brakes (and / or road, air) **[1]**; dissipation / spread thinly in surroundings **[1]**
7. a) Oppose current **[1]**; and produce heating (electrical energy is transferred to heat) **[1]**
   b) In parallel **[1]**
   c) To switch the supply to second resistor on and off **[1]**; control heating / turn heater up **[1]**
   d) A = live **[1]**; B = earth **[1]**; C = neutral **[1]**
   e) i) $resistance = \dfrac{potential\ difference}{current}$ **[1]**
         $= \dfrac{230}{4.6}$ **[1]**;
         $= 50\Omega$ **[1]**
      ii) The total (combined) current is bigger **[1]**; so the combined resistance must be smaller **[1]**
   f) power = potential difference × current **[1]**; = 9.2 × 230 **[1]**; = 2116W **[1]**
   g) energy transferred = power × time **[1]**; = 2.116 × 3 **[1]**; = 6.35kWh **[1]**
8. a) wave speed = frequency × wavelength **[1]**; = 3000 × 0.5 **[1]**; = 1500 **[1]**; m/s **[1]**
   b) Longitudinal waves do not transfer material **[1]**; sound waves are not transverse **[1]**; in transverse waves, the directions of travel and vibration are perpendicular (at right-angles) **[1]**
   c) Nuclei (of atoms) **[1]**
   d) i) Speed in space / vacuum **[1]**; can travel through space / vacuum / do not need a medium **[1]**
      ii) **Any two from:** wavelength **[1]**; frequency **[1]**; ionising ability **[1]**; penetrating ability **[1]**
   e) The range / set **[1]**; of different wavelengths / frequencies **[1]**; of the family of electromagnetic waves that travel with the same speed through space **[1]**

> Radio waves are not harmful to us because our bodies absorb them extremely weakly. They transfer very little energy to our bodies, either for heating or for ionisation.

f) i) The eyes detect / respond to visible light **[1]**
   ii) There is a warming effect **[1]**
   iii) Possible sunburn **[1]**
   iv) Ionisation effects / no noticeable sensation **[1]**
   v) There is no (measurable) response **[1]**

9. a) There is more / faster energy loss from A than B **[1]**; A has poorer insulation **[1]**
   b) They must replace the lost energy **[1]**
   c) It heats the surroundings **[1]**; spreads thinly / dissipates **[1]**
   d) Power **[1]**
   e) i) and ii) Any three sources, plus correct identification of type of resource, from: gas – non-renewable **[2]**; oil – non-renewable **[2]**; coal – non-renewable **[2]**; electricity – a mixture **[2]**; wood or other biofuel – renewable **[2]**; solar cells – renewable **[2]**; solar panels – renewable **[2]** (1 mark for source; 1 mark for correct type)
   f) Advantages: renewable **[1]**; low pollution / low CO_2 emission **[1]**;

Disadvantages: use large areas of land / not always beautiful **[1]**; power output not high **[1]**

> 25 years ago there were very few wind farms in the world. Now there are many. You need to be able to explain this change.

   g) Advantages: widely available **[1]**; easy to transport **[1]**; Disadvantages: non-renewable **[1]**; polluting / high CO_2 emission **[1]**
   h) More renewables **[1]**; primarily due to concern over climate change **[1]**

10. a) Vibrating / oscillating charged particles **[1]**; emit (electromagnetic radiation) electrons **[1]**; that can fall towards nucleus / lose energy **[1]**; within atoms, nuclei can emit gamma rays **[1]**
   b) The emission **[1]**; of ionising radiation **[1]**; from substances with unstable nuclei **[1]**
   c) The removal (or sometimes addition) of electrons from atoms when energy is absorbed **[1]**; which results in atoms becoming charged **[1]**

   d) Gamma rays **[1]**
   e) i) Alpha **[1]**
      ii) Shortest tracks **[1]**
      iii) It causes ionisation **[1]**; and passes energy to atoms **[1]**; of the surrounding substance **[1]**
   f) Beta particles are electrons **[1]**; they are extremely small and easily change direction **[1]**; when they collide / ionise / lose energy **[1]**

11. a) Use a plotting compass **[1]**; turn current on and off and watch movement of the needle **[1]**
   b) Make a coil **[1]**; with many turns / add an iron core **[1]**
   c) i) Heating **[1]**
      ii) The surroundings are heated / energy transfers thermally to the surroundings **[1]**
      iii) It can be used for heating rooms / heating spaces / in appliances **[1]**
   d) i) As the wire gets hot resistance increases (proportionally) **[1]**
      ii) The gradient **[1]**; slopes down / is negative **[1]**
   e) Less energy is wasted / a slower rate of energy waste **[1]**

Periodic Table

Key
atomic number
symbol
name
relative atomic mass

(1) 1	(2) 2	(3)	(4)	(5)	(6)	(7)	8	9	10	11	12	(3) 13	(4) 14	(5) 15	(6) 16	(7) 17	(0) 18
1 **H** hydrogen 1.0																	2 **He** helium 4.0
3 **Li** lithium 6.9	4 **Be** beryllium 9.0											5 **B** boron 10.8	6 **C** carbon 12.0	7 **N** nitrogen 14.0	8 **O** oxygen 16.0	9 **F** fluorine 19.0	10 **Ne** neon 20.2
11 **Na** sodium 23.0	12 **Mg** magnesium 24.3											13 **Al** aluminium 27.0	14 **Si** silicon 28.1	15 **P** phosphorus 31.0	16 **S** sulfur 32.1	17 **Cl** chlorine 35.5	18 **Ar** argon 39.9
19 **K** potassium 39.1	20 **Ca** calcium 40.1	21 **Sc** scandium 45.0	22 **Ti** titanium 47.9	23 **V** vanadium 50.9	24 **Cr** chromium 52.0	25 **Mn** manganese 54.9	26 **Fe** iron 55.8	27 **Co** cobalt 58.9	28 **Ni** nickel 58.7	29 **Cu** copper 63.5	30 **Zn** zinc 65.4	31 **Ga** gallium 69.7	32 **Ge** germanium 72.6	33 **As** arsenic 74.9	34 **Se** selenium 79.0	35 **Br** bromine 79.9	36 **Kr** krypton 83.8
37 **Rb** rubidium 85.5	38 **Sr** strontium 87.6	39 **Y** yttrium 88.9	40 **Zr** zirconium 91.2	41 **Nb** niobium 92.9	42 **Mo** molybdenum 95.9	43 **Tc** technetium	44 **Ru** ruthenium 101.1	45 **Rh** rhodium 102.9	46 **Pd** palladium 106.4	47 **Ag** silver 107.9	48 **Cd** cadmium 112.4	49 **In** indium 114.8	50 **Sn** tin 118.7	51 **Sb** antimony 121.8	52 **Te** tellurium 127.6	53 **I** iodine 126.9	54 **Xe** xenon 131.3
55 **Cs** caesium 132.9	56 **Ba** barium 137.3	57-71 lanthanides	72 **Hf** hafnium 178.5	73 **Ta** tantalum 180.9	74 **W** tungsten 183.8	75 **Re** rhenium 186.2	76 **Os** osmium 190.2	77 **Ir** iridium 192.2	78 **Pt** platinum 195.1	79 **Au** gold 197.0	80 **Hg** mercury 200.5	81 **Tl** thallium 204.4	82 **Pb** lead 207.2	83 **Bi** bismuth 209.0	84 **Po** polonium	85 **At** astatine	86 **Rn** radon
87 **Fr** francium	88 **Ra** radium	89-103 actinides	104 **Rf** rutherfordium	105 **Db** dubnium	106 **Sg** seaborgium	107 **Bh** bohrium	108 **Hs** hassium	109 **Mt** meitnerium	110 **Ds** darmstadtium	111 **Rg** roentgenium	112 **Cn** copernicium	114 **Fl** flerovium		116 **Lv** livermorium			

Physics Equations

You must be able to recall and apply the following equations using the appropriate SI units:

Word Equation
density (kg/m^3) = $\dfrac{\text{mass (kg)}}{\text{volume (m}^3\text{)}}$
distance travelled (m) = speed (m/s) × time (s)
acceleration (m/s^2) = $\dfrac{\text{change in velocity (m/s)}}{\text{time (s)}}$
kinetic energy (J) = 0.5 × mass (kg) × (speed $(m/s))^2$
force (N) = mass (kg) × acceleration (m/s^2)
HT momentum (kgm/s) = mass (kg) × velocity (m/s)
work done (J) = force (N) × distance (m) (along the line of action of the force)
power (W) = $\dfrac{\text{work done (J)}}{\text{time (s)}}$
force exerted by a spring (N) = extension (m) × spring constant (N/m)
gravity force (N) = mass (kg) × gravitational field strength, g (N/kg)
(in a gravity field) potential energy (J) = mass (kg) × height (m) × gravitational field strength, g (N/kg)
charge flow (C) = current (A) × time (s)
potential difference (V) = current (A) × resistance (Ω)
energy transferred (J) = charge (C) × potential difference (V)
power (W) = potential difference (V) × current (A) = (current $(A))^2$ × resistance (Ω)
energy transferred (J, kWh) = power (W, kW) × time (s, h)
wave speed (m/s) = frequency (Hz) × wavelength (m)
efficiency = $\dfrac{\text{useful output energy transfer (J)}}{\text{input energy transfer (J)}}$ × 100%

You must be able to select and apply the following equations using the appropriate SI units:

Word Equation
change in thermal energy (J) = mass (kg) × specific heat capacity (J/kg°C) × change in temperature (°C)
thermal energy for a change in state (J) = mass (kg) × specific latent heat (J/kg)
(final velocity (m/s))^2 − (initial velocity (m/s))^2 = 2 × acceleration (m/s^2) × distance (m)
energy transferred in stretching (J) = 0.5 × spring constant (N/m) × (extension (m))^2
HT force on a conductor (at right-angles to a magnetic field) carrying a current (N) = magnetic field strength (T) × current (A) × length (m)
potential difference across primary coil (V) × current in primary coil (A) = potential difference across secondary coil (V) × current in secondary coil (A)

Glossary and Index

Renewable   can be replenished naturally over the progress of time   **141, 204**

Repeatable   refers to results that will be shown again if the same experiment is performed again   **27**

Repel   push away   **164**

Reproducible   refers to results that will be found again if the same experiment is performed by another person   **27**

Resistance   opposition to current   **171**

HT Resistive force   force that opposes motion   **165**

Resolution   the ability to distinguish two objects when separated by a small distance   **16**

Respiratory   of the respiratory (breathing) system   **146**

HT Restriction enzyme   enzyme used in genetic engineering to 'cut' genes from DNA   **69**

Resultant force   the combined effect of more than one force   **164**

Retina   part of the back of the eye with many light-sensitive cells   **192**

Reverse osmosis   a water purification technology that uses a semi-permeable membrane to remove larger particles from drinking water   **147**

Reversible   can react in both directions; reactant $\rightleftharpoons$ product; description of a change for which the starting condition can be restored   **136–137, 160**

R_f value   the ratio of how far a sample moves to the movement of the mobile phase in chromatography   **89**

S

Salt   an ionic compound resulting from the neutralisation of an acid and a base   **114–115**

Sampling   surveying a number of small portions, which are representative of the total portion, to allow an estimate to be made for the whole portion   **66**

Scalar quantity   a quantity with no particular direction   **162**

Sedimentation tank   a tank that allows small particles in water to drop to the bottom to form a sediment   **147**

Selective breeding   breeding two adult organisms to get offspring with certain desired characteristics   **68**

Sensor   a component or device that responds (such as by change of resistance) to changes in the environment (such as change of temperature or brightness of light)   **173**

Series   components of a circuit that are connected one after the other in a single loop   **172, 172**

SI system   the international system of units, including the kilogram, metre and second   **162**

Simple molecule   a small, covalently-bonded molecule that only contains a few atoms, e.g. CO_2 **93**

Slope   slope or gradient (of a graph) = $\dfrac{\text{increase in } y\text{-axis quantity}}{\text{increase in } x\text{-axis quantity}}$ **162**

Solar cell   voltage-generating device, using sunlight   **205**

Solar panel   water-heating device, using sunlight   **205**

Solenoid   a coil of wire around which a magnetic field is induced by passing a current through it   **188**

Solid   a state of matter where the atoms or molecules are densely packed together, leading to a stable volume and shape   **106**

Sound wave   a kind of longitudinal wave in which the vibrations spread from a source, travel through a medium and can be detected by animal ears   **191**

Species   (chemistry) the atoms, molecules or ions undergoing a chemical process   **109, 118–119**

Specific heat capacity   energy needed per kilogram of material for each degree Celsius (or kelvin) change in temperature   **161**

Specific latent heat   energy needed per kilogram of material to change state   **160**

HT Spectator ions   ions that, although present in a reaction, do not change charge   **109**

Spring constant   ratio of force to extension for a spring   **169**

Stable   not subject to change   **196**

Stain   dye used to colour cells and cell structures   **16**

State   the condition in which a substance exists (solid, liquid or gas)   **86–87**

Statin   drug given to reduce cholesterol levels in the body   **74**

Stationary phase   the solid or liquid that the sample is added to in chromatography   **89**

Stent   a small mesh tube used to inflate blocked arteries   **74**

Step-down transformer   a transformer that provides lower voltage, but potentially higher total current, in the secondary circuit than the primary circuit   **206**

Stimulus / stimuli (pl.)   something that can elicit (give rise to) a response   **38**

HT Stoichiometry   the ratio between the relative amounts of substances taking part in a reaction or forming a compound   **109**

Stomata   pores in the underside of a leaf, which are opened and closed by guard cells, that allow gases in and out   **31**

Stopping distance   the sum of thinking distance and braking distance   **203**

HT Strong acid   an acid that fully dissociates into its component ions   **116–117**

Sublimate   change directly from solid to gas   **160**

Substrate   the molecule upon which an enzyme acts (the key)   **18**

Surface area   the area that is available for a reactant to come into contact with   **132–133**

Sustainable   refers to methods and processes with minimal impact on the environment   **65, 140–141**

Synapse   the gap between two nerve cells   **38**

Systemic / systematic error   an error that is made repeatedly, e.g. using a balance that has not been calibrated correctly   **67**

T

HT Terminal velocity   the constant velocity of a falling object once the upwards resistive force is as big as the downwards force of gravity (Forces are then balanced, so there can be no acceleration)   **165**

Thermal conductivity   a way of comparing the conduction ability of different materials   **201**

Thermal energy   energy due to temperature difference   **167**

Notes

Notes

Collins

OCR Gateway
GCSE 9-1
Combined Science
Higher

Workbook

Fran Walsh, Eliot Attridge and Trevor Baker

Revision Tips

Rethink Revision

Have you ever taken part in a quiz and thought *'I know this!'* but, despite frantically racking your brain, you just couldn't come up with the answer?

It's very frustrating when this happens but, in a fun situation, it doesn't really matter. However, in your GCSE exams, it will be essential that you can recall the relevant information quickly when you need to.

Most students think that revision is about making sure you *know* stuff. Of course, this is important, but it is also about becoming confident that you can *retain* that *stuff* over time and **recall** it quickly when needed.

Revision That Really Works

Experts have discovered that there are two techniques that help with all of these things and consistently produce better results in exams compared to other revision techniques.

Applying these techniques to your GCSE revision will ensure you get better results in your exams and will have all the relevant knowledge at your fingertips when you start studying for further qualifications, like AS and A Levels, or begin work.

It really isn't rocket science either – you simply need to:

- **test yourself** on each topic as many times as possible
- **leave a gap** between the test sessions.

Three Essential Revision Tips

1. **Use Your Time Wisely**

   - Allow yourself plenty of time.
   - Try to start revising at least six months before your exams – it's more effective and less stressful.
   - Your revision time is precious so use it wisely – using the techniques described on this page will ensure you revise effectively and efficiently and get the best results.
   - Don't waste time re-reading the same information over and over again – it's time-consuming and not effective!

2. **Make a Plan**

   - Identify all the topics you need to revise (this Complete Revision & Practice book will help you).
   - Plan at least five sessions for each topic.
   - One hour should be ample time to test yourself on the key ideas for a topic.
   - Spread out the practice sessions for each topic – the optimum time to leave between each session is about one month but, if this isn't possible, just make the gaps as big as realistically possible.

3. **Test Yourself**

   - Methods for testing yourself include: quizzes, practice questions, flashcards, past papers, explaining a topic to someone else, etc.
   - This Complete Revision & Practice book provides seven practice opportunities per topic.
   - Don't worry if you get an answer wrong – provided you check what the correct answer is, you are more likely to get the same or similar questions right in future!

Visit our website to download your free flashcards, for more information about the benefits of these techniques, and for further guidance on how to plan ahead and make them work for you.

collins.co.uk/collinsGCSErevision

Contents

Contents

Contents

Cell Structures

1 **a)** Draw a line from each cellular structure to its function.

Structure	Function
Nucleus	Contains chlorophyll
Cell membrane	Contains receptor molecules
Chloroplast	Contains enzymes for respiration
Mitochondrion	Contains genetic material

[3]

b) Which of the structures above is only found in plant cells?

.... [1]

2 Which of the following are features of an electron microscope? Tick (✓) **two** boxes.

Higher resolution than light microscope	
Does not require specimens to be stained	
Uses more powerful lenses than a light microscope	
Uses magnets to focus an electron beam	

[2]

3 Jade wants to look at some blood using a light microscope.
She pricks her finger and places a drop of blood on a glass slide.

a) Once the blood has dried, what can Jade do to make the cells easier to see under the microscope?

.... [1]

b) The diagram shows what Jade can see under the microscope.

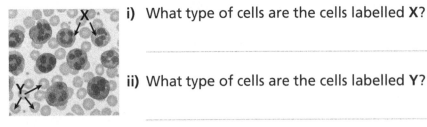

i) What type of cells are the cells labelled **X**?

.... [1]

ii) What type of cells are the cells labelled **Y**?

.... [1]

c) Jade is using a microscope with a ×4 magnification eyepiece and a ×100 magnification objective lens.

What is the total magnification she is using?

.... [1]

Total Marks .... / 10

What Happens in Cells

1 **a)** Protease enzyme digests proteins. What does carbohydrase enzyme digest?

_____ **[1]**

b) An enzyme reaction can be summarised as: enzyme + substrate ⟶ products (+ enzyme)

Complete the equation to show the reaction for lipase enzyme.

lipase + _____  _____ + _____ (+ lipase) **[3]**

c) Enzymes are denatured by high temperatures.

Suggest **one** other factor that may denature enzymes.

_____ **[1]**

2 The molecule below is made of nucleic acids and carries genetic information.

a) What is the name of this molecule?

_____ **[1]**

b) In which part of the cell is the molecule found?

_____ **[1]**

c) The two strands of nucleic acid are coiled around each other.

What is the name of this coiled structure?

_____ **[1]**

3 The enzyme amylase breaks down starch. It is present in saliva.

Explain why amylase stops working when it reaches the stomach.

_____

_____ **[2]**

Total Marks _____ / 10

Respiration

1 The table below shows the percentage of a person's blood that is delivered to different areas of the body during exercise and at rest.

Body Parts	Percentage of Blood Delivered to Each Part	
	At Rest	During Exercise
Muscles	15	65
Brain	17	5
Liver	30	8
Other	38	22

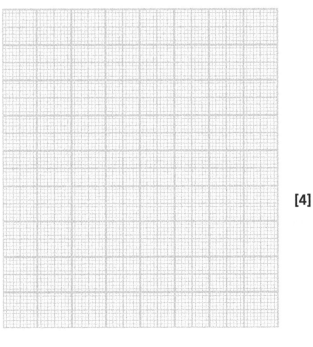

a) Plot these results on the graph paper. [4]

b) Suggest why there is more blood flowing to the muscles during exercise than at rest.

_____

_____

_____ [3]

2 The diagram shows a yeast cell.

a) To what group of microorganisms does yeast belong?

_____

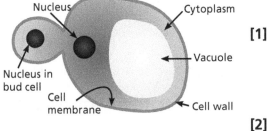

[1]

b) Which **two** structures are present in both yeast and plants cells, but not animal cells?

_____ [2]

c) Yeast can respire in the absence of oxygen. Write the word equation for this reaction.

_____ [3]

3 Most living things use oxygen when they respire. This is known as aerobic respiration.

a) Write the word equation for aerobic respiration.

_____ [2]

b) How are the reactants in this equation delivered to body cells?

_____ [1]

Total Marks _____ / 16

Photosynthesis

1 The following questions are about plants and food:

a) What is the name of the process plants use to make food?

.... [1]

b) What is the name of the food molecule made by plants?

.... [1]

c) What is this food molecule used for?

....

.... [2]

d) Explain why the concentration of carbon dioxide in a greenhouse may be higher at night than during the day.

....

....

.... [2]

2 **a)** Complete the word equation for photosynthesis.

$$\text{carbon dioxide} + \text{..........} \xrightarrow[\text{chlorophyll}]{\text{light energy}} \text{glucose} + \text{..........}$$

[2]

b) Complete the symbol equation for photosynthesis.

$$6CO_2 + \text{..........} \xrightarrow[\text{chlorophyll}]{\text{light energy}} C_6H_{12}O_6 + \text{..........}$$

[2]

c) The chlorophyll needed for photosynthesis is found inside which subcellular structures?

.... [1]

d) Photosynthesis is an endothermic reaction. What does this mean?

.... [1]

e) Apart from the amount of light, state **two** other factors that limit photosynthesis.

.... [2]

Total Marks .... / 14

Supplying the Cell

1 The diagram below shows the stages in the cell cycle.

| A | B | C | D | E |

a) Put the stages in the correct order. The first one has been done for you.

____B____ _____ _____ _____ _____ [1]

b) What is the name of the type of cell division shown in the diagram?

_____ [1]

c) Apart from growth, state **one** other reason for this type of cell division.

_____ [1]

2 The boxes below represent the concentrations of salts inside and outside a cell.

| Low salt concentration | | High salt concentration |

Draw three arrows on the diagram and label them to show the direction of movement for:

a) diffusion　　　　b) active transport　　　　c) movement of water by osmosis.　　[3]

3 A stem cell is a cell that can differentiate into many different types of cell.

a) Name **two** places from where human stem cells can be obtained.

_____ [2]

b) The diagram shows two cells that have differentiated to become specialised cells.

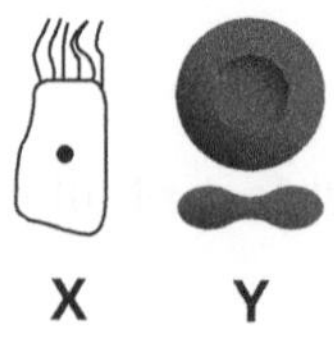

Name each cell and describe how it is adapted to its function.

X _____

_____

Y _____

_____ [6]

Total Marks _____ / 14

The Challenges of Size

1 The diagram shows two cells from different parts of a plant.

   a) What is the name of cell **Y**?

      .... [1]

   b) In which part of a plant would you find cell **Z**?

      .... [1]

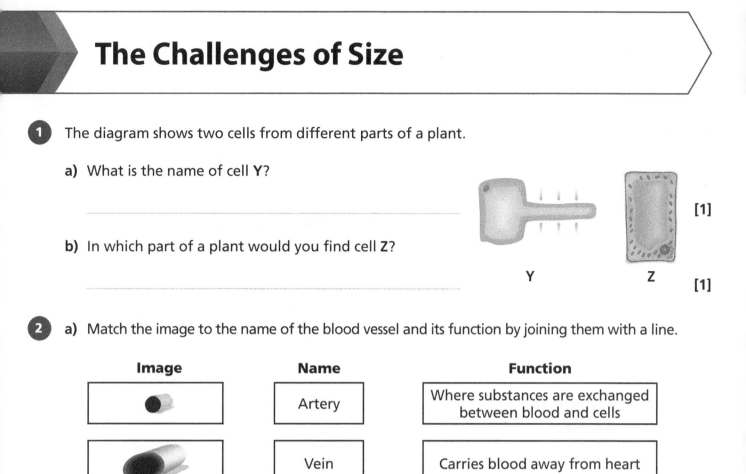

Y   Z

2 **a)** Match the image to the name of the blood vessel and its function by joining them with a line.

Image	Name	Function
●	Artery	Where substances are exchanged between blood and cells
●	Vein	Carries blood away from heart
●	Capillary	Carries blood back to heart

[6]

   b) Which blood vessels carry blood under low pressure?

      .... [1]

   c) Which blood vessels have valves?

      .... [1]

   d) The diagram shows a network of capillaries in a muscle.

      Tick (✓) the boxes next to the two substances that are most likely to be found in the blood flowing through an artery.

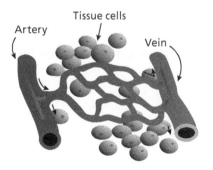

Tissue cells

Artery

Vein

Carbon dioxide	
Glucose	
Urea	
Starch	
Oxygen	

[2]

Total Marks .... / 12

The Heart and Blood Cells

1 The diagram shows a human heart.

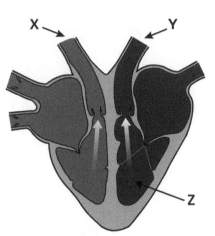

The bottom chambers are contracting, forcing blood through the blood vessels labelled **X** and **Y**.

a) What name is given to the bottom chambers of the heart?

.... **[1]**

b) Are blood vessels **X** and **Y** arteries or veins?

Explain your answer.

....

.... **[2]**

c) Where is blood flowing to when it flows through blood vessel **X**?

.... **[1]**

d) How is the chamber labelled **Z** adapted to its function?

....

....

.... **[3]**

2 Name **three** substances carried by plasma.

.... **[3]**

Total Marks .... / 10

Plants, Water and Minerals

1 The diagram below shows two similar plants in different conditions.

Plant X	Plant Y
Plant cells short of water so the plant wilts	H_2O H_2O H_2O H_2O H_2O Plant cells full of water so the plant stays erect
Not enough water in soil	Plenty of water in soil

Plants lose water through pores in the leaf called stomata.

a) In which plant will the stomata be closed?

.... [1]

b) Explain why plant **X** will not be photosynthesising.

.... [2]

c) Which terms describes the cells of plant **Y**?
    Tick (✓) the correct answer.

Turgid	
Flaccid	
Wilting	

[1]

d) Water moves from the roots to the leaves before being lost from the leaves.

    What is this process called?

.... [1]

e) How do the following factors affect water loss from a plant's leaves?

    i)   Wind speed .... [1]

    ii) Temperature .... [1]

f) The diagram shows a stoma.

    What is the role of the cell labelled **A**?

.... [1]

g) Explain why it is usual to find more stomata on the lower surface of a leaf compared with the upper surface.

....

.... [2]

Total Marks .... / 10

Coordination and Control

1 Match the part of the nervous system to its function by drawing a line between them.

Nervous System Part   **Function**

| Sensory neurone | | Carries impulses from relay neurone to effector |

| Receptor | | Carries impulses from receptor to relay neurone |

| Brain | | Detects a stimulus |

| Motor neurone | | Responsible for coordinating a response | **[3]**

2 The diagram below shows a neurone.

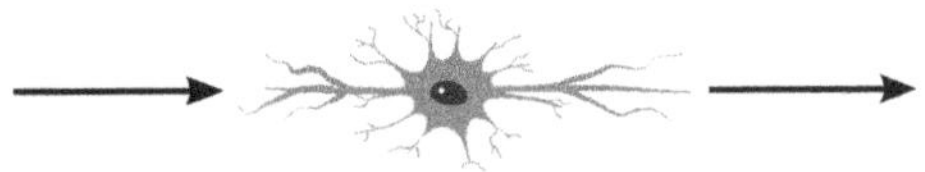

Impulse travels towards cell body and then away from cell body

What type of neurone is it?

.... **[1]**

3 The diagram shows the junction between two nerve cells, **A** and **B**.
The arrow shows the direction of the nerve impulse.

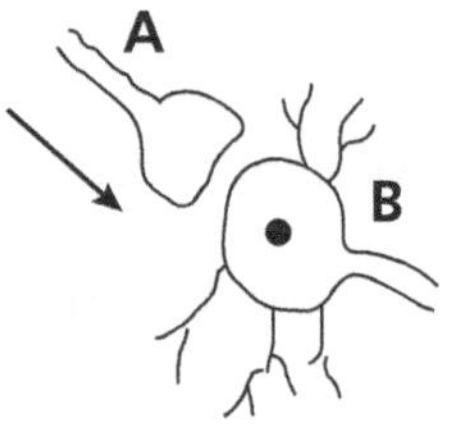

Complete the passage below about transmission of nerve impulses by
filling in the gaps.

A receptor receives a .... and generates an .... impulse in

the .... neurone. The impulse travels along the neurone until it reaches a

.... The sensory neurone releases a chemical called a ....

This crosses the gap and binds to .... on the next neurone. This generates

another electrical impulse. **[6]**

4 Which statement is true for a nervous response? Tick (✓) one box only.

A very quick response	
A generalised response	
Carried by chemical impulses	
Long-lasting effect	

[1]

Total Marks .... / 11

The Endocrine System

1 The diagram shows the major glands in the body that produce hormones.

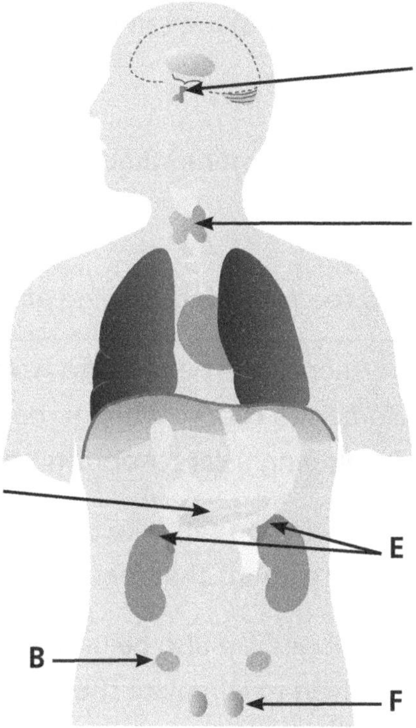

a) What is the name of gland **C**?

_____ [1]

b) Name the **two** hormones produced by gland **A**.

_____ [2]

c) Which hormone causes gland **D** to produce thyroxine?

_____ [1]

d) The glands labelled **E** are found on top of which organs?

_____ [1]

e) The organs labelled **F** produce the male sex hormone.
What are the organs called?

_____ [1]

f) How do hormones travel to their site of action?

_____ [1]

2 Which of the following statements are true about hormones? Tick (✓) two boxes.

They are fast acting	
They are chemicals	
They are proteins	
Their response is long lasting	

[2]

3 The flow diagram is about control of blood glucose levels.

a) Process **A** happens in which organ?

_____ [1]

b) Process **B** happens in which organ?

_____ [1]

Blood glucose levels rise

↓

A Insulin produced

↓

B Glucose converted to glycogen

↓

C Increased permeability of membrane to glucose

c) When blood glucose levels return to normal, insulin production stops. What is this an example of?

_____ [1]

Total Marks _____ / 12

Hormones and Their Uses

1 A woman's fertility can be controlled by drugs made from female sex hormones.

a) What hormone can be combined with oestrogen in the female contraceptive pill?

.... [1]

b) What hormone stimulates the release of the egg?

.... [1]

2 Read the following passage about the use of hormones to control fertility.

> Hormones can be given to women to control their fertility. The contraceptive pill allows women to choose when to have children. They may wish to delay having children while developing a career or until they can afford to have children.
>
> Not everyone can take the pill. It can cause side effects in some people and there are some religions that feel it is wrong for humans to control their fertility.
>
> Hormones can also be used to help women who are struggling to conceive. They can be given to stimulate the release of eggs, which can be fertilised naturally or by in vitro fertilisation (IVF).
>
> IVF is expensive and has less than a 50% success rate but does allow infertile couples to have their own children rather than adopt. Sometimes two or more embryos are fertilised successfully, resulting in multiple births. Embryos produced by IVF can be screened for genetic diseases before being implanted in the womb. Embryos not implanted can be used for research or are destroyed.

a) Which **two** hormones might be used in fertility treatment?

.... [2]

b) Give **two** arguments against taking the contraceptive pill.

.... [2]

c) Discuss the advantages and disadvantages of IVF.

....

....

....

....

.... [6]

Total Marks .... / 12

Maintaining Internal Environments

1 Enzymes have an optimum temperature at which they work.

Explain what happens to the enzymes if the temperature of the body:

a) increases _____ [1]

b) decreases _____ [1]

2 Describe the similarities and differences between type 1 and 2 diabetes.

_____

_____

_____

_____ [3]

3 The following table contains information about control of blood glucose levels.

Fill in the missing information in boxes **a)** to **d)**.

Gland	Hormone Produced	Transported by	Target Organ
Pancreas	Insulin	a)	b)
c)	Glucagon	Blood plasma	d)

[4]

4 The diagram below shows the digestive system.

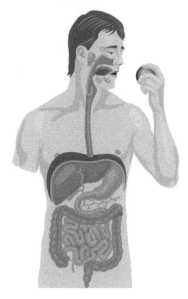

On the diagram:

a) Label the organ where glycogen is stored with an **X**. [1]

b) Label the organ where insulin is produced with a **Z**. [1]

Total Marks _____ / 11

Recycling

1 **a)** In the carbon cycle, carbon is returned to the atmosphere by the respiration of **three** groups of organisms.

What are the **three** groups?

....

....

.... **[3]**

b) Which molecule do plants convert atmospheric carbon dioxide into during photosynthesis?

.... **[1]**

2 **a)** Some plants have nitrogen fixing bacteria in their root nodules.
These bacteria can turn nitrogen into nitrates.

What natural process also fixes nitrogen?

.... **[1]**

b) Plants need nitrogen to make certain kinds of molecules.

What are these molecules?

.... **[1]**

3 The flow diagram below shows part of the nitrogen cycle.

Nitrogen in plants **A** Nitrogen in animals **B** Ammonium compounds

a) What process is shown by the letter **A**?
Underline the correct answer from the list below.

lightning   decomposition   feeding   nitrogen fixation   **[1]**

b) What process is shown by letter **B**?
Underline the correct answer from the list below.

lightning   decomposition   feeding   nitrogen fixation   **[1]**

Total Marks .... / 8

Interdependence

1 The diagram shows a farmland food web.

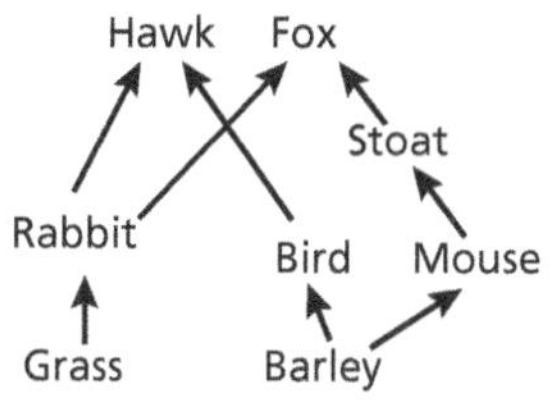

a) Name **two** animals that are in competition with each other for food.

.... **[1]**

b) Name **one** primary consumer from the web.

.... **[1]**

c) Name **one** producer from the web.

.... **[1]**

Hawk Fox

Stoat

Rabbit Bird Mouse

Grass Barley

2 The graph shows the population of foxes in a woodland over time.

a) Suggest **three** reasons for the decrease in fox numbers at point **A**.

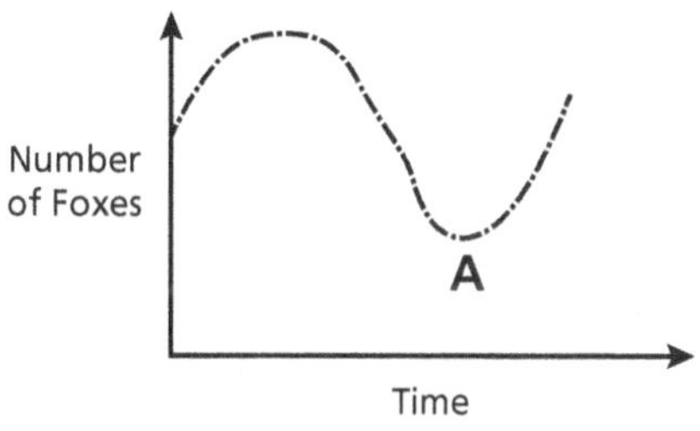

....

....

.... **[3]**

Number of Foxes

A

Time

b) Give **three** resources for which foxes will compete.

.... **[3]**

c) The trees in the woodland will also be in competition with each other. Give **three** resources for which trees will compete.

.... **[3]**

3 The image shows a buffalo and some oxpecker birds. The birds feed on tiny parasites on the buffalo's skin.

a) What sort of relationship is this?

....

.... **[1]**

b) Marcus says that the oxpecker bird is a parasite.

Why is Marcus wrong?

....

.... **[1]**

Total Marks .... / 14

Genes

1 Complete the following sentences by filling in the missing words.

An inherited characteristic is controlled by ...., which are found on structures

called chromosomes. Chromosomes are made of .... which is coiled to form a

.... Body cells have pairs of chromosomes and are said to be

diploid, while gametes have single chromosomes and are said to be .... **[4]**

2 Chelsea is describing herself on a dating website.

I am 27 years old, 150cm tall and quite thin. I have blonde hair and blue eyes. I have my ears and eyebrows pierced. I work in a sports shop and am a very good swimmer.

a) Which **two** characteristics are inherited?

.... **[2]**

b) Name **one** characteristic that is most likely due to a combination of inherited and environmental factors.

.... **[1]**

3 Match the pairs of alleles to the correct description by drawing a line between them.

tt		Heterozygous

Tt		Homozygous dominant

| TT | | Homozygous recessive | **[2]**
|----|--|----------------------|

4 A gardener is investigating the range of heights of sweet pea plants.
She measured the height of eight plants in a population.
Her results, in cm, are given below.

125   152   160   139   119   167   158   158

Calculate the mean height of the plants to 1 decimal place.

Mean = ... cm **[3]**

Total Marks .... / 12

Genetics and Reproduction

 1 A scientist carried out many investigations into breeding pea plants.

He concluded that the height of the pea plants was controlled by 'determiners' (which we now know are genes) and that plants are always tall or short.

The scientist crossed two pure bred pea plants – one tall and one short and found that all the offspring were tall.

a) Which is the dominant trait?

_____ **[1]**

b) The scientist then crossed two of the offspring and found that in the next generation the ratio of tall to small plants was 3 : 1.

Draw a genetic cross to show why this happened. **[2]**

2 A camel body cell has 70 chromosomes.

a) Following mitosis, how many chromosomes will each resulting cell have? _____ **[1]**

b) Following meiosis, how many chromosomes will each resulting cell have? _____ **[1]**

3 The 23rd pair of chromosomes in humans is responsible for determining sex.

Draw a genetic diagram to show how sex is determined. **[3]**

Total Marks _____ / 8

Natural Selection and Evolution

1 The phylogenic tree shows the classification of some of the big cats together with their Latin names.

a) Jordan says the snow leopard will be more closely related to the clouded leopard than the tiger, because the tiger is not a leopard.

Explain why Jordan is **not** correct.

....

....

....

| Lion *Panthera leo* |
| Jaguar *Panthera onca* |
| Leopard *Panthera pardus* |
| Tiger *Panthera tigris* |
| Snow leopard *Panthera uncia* |
| Clouded leopard *Neofelis nebulosa* |

[2]

b) Scientists use fossil evidence to help classify organisms.

What other technique can they use that helps to show the evolutionary relationships between animals?

.... [1]

2 Put a circle around the best word from each pair in bold to complete the passage about natural selection.

Individual organisms of a species may show a wide range of **randomness / variation** because of differences in their genes.

Occasionally, a **mutation / aberration** occurs that gives an individual organism an advantage over other organisms.

Individuals with characteristics most suited to the **environment / population** are more likely to **mutate / survive** and breed successfully.

The genes that have helped these individuals to survive are then passed on to their **generation / offspring**. [5]

3 Antibiotic resistance in bacteria provides evidence for evolution.

Explain how a population of bacteria with resistance to the antibiotic Penicillin would develop.

....

....

....

....

....

[5]

Total Marks .... / 13

Monitoring and Maintaining the Environment

1 Recent estimates suggest that if deforestation continues at its present rate, the world's rainforests will have vanished by the year 2150.

Explain why this is a problem.

....

....

....

....

[4]

2 Read the passage below about Kasanka in Zambia and answer the questions that follow.

> Thirty years ago, Kasanka was threatened by rampant poaching. There was no tourism, little wildlife and little community development.
>
> A British expatriate, called David Lloyd, visited the area and was so concerned that he applied for funding and permission to develop the area. He built tourist camps, roads and bridges. He also set up Kasanka National Park and Trust to raise funds for community-based projects and provide a safe habitat for wildlife.
>
> Income from tourism goes towards park management and community development. The community is involved in developing local schools, healthcare and horticulture in the area. The local communities also sell local produce to tourists.
>
> The trust promotes wildlife conservation, helps to enforce laws against poaching and helps to maintain habitats. Today, wildlife and the local community are both flourishing.

a) Suggested **one** way that tourism has benefited the local community.

....

.... **[1]**

b) How has the local environment and wildlife benefited from tourism?

....

....

.... **[2]**

Total Marks .... / 7

Investigations

1 A local wildlife group wanted to find out about the number of hedgehogs in an area.
They spent a night in the area and counted and marked each hedgehog they found.
A week later, they returned to the area and repeated the exercise.
This time they counted the hedgehogs and noted if they were marked or not.
On the first night, 12 hedgehogs were counted and marked.
On the second night, 10 hedgehogs were counted, 6 of which were marked.

a) What is the name of the technique used by the wildlife group?

.... **[1]**

b) Calculate the number of hedgehogs in the area. Show your working.

Answer: .... **[2]**

2 Some students investigated the effect of different concentrations of acid rain on the growth of broad bean plants.
They germinated 50 bean plants and selected five plants that had roots and shoots of the same length.
The students planted the five plants in different pots and watered each pot weekly with a different concentration of acid rain.
After six weeks they measured the height of the bean plants.
Their results are shown in the table.

pH of Rainwater	7.0	6.5	6.0	5.5	5.0	4.5
Height of Bean Plant (cm)	47	48	47	42	20	8

a) Suggest **two** variables that needed to be controlled when planting the bean plants in pots.

.... **[2]**

b) Which of these conclusions could the students draw from the results?

 A The plants with the highest pH grew best.

 B Increasing the pH resulted in taller plants.

 C When the pH gets too acid, plants stop growing.

 D Bean plants are not affected until the pH drops to 5.5.

Your answer ☐ **[1]**

c) Describe **one** way by which the students could improve the reliability of their results.

.... **[2]**

Total Marks .... / 8

Feeding the Human Race

1 Genetic engineering involves modifying the genes of an animal or plant to introduce desirable characteristics. The diagram shows part of the process.

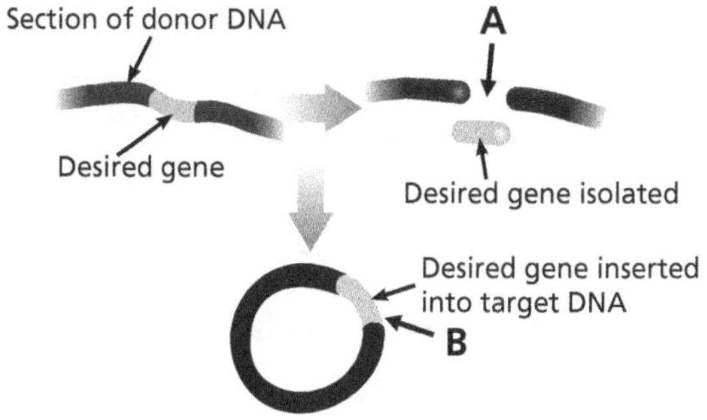

a) What type of enzyme is used at point **A**?

.... [1]

b) What type of enzyme is used at point **B**?

.... [1]

Section of donor DNA
A
Desired gene
Desired gene isolated
Desired gene inserted into target DNA
B

c) The desired gene is often inserted into bacterial hosts. Explain why bacteria are used.

....

.... [2]

d) Suggest why each of the following traits would be desirable

i) Maize that produces an insecticide in its leaves.

.... [1]

ii) Rice that contains high levels of vitamins.

.... [1]

2 Aphids (greenfly) cause considerable damage, particularly to wheat crops.
One method of control is to spray crops with insecticides. However, this can lead to loss of biodiversity. Scientists have discovered that aphids produce an 'alarm' odour called β-farnesene, which they use to alert each other to danger. They are now working to produce genetically engineered wheat plants that will release β-farnesene.

a) Why does the use of insecticides reduce biodiversity?

.... [1]

b) Why do scientists want to produce wheat that releases β-farnesene?

.... [1]

c) Explain how scientists could produce this wheat using gene technology.

....

....

....

.... [5]

Total Marks .... / 13

Monitoring and Maintaining Health

1 Which of the following diseases are communicable diseases?
Tick (✓) **three** correct boxes.

AIDS		Haemophilia	
Measles		Cholera	
Diabetes		Liver disease	
Cancer		Goitre	

[3]

2 Match each microorganism to its size and the disease that it causes by drawing lines between the boxes.

Type of Microorganism	Size	Disease
Virus	10µm	Thrush
Bacteria	1µm	Malaria
Protozoa	100µm	Tuberculosis
Yeast	0.1µm	Chickenpox

[4]

3 **a)** Name the virus that leads to AIDS.

.... [1]

b) Give **three** ways in which the virus can be transmitted from one person to another.

....

....

.... [3]

c) Antibiotics have no effect on viruses. However, sometimes people with AIDS are treated with antibiotics.

Suggest why.

....

....

.... [2]

Total Marks .... / 13

Prevention and Treatment of Disease

1 The human body has a number of mechanisms to prevent pathogens from entering.

   a) Describe the role of platelets in defence against disease.

   ....

   .... **[2]**

   b) The respiratory system produces mucus.

   What is the role of mucus in defence against disease?

   .... **[1]**

2 **a)** Name **two** tests that may be carried out on new drugs before clinical trials.

   .... **[2]**

   b) In clinical trials, some patients are given the drug, and others are given a sugar pill that contains no drugs.

   Explain why.

   ....

   .... **[2]**

3 Animals have physical barriers, such as skin and nose hairs, which stop pathogens entering their bodies.

Plants also have physical barriers.

   a) Name **two** physical barriers that plants have to prevent entry of pathogens.

   ....

   .... **[2]**

   b) Plant pathogens can be spread by direct or indirect contact.

   Give **two** examples of how plant pathogens can be spread by indirect contact.

   ....

   ....

   .... **[2]**

Total Marks .... / 11

Non-Communicable Diseases

1 By studying the human genome, scientists are able to predict the likelihood of an individual having a particular disease.
Potentially, this could be used to predict the chances of someone getting cancer.

Suggest **one** advantage and **one** disadvantage of being able to use the human genome in this way.

Advantage: ....

Disadvantage: .... **[2]**

2 Body mass index (BMI) is a simple measurement that is commonly used to classify a person as underweight, normal weight, overweight, or obese.
It is calculated using the following equation:

$$BMI = \frac{weight\,(kg)}{height^2\,(m)}$$

A BMI greater than or equal to 25 is overweight.
A BMI greater than or equal to 30 is obese.

Karanjit is 1.62 metres tall and weighs 74kg

a) Calculate Karanjit's BMI to one decimal place.

.... **[2]**

b) Name **two** health risks associated with obesity.

.... **[2]**

3 Which statement best describes why smoking increases the risk of coronary heart disease?

    A It increases the ability of the blood to carry oxygen, which puts extra strain on the circulatory system.

    B It reduces the ability of the blood to carry oxygen, which puts extra strain on the circulatory system.

    C It reduces the ability of the blood to carry carbon dioxide, which puts extra strain on the circulatory system.

    D It increases the ability of the blood to carry carbon dioxide, which puts extra strain on the circulatory system.

Your answer ☐ **[1]**

Total Marks .... / 7

Particle Model and Atomic Structure

1 The smallest particle that retains the properties of the element is the:

A neutron   **B** proton   **C** atom   **D** Higgs boson

Your answer ☐ **[1]**

2 Which of the following correctly explains why an atom has no charge?

A The number of neutrons equals the number of protons.

B The number of protons equals the number of electrons.

C The number of neutrons equals the number of electrons.

D The total number of protons and neutrons equals the number of electrons.

Your answer ☐ **[1]**

3 This table about the three main isotopes of carbon, is missing some key information.

Isotope	Symbol	Mass Number	Atomic Number	Protons	Neutrons	Electrons
Carbon-12	$^{12}_{6}C$	12	6	**B**	6	**D**
A	$^{13}_{6}C$	13	6	6	7	6
Carbon-14	$^{14}_{6}C$	14	6	6	**C**	6

a) Fill in the missing sections of the table. **[4]**

b) What is the name of the particle whose number defines an element? **[1]**

....

c) Palladium has an atomic mass of 106 and atomic number of 46.

How many **neutrons** are in the nucleus? ☐ **[1]**

4 Look at the diagram of a gas. Particles are drawn as circles.

Explain **three** limitations of this model.

....

....

[3]

Total Marks _____ / 11

Purity and Separating Mixtures

1 Glucose, $C_6H_{12}O_6$, can be bought at the following price points:

- Supermarket: £3/kg
- Industrial chemical supplier: £6/kg; £18/kg; £114/kg

Suggest why the chemical supplier can charge £114/kg for glucose.

....

.... **[2]**

2 Which of the following is the empirical formula of glucose, $C_6H_{12}O_6$?

A $C_6H_{12}O_6$ **B** $C_2H_3O_2$ **C** CHO   **D** CH_2O

Your answer ☐ **[1]**

3 Calculate the relative formula mass of $Mg(OH)_2$. Show your working.

relative formula mass = ... **[2]**

4 What is the empirical formula of a chemical containing 60g of carbon, 12g of hydrogen and 16g of oxygen? Show your working.

empirical formula = ... **[2]**

5 Sandy wants to extract the pigment in ink so that she can transport it easily on a camping expedition.

Explain how she can separate the pigment from the ink.

....

....

.... **[3]**

Total Marks .... / 10

Bonding

1 Tungsten is a metal. Which of the following is **not** a reason why it is a metal?

A It conducts electricity easily.

B It is malleable.

C It is a good conductor of heat.

D It has the symbol W.

Your answer ☐ **[1]**

2 Which of the following shows the **electronic structure** for magnesium?

A 2.8.2 **B** 2.8.1 **C** 2.8.8.2 **D** 2.8.8.1

Your answer ☐ **[1]**

3 Explain the differences between a **covalent bond** and an **ionic bond**.

....

....

....

....

[4]

4 Sodium, Na, is a metal. It has the atomic number 11.

Which of the following can you **not** deduce using the atomic number?

A The electronic structure. **B** The period sodium is in.

C The number of isotopes. **D** The group sodium is in.

Your answer ☐ **[1]**

5 Draw the electronic structure for the element chlorine, Cl.

☐

[1]

Total Marks .... / 8

Models of Bonding

Methane has the formula CH_4.

a) Draw a dot and cross diagram and the displayed formula for methane in the spaces below.

Dot and Cross Diagram   **Displayed Formula**   [2]

b) Give **one** advantage that ball and stick models have over displayed formulae.

.... [1]

2 Aluminium forms an ion.

Draw a dot and cross diagram to show an aluminium cation.

[1]

3 Hydrogen sulfide, H_2S, is a covalent compound.

a) Draw a dot and cross diagram to show a molecule of H_2S.

[1]

b) How many electrons are there in a **single** covalent bond?

.... [1]

Total Marks .... / 6

Properties of Materials

1 Which of the following best describes an **allotrope**?

A   An element with the same number of protons and neutrons.

B   An element with the same number of protons but a different number of neutrons.

C   An element that can form different physical forms.

D   An element that can form long-chained molecules.

Your answer ☐ **[1]**

2 Graphite melts at 3600°C. Water melts at 0°C.

Explain why graphite and water melt at such widely different temperatures.

....

....

....

....

[4]

3 Which of the following is **not** an allotrope of carbon?

A   glucose   B   diamond   C   buckyballs   D   graphene

Your answer ☐ **[1]**

4 Why is diamond such a hard material?

....

....

[2]

5 a) Draw the structure of diamond in the space alongside.   **[2]**

b) Explain why diamond does **not** conduct electricity.

....

....

....

....

[2]

Total Marks _____ / 12

Introducing Chemical Reactions

1. Write the number of atoms of each element for each formula below:

a) $BeAl_2O_4$

.... [1]

b) Al_2O_5Si

.... [1]

c) $Ba(PO_3)_2$

.... [1]

d) $CH_3(CH_2)_{16}COOH$

.... [1]

2. Copper oxide reacts with sulfuric acid.

Which of the following is the correct **word** equation for this reaction?

A copper oxide + sulfuric acid → copper sulfate

B copper oxide + sulfuric acid → copper sulfate + water

C copper oxide + sulfuric acid → copper + hydrogen sulfate

D copper oxide + sulfuric acid → copper sulfate + carbon dioxide + water

Your answer ☐ [1]

3. Magnesium carbonate reacts with nitric acid.

magnesium carbonate + nitric acid → magnesium nitrate + carbon dioxide + water

Write the **balanced symbol** equation for this reaction.

.... [2]

Total Marks .... / 7

Chemical Equations

1 This question is about the four ions: **iron(II)**, **magnesium**, **sulfate** and **hydrogen carbonate**.

Which of the following correctly shows the charges on each ion?

	Iron(II)	Magnesium	Sulfate	Hydrogen Carbonate
A	3+	2+	2+	2–
B	2+	3+	2–	1–
C	2+	2+	2–	1–
D	2+	2+	2–	1+

Your answer ☐ **[1]**

2 Give the definition of the term **stoichiometry**.

....

.... **[2]**

3 Write the **half equation** for the following:

a) Bromine ions to bromine gas.

.... **[1]**

b) Hydrogen gas to hydrogen ions.

.... **[1]**

c) Iron(III) ions to elemental iron.

.... **[1]**

4 Look at the following reaction:

sodium carbonate (aq) + copper sulfate (aq) → copper carbonate (s) + sodium sulfate (aq)

a) What are **spectator ions**?

.... **[1]**

b) Which are the spectator ions in this reaction?

.... **[1]**

c) Write the **net ionic** equation for this reaction.

.... **[2]**

Total Marks .... / 10

Moles and Mass

1 Which of the following is the best explanation of the chemical mole?

    A The amount of a chemical substance that is the same as in 12g of the element carbon.

    B The amount of a chemical substance that is the same as in 12g of the element carbon-12.

    C The amount of a chemical substance that is the same as in the element carbon.

    D The amount of a chemical substance that is the same as in 12g of the element hydrogen.

Your answer ☐ [1]

2 Peter wants to measure out 3 moles of carbon atoms.

Given that the A_r for carbon-12 is 12, which of the following shows the mass of carbon he should measure out?

    A 12g   **B** 2.4g   **C** 36g   **D** 3.6g

Your answer ☐ [1]

3 Calculate the **relative formula mass** of the mineral chrysoberyl, $BeAl_2O_4$.
Show your working.

relative formula mass = ... [3]

4 How many grams of water will be produced when 5 moles of hydrogen is completely combusted in air?
Show your working.

amount of water = ... [3]

5 The A_r of rubidium is 85. What is the mass of a single Rb atom?

mass = ... [2]

Total Marks .... / 10

Energetics

1 Hannah is going to a concert. On the way there, she passes a stall selling glow stick jewellery. She buys a glow stick bracelet.

Before the bracelet will glow, the two chemicals inside the bracelet have to come into contact with each other.

The chemicals, luminol and hydrogen peroxide, react and produce a chemical that emits light.

Explain why this reaction is classed as being **exothermic**.

....

....

.... **[3]**

2 Which of the following shows a reaction profile for an **endothermic** reaction?

A   B   C   D

Reactants → Product (A), Reactants → Product (B), Product → Reactants (C), Products → Reactants (D)

Your answer ☐ **[1]**

3 Describe the features of an **endothermic** reaction.

....

.... **[2]**

4 A series of reactions was carried out and the overall bond energies are recorded below.

For each, state whether the reaction is **exothermic** or **endothermic**.

a) −4kJ/mol **[1]**

b) +134kJ/mol **[1]**

c) −1kJ/mol **[1]**

d) +8kJ/mol **[1]**

Total Marks .... / 10

Types of Chemical Reactions

1 Draw a (circle) around the substance that is being **reduced** in each reaction.

    a) lithium + chlorine → lithium chloride   **[1]**

    b) beryllium + oxygen → beryllium oxide   **[1]**

    c) potassium + bromine → potassium bromide   **[1]**

    d) lead oxide + hydrogen → lead + water   **[1]**

2 Look at the following reaction:

    iron(III) oxide (aq) + carbon monoxide (g) → iron (l) + carbon dioxide (g)

    a) Which species is being **reduced**?

    .... **[1]**

    b) Which species is being **oxidised**?

    .... **[1]**

    c) Write the **balanced symbol** equation for the reaction.

    .... **[2]**

3 Underline the **acid** in each of the following reactions.

    a) $MgCO_3(aq) + 2HCl(aq) \rightarrow MgCl_2(aq) + CO_2 (g) + H_2O(l)$ **[1]**

    b) $HNO_3(aq) + NaOH(aq) \rightarrow NaNO_3(aq) + H_2O(l)$ **[1]**

    c) $2CH_3CH_2COOH(aq) + 2K(s) \rightarrow H_2(g) + 2CH_3CH_2COOK(aq)$ **[1]**

    d) $2HBr(aq) + Mg(s) \rightarrow MgBr_2(aq) + H_2(g)$ **[1]**

4 Which of the following represents a **low** pH?

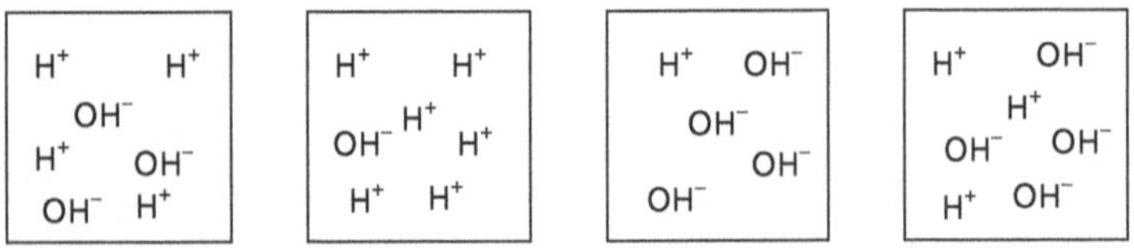

A	B	C	D

H^+ H^+
  OH^-
H^+ OH^-
OH^- H^+
	H^+ H^+
	OH^- H^+ H^+
	H^+ H^+
		H^+ OH^-
		  OH^-
		     OH^-
		OH^-
			H^+ OH^-
			   H^+
			OH^- OH^-
			H^+ OH^-

    Your answer ☐ **[1]**

Total Marks .... / 13

pH, Acids and Neutralisation

1. Trevor is measuring the pH of a local river.
He uses universal indicator.

Explain why using universal indicator is **less** accurate than using a data logger with a pH probe.

....

....

.... [3]

2. Read the following statements about acids and alkalis.

Which statements are **true**? Write the letters below.

A Strong acids have a high pH.

B Strong acids dissociate easily.

C A dilute acid has more H^+ ions per unit volume than a concentrated acid.

D A dilute acid has fewer H^+ ions per unit volume than a concentrated acid.

E Weak acids form equilibria.

.... [3]

3. Look at the following representations of acids:

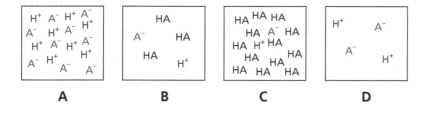

A   B   C   D

a) Which diagram represents a **concentrated, weak** acid?... [1]

b) Which diagram represents a **dilute, strong** acid? .... [1]

c) Which diagram represents a **concentrated, strong** acid? .... [1]

d) Which diagram represents a **dilute, weak** acid? .... [1]

Total Marks .... / 10

Electrolysis

1 When a metal is reacted with an acid it forms a metal salt, plus hydrogen gas.
A reactivity series is shown on the right.

Look at the following reactions:

W sodium + sulfuric acid ➜ sodium sulfate + hydrogen

X calcium + sulfuric acid ➜ calcium sulfate + hydrogen

Y potassium + sulfuric acid ➜ potassium sulfate + hydrogen

Z magnesium + sulfuric acid ➜ magnesium sulfate + hydrogen

Potassium
Sodium
Calcium
Magnesium
Aluminium
Carbon
Zinc
Iron
Tin
Lead
Hydrogen
Copper
Silver
Gold
Platinum

a) Which of the following shows the **correct order** of reactivity for these reactions, least to most reactive?

A WXYZ

B ZWXY

C ZXWY

D YWXZ

Your answer ☐ **[1]**

b) Element **V** can displace zinc from zinc oxide. It cannot be displaced by carbon.

Suggest where element **V** would appear in the reactivity series.

.... **[1]**

2 Explain why some metals have to be extracted from their ores by electrolysis.

....

....

....

.... **[3]**

Total Marks .... / 5

Predicting Chemical Reactions

1 Which of the following correctly explains the reactivity of **Group 1** metals?

   A As the outermost electron is closest to the nucleus it is easier to remove that electron.

   B As the outermost electron is furthest from the nucleus it is easier to remove that electron.

   C The outermost electrons are closest to the nucleus and are easier to remove.

   D The outermost electrons are closer to the nucleus and are harder to remove.

Your answer   ☐   [1]

2 Argon is a noble gas. It has the atomic number 18.

Which of the following correctly shows the electronic structure of argon?

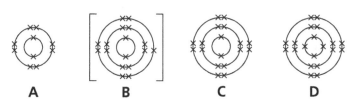

   A   **B**   **C**   **D**

Your answer   ☐   [1]

3 Describe how you would test if the gas produced in a reaction is oxygen.

....

.... **[2]**

4 Fluorine and oxygen are next to each other in the periodic table in Period 2.

Describe the similarities and differences between the reactivity of fluorine and oxygen.

....

....

....

....

.... **[3]**

Total Marks .... / 7

Controlling Chemical Reactions

1 Which of the following shows the correct units for measuring the rate of reaction?

A g/s   **B** g/m   **C** m/s   **D** cm²/s

Your answer ☐ **[1]**

2 Which of the following correctly shows the reason why a chemical reaction stops?

A The maximum quantity of reactant has been made.

B The minimum quantity of reactant has been made.

C A reactant is used up.

D A product is used up.

Your answer ☐ **[1]**

3 Give **two** ways in which the rate of a reaction can be changed.

....

.... **[2]**

4 Look at the following diagrams.

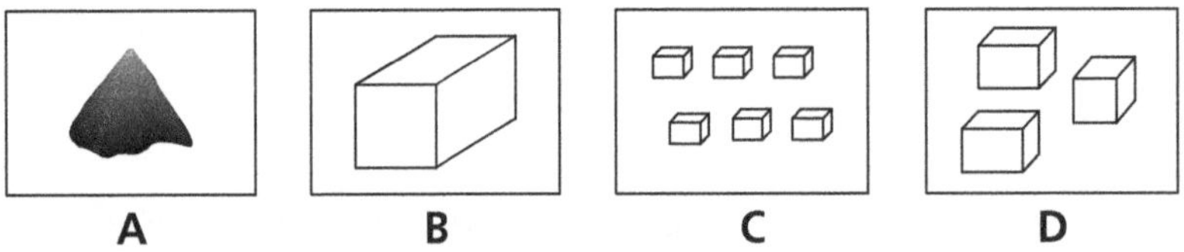

The diagrams all show the same mass of calcium carbonate.
Calcium carbonate reacts with nitric acid.

a) Which diagram shows the calcium carbonate that would react the **slowest** with nitric acid?

Your answer ☐ **[1]**

b) Describe how surface area affects the rate of reaction.

....

.... **[2]**

Total Marks .... / 7

Catalysts and Activation Energy

1 Fraser investigated the reaction between copper carbonate and nitric acid.
He used the same amount of copper carbonate and nitric acid each time.
He changed the temperature.
He measured the amount of carbon dioxide produced and drew a graph.

a) What does a **steeper** line on the graph indicate?

_____ **[1]**

b) In Fraser's experiment, which **temperature** led to the quickest CO_2 production?

_____ **[1]**

c) What was the **maximum** amount of CO_2 produced?

_____ **[1]**

d) At what **time** did the reaction end for the experiment at 30°C?

_____ **[1]**

e) Predict how long it would take for the reaction to finish if the temperature was 60°C.

_____ **[1]**

2 Explain how a catalyst affects the rate of reaction.

_____

_____

_____

_____

_____ **[4]**

Total Marks _____ / 9

Equilibria

1 What is meant by the term **irreversible** reaction?

_____ [1]

2 Which of the following reactions are classed as being **reversible**? Write down the correct letters.

A $CH_4(g) + 2O_2(g) \rightarrow CO_2(g) + 2H_2O(l)$

B $2Li(s) + 2HCl(aq) \rightarrow 2LiCl(aq) + H_2(g)$

C $H_2(g) + I_2(g) \rightleftharpoons 2HI(g)$

D $NH_4Cl(s) \rightleftharpoons NH_3(g) + HCl(g)$

_____ [1]

3 The graph on the right shows a dynamic equilibrium.

Which of the following shows the correct labels for the graph?

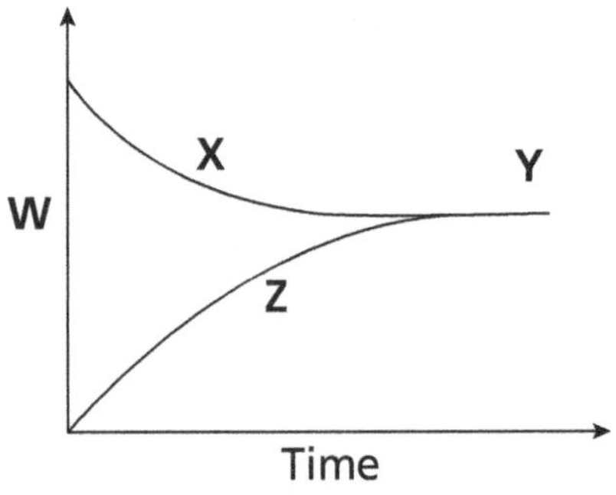

A W = forward reaction   X = reaction rate   Y = equilibrium   Z = backward reaction

B Y = forward reaction   Z = reaction rate   W = equilibrium   X = backward reaction

C X = forward reaction   W = reaction rate   Y = equilibrium   Z = backward reaction

D X = forward reaction   Y = reaction rate   W = equilibrium   Z = backward reaction

Your answer ☐ [1]

4 Look at the following reaction:

$H_2(g) + F_2(g) \rightleftharpoons 2HF(g)$

The pressure of the system is **decreased**.

How will this affect the position of the equilibrium?

_____ [1]

Total Marks _____ / 4

Improving Processes and Products

1   What is the name given to a rock containing minerals of metal compounds?

.... **[1]**

2   Which of the following is the element that most metals on the planet are bonded to?

A   sulfur   **B**   silica   **C**   oxygen   **D**   carbon

Your answer ☐   **[1]**

3   Andy is carrying out the experiment shown in the diagram below.

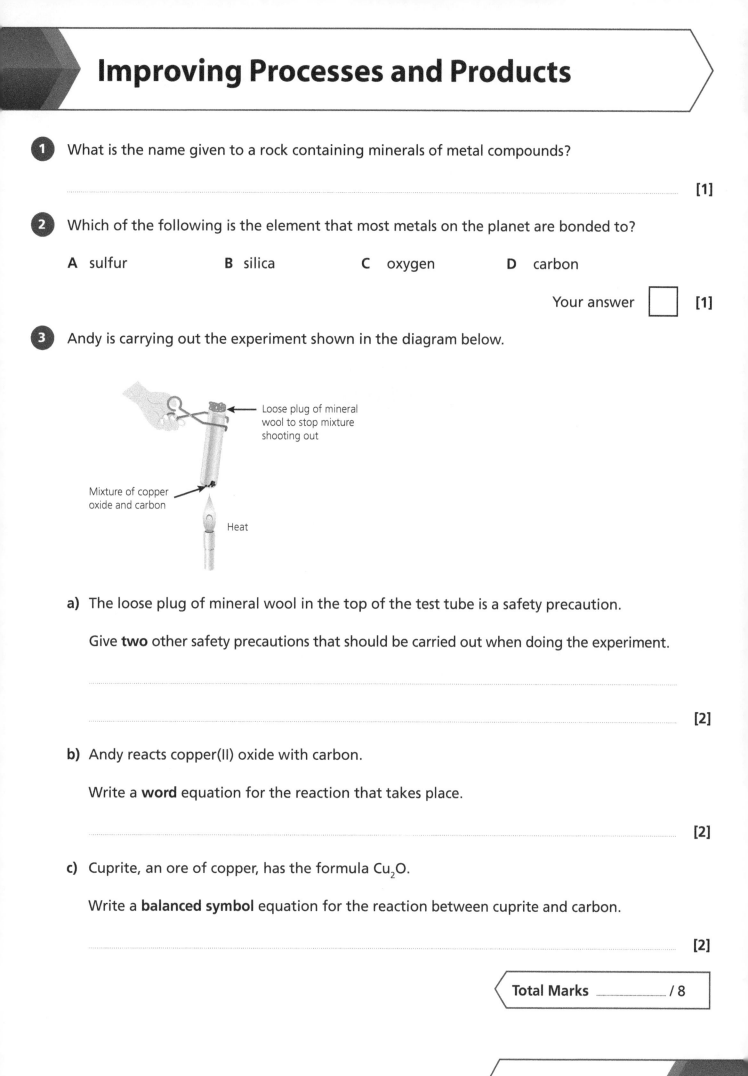

Loose plug of mineral wool to stop mixture shooting out

Mixture of copper oxide and carbon

Heat

a) The loose plug of mineral wool in the top of the test tube is a safety precaution.

Give **two** other safety precautions that should be carried out when doing the experiment.

....

.... **[2]**

b) Andy reacts copper(II) oxide with carbon.

Write a **word** equation for the reaction that takes place.

.... **[2]**

c) Cuprite, an ore of copper, has the formula Cu_2O.

Write a **balanced symbol** equation for the reaction between cuprite and carbon.

.... **[2]**

Total Marks .... / 8

Life Cycle Assessments and Recycling

1 Lochmara Lodge Cafe prides itself on being environmentally friendly.
Inside the cafe the lights have lampshades made out of used computer circuit boards.

The lampshades are an example of recycling.

Give **three** reasons why recycling is beneficial.

....

....

.... **[3]**

2 Some car parts are made out of carbon fibre instead of plastic.
When designing the car, designers had to take into account a life cycle assessment.

a) Give the stages involved in a **life cycle assessment**.

....

....

....

.... **[4]**

b) Which is more sustainable, carbon fibre or plastic?
Explain your answer.

....

.... **[2]**

Total Marks .... / 9

Crude Oil

1 The diagram shows a fractional distillation column.

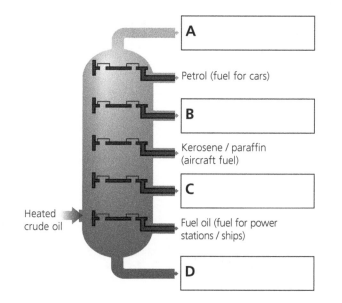

A

Petrol (fuel for cars)

B

Kerosene / paraffin
(aircraft fuel)

C

Heated
crude oil

Fuel oil (fuel for power
stations / ships)

D

The diagram is missing the following labels:

Bitumen   Diesel   Naphtha   Refinery gases

Write each label in the correct place on the diagram.   **[4]**

2 Explain why short chain hydrocarbons have a low boiling point and long chain hydrocarbons have a high boiling point.

....

....   **[2]**

3 Not all of the products from fractional distillation are useful.
To make them useful, they have to be cracked.

What is meant by the term **cracking**?

....   **[1]**

4 The hydrocarbons in crude oil are mainly alkanes.

Which of the following gives the correct formula for alkanes?

A C_2H_n **B** $C_{2n}H_{n+2}$ **C** C_nH_{2n+2} **D** C_nH_{2n}

Your answer ☐ **[1]**

Total Marks .... / 8

Interpreting and Interacting with Earth's Systems

1 The atmosphere of Earth currently is made up of the following gases.

Gas	Proportion of the Atmosphere (%)
Nitrogen	78
Oxygen	21
Carbon dioxide	0.04
Other gases	1

a) The numbers in the table do not add up to 100%.

Suggest why this is the case.

.... **[1]**

b) Scientists have created models of the atmosphere of early Earth.
One of the models is shown below. It is in the **incorrect** order.

Put these statements into the correct order.

A These newly formed oceans removed some carbon dioxide by dissolving the gas.

B Primitive plants that could photosynthesise removed carbon dioxide from the atmosphere, and added oxygen.

C Levels of nitrogen in the atmosphere increased as nitrifying bacteria released nitrogen.

D A hot volcanic Earth released ammonia, carbon dioxide and water vapour.

E The Earth cooled and water vapour condensed into liquid water.

The **correct** order is:

[2]

2 What part of the electromagnetic spectrum is affected by greenhouse gases?

A gamma   **B** ultraviolet   **C** infrared   **D** heat

Your answer [ ] **[1]**

Total Marks .... / 4

Air Pollution and Potable Water

1 Draw a line from each air pollutant to the effect it produces.

Carbon monoxide		coats surfaces with soot
Sulfur dioxide		kills plants and aquatic life
Nitrogen oxides		forms photochemical smogs
Particulates		air poisoning

[3]

2 Clean drinking water is essential for everyday life.
Look at the flowchart.
It shows the treatment of sewage water leading to fresh drinking water.

Water filtration
(A)
Sedimentation tank
(C) (B)
Ultra filtration
(D)
Reservoir

a) What is **removed** from the water treatment system at **A** and **B**?

A ....

B .... [2]

b) What is **added** to the water treatment system at **C** and **D**?

C ....

D .... [2]

3 Draw a diagram to show how **thermal desalination** takes place.

[2]

Total Marks .... / 9

Matter, Models and Density

1 Describe the forces between the molecules in a gas.

.... [1]

2 State the evidence that led to Rutherford claiming that the atomic nucleus was positively charged.

....

.... [2]

3 Explain why solid aluminium has a higher density than molten aluminium.

....

.... [2]

4 The density of copper is $8.96 \times 10^3 \text{kg/m}^3$.

Calculate the mass of a 2m^3 cube of copper.

mass = ... [2]

5 Name the particle that J. J. Thomson discovered.   Answer: .... [1]

6 State how the Bohr model of the atom differed from Rutherford's model.

....

.... [2]

7 1kg of steam condenses to water at 100°C.

In the boxes below, sketch how the molecules might appear in the two states.

Steam (gas)   Water (liquid)   [2]

Total Marks .... / 12

Temperature and State

Refer to the Data Sheet on page 456 when answering these questions.

1 A small block of aluminium is given just sufficient energy to melt.

Identify **one** factor that remains unchanged when it melts.

.... **[1]**

2 What type of change is a change of state?

.... **[2]**

3 State **two** factors that are linked to the internal energy of a solid or liquid.

....

.... **[2]**

4 **a)** The latent heat of vaporisation of water is 2.26×10^6 J/kg.

Calculate how much energy is released if 0.1kg of steam condenses to water at 100°C.

Answer: .... **[2]**

b) The specific heat capacity of water is 4200J/kg/°C.

Calculate how much energy is released when 0.1kg of water at 100°C cools to 35°C.

Answer: .... **[2]**

c) Use your answers to parts **a)** and **b)** to explain why steam burns are more serious than burns from boiling water.

....

....

....

.... **[4]**

Total Marks .... / 13

Journeys

Refer to the Data Sheet on page 456 when answering these questions.

1 Describe the difference between **speed** and **velocity**.

....

.... **[2]**

2 A car travels at an average speed of 30m/s.

Calculate how far it travels in 40 seconds.

Answer: .... **[2]**

3 A car travelling at 30m/s increased its speed to 70m/s in 8 seconds.

Calculate the acceleration.

Answer: .... **[2]**

4 A moon is orbiting a planet in a circular orbit.

State which of the following are changing: **acceleration, velocity** or **speed**.

Answer: .... **[1]**

5 A 50kg athlete is running at 8m/s.

Calculate how much kinetic energy they have.

Answer: .... **[3]**

Total Marks .... / 10

Forces

1 What is the difference between a **gravitational** force and an **electrostatic** force?

....

.... **[1]**

2 State Newton's **first** law of motion.

....

.... **[2]**

3 The Earth makes a complete orbit of the Sun each year.

Describe the forces acting on the two bodies while this is happening and compare their magnitude.

....

....

.... **[3]**

4 A skydiver falling from an aircraft reaches terminal velocity.

Explain, in terms of forces, why the skydiver is not accelerating.

....

....

.... **[3]**

5 Explain what is meant by **inertia**.

.... **[2]**

6 Explain why a satellite that is in a circular orbit around the Earth has a **constant speed** but **changing velocity**.

....

.... **[2]**

Total Marks _____ / 13

Force, Energy and Power

1 A motorcyclist and their bike have a total mass of 310kg.

Calculate their momentum when travelling at 20m/s.

Answer: _____ **[2]**

2 When a toy water rocket at rest is pumped up, water is propelled from the bottom and it takes off.

Explain why this happens in terms of **momentum**.

_____

_____

_____

_____ **[4]**

3 Calculate how much gravitational potential energy a 60kg person gains if they climb to the top of a set of steps 12m high. (Take g to equal 10m/s².)

Answer: _____ **[2]**

4 A machine can lift a 150kg mass 12m vertically in 90 seconds.

Calculate the power of the machine. (Take g to equal 10m/s².)

Answer: _____ **[2]**

5 Explain why a person might find that pushing a piano is difficult, despite it having wheels.

_____

_____ **[2]**

Total Marks _____ / 12

Changes of Shape

Refer to the Data Sheet on page 456 when answering these questions.

1 What is meant by a **plastic** material?

....

.... **[1]**

2 Explain why the value for gravitational field strength is larger on Jupiter than it is on Mars.

.... **[1]**

3 A spring extends 2cm when a force of 1N is applied.

Calculate the value of the spring constant.

Answer: .... **[2]**

4 A student added weights to a spring, recording the extension for each weight.
A graph was plotted of the results, as shown below.

Explain the shape of the graph.

....

....

....

.... **[3]**

5 A spring with spring constant 80N/m is stretched 30cm.

Calculate how much energy is transferred.

Answer: .... **[2]**

Total Marks .... / 9

Electric Charge

1 What name is given to the space around a charged object in which another charged object would experience a force?

Answer: .... **[1]**

2 Apart from giving a reading on an ammeter, state the **two** other detectable effects that a flow of charge can produce.

....

.... **[2]**

3 Explain why a plastic comb is an insulator, despite containing electrons.

.... **[2]**

4 A battery produces a current of 5A for 10 seconds.

Calculate how much charge flows.

Answer: .... **[2]**

5 A teacher rubs a plastic rod with a woollen cloth. Both the rod and cloth are insulators.

a) Explain why the plastic rod becomes positively charged and the woollen cloth becomes negatively charged.

....

....

.... **[3]**

b) A student holds the positive rod over a discharged electroscope.

Explain why the gold leaf inside the electroscope rises.

....

.... **[3]**

Total Marks .... / 13

Circuits

1 State what the gradient of a **potential difference–current** (*V–I*) graph represents for a component.

Answer: .... **[1]**

2 A student plotted the graph below for a component in a circuit after varying the potential difference across it and measuring the current.

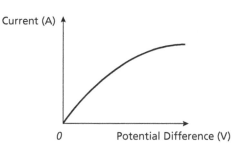

a) Suggest what the component was most likely to be.

Answer: .... **[1]**

b) Explain why the graph in part **a)** has that shape for the component.

....

....

....

....

....

.... **[4]**

3 A student obtained the graph below by varying the potential difference across a component and measuring the current.

a) Name the component.

Answer: .... **[1]**

b) State what the graph indicates about the current direction.

.... **[1]**

Total Marks .... / 8

Resistors and Energy Transfers

1. Calculate the value of the potential difference across **V** in the circuit below.

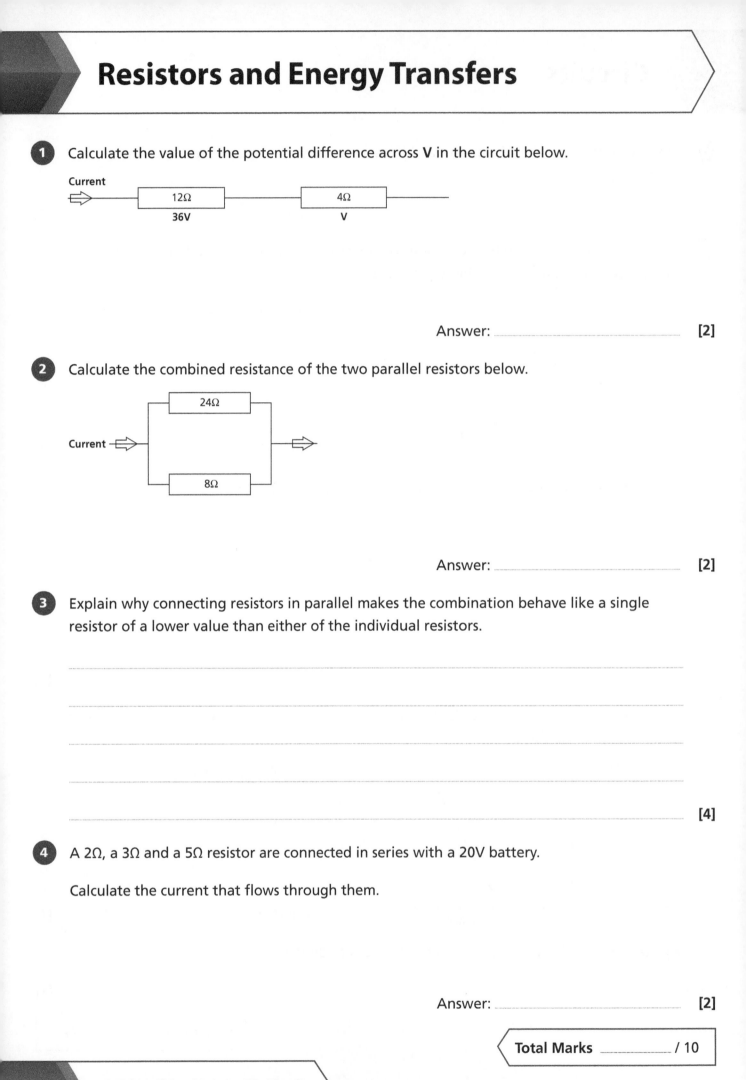

Answer: .... [2]

2. Calculate the combined resistance of the two parallel resistors below.

Answer: .... [2]

3. Explain why connecting resistors in parallel makes the combination behave like a single resistor of a lower value than either of the individual resistors.

....

....

....

....

.... [4]

4. A 2Ω, a 3Ω and a 5Ω resistor are connected in series with a 20V battery.

Calculate the current that flows through them.

Answer: .... [2]

Total Marks .... / 10

Magnetic Fields and Motors

Refer to the Data Sheet on page 456 when answering these questions.

1 A compass is placed near a coil of wire that is connected via a closed switch to a 12V battery. After 15 seconds, the switch is opened.
The compass needle moves and points in a different direction.

Suggest a reason for this.

....

....

....

[2]

2 A student investigates the magnetic field due to a coil placed on a piece of paper and connected to a 6V battery.
They sprinkle iron filings around the coil.

a) Suggest how they would identify the **strongest** region of the magnetic field.

....

....

[1]

b) What difference would the student have noticed if the same experiment had been carried out with a 20V battery rather than a 6V one? Explain your answer.

....

....

[3]

3 A wire of length 0.5m is placed in a magnetic field of magnetic flux density 2.4×10^{-2}T.

Calculate the **force** that acts on the wire if a current of 3A flows in it.

Answer: .... **[2]**

Total Marks .... / 8

Wave Behaviour

1 What type of wave is a **sound** wave? Answer: _____ **[1]**

2 A student uses a 0.9 m long ripple tank to measure the velocity of a water wave.
They find out that the wave travels the length of the tank in 3 seconds.
The distance between waves was 0.01 m.

   a) Calculate the velocity of the wave.

velocity = _____ **[2]**

   b) What would be the frequency of this wave?

frequency = _____ **[2]**

3 A student stands a distance from a wall and claps their hands.
When they hear the echo, they clap again.
They keep this going in a regular rhythm while another student measures the time between claps.

   a) Suggest how the speed of sound could be calculated using this data.

   _____

   _____ **[2]**

   b) Explain why it would have been better to measure the time for 10 claps.

   _____

   _____

   _____ **[2]**

4 The diagram below shows the refraction of water waves as they pass into shallow water.

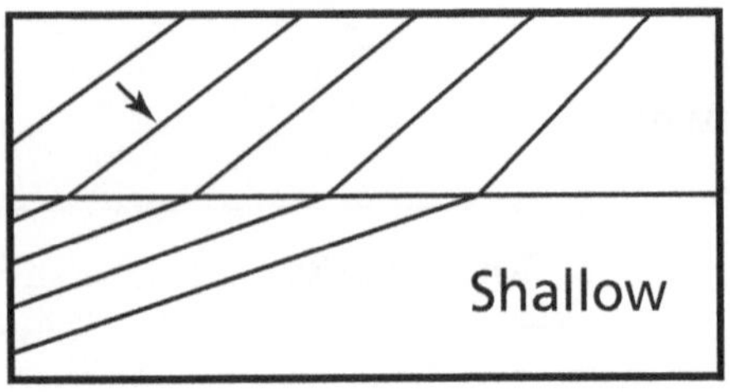

Shallow

What **two** wave measurements have changed during refraction?

_____

_____ **[2]**

Total Marks _____ / 11

Electromagnetic Radiation

1 A pebble is dropped into a ripple tank, producing a water wave. This can model an electromagnetic wave.

State what property of the wave cannot now be changed.

Answer: _____ [1]

2 A dentist will leave the room when a patient is having an X-ray.

Suggest the reason for this.

_____ [2]

3 Explain how X-rays are used at a hospital to check for fractures in the body.

_____

_____

_____ [3]

4 A radio station broadcasts by using radio waves of wavelength 1500m.
The speed of electromagnetic radiation is 3×10^8m/s.

Calculate the **frequency** of the station.

Answer: _____ [3]

5 Multiple sclerosis is a condition that leads to poor circulation and lower temperatures in the extremes of the body, such as the hands.

Identify the part of the electromagnetic spectrum that could be used by a scanner to detect this condition.

Answer: _____ [1]

6 X-rays and gamma rays are forms of **ionising radiation**.

Describe how they can affect the body.

_____

_____ [2]

Total Marks _____ / 12

Nuclei of Atoms

1 Name the particles that are found in an atomic **nucleus**.

.... [2]

2 Uranium atoms, atomic number 92, can be found with different numbers of neutrons.

What word describes these different forms of the same element?

Answer: .... [1]

3 Gamma rays are known as ionising radiation.

Describe how they can affect atoms.

....

.... [2]

4 A teacher puts different absorbers between a radioactive source and a Geiger counter.
When paper is used, the count rate on the Geiger counter doesn't change.
However, the count rate drops when aluminium is used and drops further when lead is used.

Name the types of radiation that the source emitted.

.... [2]

5 State the values for **X** and **Y** in the equation for radioactive decay below.

$^{235}_{92}U \rightarrow {}^{Y}_{X}Th + {}^{4}_{2}He$

.... [2]

6 Two nuclei of carbon have different numbers of neutrons.

$^{12}_{6}C$ $^{14}_{6}C$

a) Describe the structure of each nucleus.

....

.... [2]

b) State the charge on each nucleus.

.... [1]

Total Marks .... / 12

Half-Life

1 Explain what is meant by the term **'half-life'**.

_____

_____ **[1]**

2 A sample of radioactive iodine, with a mass of 64g and a half-life of 13 hours, was left to decay.

a) Calculate the amount of iodine that was left after 52 hours.

Answer: _____ **[3]**

b) Calculate the fraction of the starting nuclei that remained after 52 hours.

Answer: _____ **[1]**

3 Explain what is meant by the term **'irradiation'**.

_____ **[1]**

4 Americium-241 is used in smoke detectors. It emits alpha particles.
The equation for the radioactive decay is shown below.

$$^{241}_{95}\text{Am} \rightarrow \, ^{237}_{x}\text{Np} + \, ^{y}_{2}\text{He}$$

a) What are the values of x and y?

$x =$ _____ $y =$ _____ **[2]**

b) Americium-241 has a half life of 432 years, so will not need to be replaced in the smoke detector.

Suggest why this is an advantage.

_____

_____ **[2]**

Total Marks _____ / 10

Systems and Transfers

Refer to the Data Sheet on page 456 when answering these questions.

1 A hot saucepan is dropped into a sink full of cold water.

    a) Explain why the saucepan cools down.

    ....

    .... **[2]**

    b) Describe what happens to the cold water as a result.

    .... **[1]**

2 According to the principle of conservation of energy: 'the total energy of the system remains the same, whatever happens inside the system'.

    State **two** things that could change the energy of the system.

    ....

    .... **[2]**

3 Calculate the amount of heat required from an electric kettle to raise the temperature of 1.2kg of water from 20°C to boiling point at 100°C.
(Take the specific heat capacity of water to be 4200J/kg/°C.)

                                       Answer: .... **[2]**

4 A farm has a hydroelectric system to produce electrical power from a generator.
20kg of water falls 15m down a vertical hillside every second.

    a) State the main energy changes that take place in the system.

    ....

    .... **[3]**

    b) In reality, energy will be lost from the system. State what form that energy is likely to take.

                                         Answer: .... **[1]**

Total Marks .... / 11

Energy, Power and Efficiency

1. When charging a mobile phone from the mains, not all the electrical energy is transferred through the transformer (charger).

   Describe how energy is 'lost'.

   .... **[1]**

2. Energy used up by large electrical items is measured in kilowatt-hours. What is a kilowatt-hour?

   .... **[2]**

3. State **two** ways in which the heating of a house can be improved by reducing heat loss.

   ....

   .... **[2]**

4. A 250V mains fan draws a current of 2A from the power supply.

   a) Calculate the power of the fan.

      Answer: .... **[2]**

   b) Calculate how much energy the fan will draw from the mains power supply if it is on for 10 minutes.

      Answer: .... **[2]**

5. A 250V mains kettle draws a current of 8A from a power supply to heat 2kg of water. It takes 400 seconds to raise the temperature of the water by 70°C in order for it to boil. (The specific heat capacity of water = 4200J/kg/°C.)

   a) Calculate the power of the kettle.

      Answer: .... **[2]**

   b) Calculate how much energy it would supply in 400 seconds.

      Answer: .... **[2]**

Total Marks .... / 13

Physics on the Road

1 State **two** distances that determine the total stopping distance of a car when the driver needs to brake suddenly.

_____ **[2]**

2 **a)** A car travelling at a velocity of 25m/s is brought to rest in 10 seconds.

Calculate the **deceleration**.

Answer: _____ **[2]**

b) The car has a mass of 1500kg. Calculate what **force** would have acted to cause the deceleration.

Answer: _____ **[2]**

3 Two students carry out an experiment to measure their reaction time.
They use a metre ruler and a stopwatch.
They found the following equation useful:

$$\text{reaction time} = \frac{\sqrt{2 \times \text{distance (metres)}}}{g}$$

a) Describe how they could carry out the experiment.

_____

_____

_____

_____

_____

_____ **[6]**

b) Suggest what would be a sensible result for their reaction time.

_____ **[1]**

Total Marks _____ / 13

Energy for the World

1. Explain what is meant by a **non-renewable** energy source.

   ....

   .... [2]

2. A steam engine uses coal as a fuel.

   a) Describe the energy changes that take place in the engine as it moves along the track.

   ....

   ....

   .... [4]

   b) Alcohol is a renewable bio-fuel that can be produced from crops on farms.
   It could be used to run steam engines by burning it in a boiler.

   Suggest **two** reasons why this wouldn't be a good idea.

   ....

   ....

   .... [2]

3. Nuclear power stations use uranium-235 as a fuel.
   Uranium-235 has a long half-life of 7×10^8 years.

   a) Suggest why uranium is considered to be a **non-renewable** fuel.

   ....

   .... [1]

   b) State **one** advantage and **one** disadvantage of using nuclear power.

   ....

   ....

   .... [2]

   Total Marks .... / 11

Energy at Home

Refer to the Data Sheet on page 456 when answering these questions.

1 What is the difference between an **a.c.** supply and a **d.c.** supply?

_____

_____ **[2]**

2 State the **potential difference** and **frequency** of the mains electricity supply.

_____ **[2]**

3 A teacher sets up a model electricity transmission system.
A diagram of the apparatus, showing the number of turns in each coil of the transformer, is shown below.

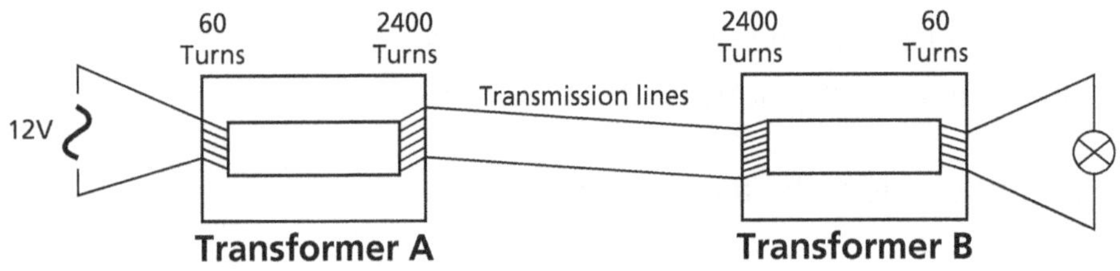

a) State what type of transformer is represented by **Transformer B**.

_____ **[1]**

b) Calculate the **potential difference** across the transmission lines.

Answer: _____ **[2]**

c) Calculate the **current** in the transmission lines if the 12V a.c. power supply delivered a current of 2A.

Answer: _____ **[3]**

Total Marks _____ / 10

GCSE (9–1)
Combined Science (Biology)

Paper 1 (Higher Tier)

H

You may use:

- a scientific calculator
- a ruler.

Time allowed: 1 hour 10 minutes

Instructions

- Use black ink. You may use a HB pencil for graphs and diagrams.
- Answer **all** the questions.
- Write your answer to each question in the space provided.
- Additional paper may be used if required.

Information

- The total mark for this paper is **60**.
- The marks for each question are shown in brackets [ ].
- Quality of extended response will be assessed in questions marked with an asterisk (*).

SECTION A

Answer **all** the questions.

You should spend a maximum of 20 minutes on this section.

1   The diagram shows a light microscope.

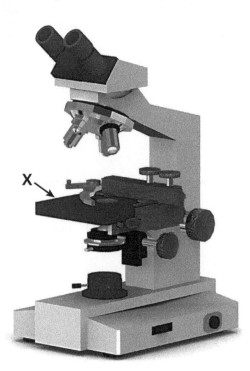

What is **X**?

A   lens

B   stage

C   spine

D   mirror   Your answer [ ]   **[1]**

2   Which of the following statements is true for **mitosis**?

A   Mitosis produces cells that are genetically identical.

B   Mitosis is an example of sexual reproduction.

C   Gametes are formed by mitosis.

D   Cells resulting from mitosis are haploid in number.   Your answer [ ]   **[1]**

3   How many strands of nucleic acid form the helical structure of DNA?

   A 1

   B 2

   C 3

   D 4   Your answer ☐ **[1]**

4   The following statements are about photosynthesis.

   Which statement is **true**?

   A The products of photosynthesis are oxygen and glucose.

   B The main site of photosynthesis in plants is the mesophyll
        cells of the leaf.

   C Photosynthesis is an exothermic reaction.

   D Chlorophyll traps heat energy from the sun for use in
        photosynthesis.   Your answer ☐ **[1]**

5   The diagram shows a plant cell.

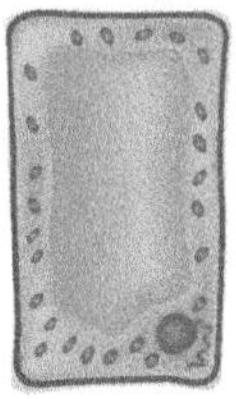

   What is the name of this cell?

   A root hair cell

   B epithelial cell

   C mesophyll cell

   D palisade cell   Your answer ☐ **[1]**

Turn over

6   Where is meristematic tissue found?

    A central nervous system

    B peripheral nervous system

    C bone marrow

    D plant roots and shoot tips   Your answer ☐ **[1]**

7   The diagram shows an egg cell.
An egg cell measures 80 μm in diameter from point A to point B.
A student measured the diameter of the cell shown in the picture and found it was 32 mm.

How many times larger is the diagram compared to a real cell?

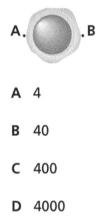

A.   .B

    A 4

    B 40

    C 400

    D 4000   Your answer ☐ **[1]**

8   The diagram shows a working muscle cell.

A   B   C

Oxygen

Glucose

Energy

Carbon dioxide

Water

Nucleus

D

What letter is most likely to represent plasma?   Your answer ☐ **[1]**

9   Martha is about to run her first marathon.

Which of the following describes the likely sequence of events over the course of the race?

   A   Her glucose levels will drop. Glycogen will be produced by the pancreas.
   Glucagon will be converted into glucose.

   B   Her glucose levels will drop. Glucagon will be produced by the pancreas.
   Glycogen will be converted into glucose.

   C   Her glucose levels will drop. Glycogen will be produced by the liver.
   Glucagon will be converted into glucose.

   D   Her glucose levels will drop. Glucagon will be produced by the
   liver. Glycogen will be converted into glucose.   Your answer [ ]   **[1]**

10   Which of the following best describes the interaction of hormones in the menstrual cycle?

   A   Oestrogen, produced in the ovaries, stimulates FSH production.

   B   FSH, produced in the pituitary, causes the ovaries to produce oestrogen.

   C   Progesterone, produced in the pituitary, inhibits FSH and LH.

   D   LH, produced in the pituitary, inhibits the release of an egg.   Your answer [ ]   **[1]**

SECTION B

Answer **all** the questions.

11 Catalase is an enzyme that breaks down hydrogen peroxide to release oxygen gas.
Catalase is present in liver.
Some students wanted to investigate the effect of pH on catalase activity.
They took five boiling tubes and placed a piece of liver in each.
They then added to each tube hydrogen peroxide that had been buffered to a different pH.

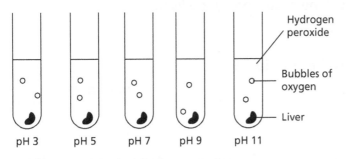

pH 3   pH 5   pH 7   pH 9   pH 11

The students counted the number of bubbles produced in five minutes.

(a) Why is counting bubbles **not** an accurate way of measuring the amount of oxygen produced?

[2]

(b) Another group of students carried out the same experiment but decided to collect the oxygen using a gas syringe.

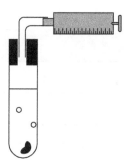

Explain why this will give a more accurate measure.

....

.... **[1]**

(c) On the axis below, sketch the graph you would expect to get from the results.

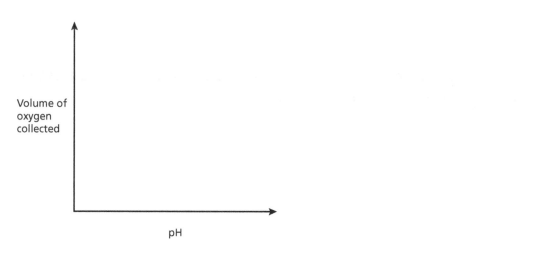

[1]

(d) The lock and key model demonstrates how enzymes work.

Explain why this will give a more accurate measure.

....

....

....

.... **[3]**

12 Some students wanted to investigate the link between surface area to volume ratio (SA : V) and rate of diffusion.

They found the following information in a textbook.

Side length of cube	Surface area	Volume	SA : V ratio
2 cm	2 × 2 × 6 = 24 cm²	2 × 2 × 2 = 8 cm³	24 to 8 = 3 : 1
4 cm			
6 cm	6 × 6 × 6 = 216 cm²	6 × 6 × 6 = 216 cm³	216 to 216 = 1 : 1
8 cm	8 × 8 × 6 = 384 cm²	8 × 8 × 8 = 512 cm³	384 to 512 = 0.75 : 1

(a) Complete the table for the cube with a side length of 4 cm. [3]

(b) In which cube will diffusion occur fastest?
Explain your answer.

....

.... [2]

(c) Explain why large multicellular organisms require specialised systems, such as the respiratory system, but unicellular organisms do not.

....

....

....

.... [2]

13 Adam, Ben and Tommy want to find out who has the quickest reactions.

They use a computer program.

When they see the shape change from green to red, they have to press the mouse button as quickly as they can. The computer calculates their reaction times and puts them in a table.

User name	1st attempt	2nd attempt	3rd attempt	Mean average
Adam	0.52	0.49	0.46	0.49
Ben	0.66	0.62	0.61	
Tommy	0.45	0.42	0.42	0.43

(a) Calculate the mean for Ben's results.

mean = ... **[3]**

(b) In this experiment, the pupils are making a voluntary response.

Write down the stages of the pathway represented by **A**, **B** and **C**.

receptor ⟶ **A** .... ⟶ relay neurone ⟶ **B** .... ⟶ **C** .... ⟶ response

A ....

B ....

C .... **[3]**

Turn over

(c) Another group of students wanted to find out the effect of caffeine on reaction times.
They divided the class into five groups.
The first group drank 100 cm³ of caffeine-free cola. The other four groups each drank different volumes of cola containing caffeine.
After 10 minutes, the students in each group tested their reaction times on the computer and found the average time.

The results are shown in the table.

Group	1	2	3	4	5
Amount of cola with caffeine (cm³)	0	100	150	200	250
Average reaction time (s)	0.52	0.48	0.47	0.40	0.37

(i) Plot a line graph of the results. **[3]**

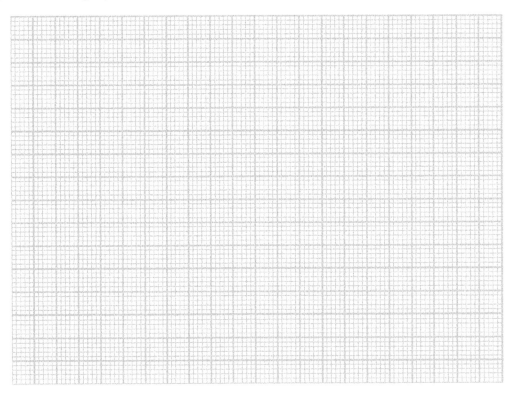

(ii) One student suggests that Group 1 should have drunk 100 cm³ of water instead of caffeine-free cola.

Do you agree with this? Give a reason for your answer.

....

....

.... **[1]**

14 Some children are born with a hole in the heart.

When this happens, there is a hole at point **X** on the diagram, between the left and right side of the heart.

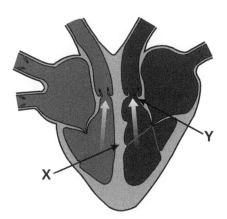

Dionne is 13 years old. She has had a hole in the heart since she was born. She easily gets short of breath and finds physical exercise very tiring.

(a) Explain why she has these symptoms.

....

....

....

....

....   **[2]**

(b) Aortic valve stenosis is when the valve labelled **Y** is too narrow.

What will be the likely effect of this condition on the heart?

....

....

....   **[1]**

Turn over

(c) In humans, blood is used to transport substances around the body.

   (i) Explain how oxygen is transported by the blood.

   ....

   ....

   .... **[2]**

   (ii) Carbon dioxide is transported by blood plasma.

   Name **two** other substances transported by plasma.

   .... **[2]**

15 **(a)** The symbol equation for aerobic respiration is shown below.

Balance the equation by writing the correct numbers in the spaces provided.

$C_6H_{12}O_6$ + _____ O_2 → _____ CO_2 + _____ H_2O **[1]**

(b) Yeast respires anaerobically to produce alcohol.
Colin wanted to make some homemade pear wine.
He mashed some pears, added water, yeast and sugar and left the mixture for a week.

(i) Why did Colin add sugar?

_____

_____ **[1]**

(ii) After a week, Colin noticed a lot of bubbles in the mixture.

Explain why there were bubbles.

_____

_____ **[1]**

(c) Describe the differences between aerobic and anaerobic respiration in humans.

_____

_____

_____

_____ **[4]**

Turn over

16   The photograph shows a water lettuce, which is a floating water plant.
It is not anchored to the ground, but has hair-like roots.

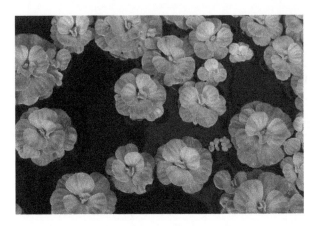

(a) Explain how water moves into the water lettuce.

....

....

....

.... **[3]**

(b) Water lettuce is a freshwater plant.

What would happen to the net movement of water if the water lettuce was placed
in sea water?
Give a reason for your answer.

....

....

....

.... **[2]**

(c) The water lettuce produces glucose during photosynthesis.
Increasing the amount of light available was found to increase the rate of photosynthesis to a certain point.
However, increasing the light above this point no longer increased the rate of photosynthesis.
What was limiting photosynthesis?

[1]

(d)* Explain how glucose and minerals are transported through a plant.

[6]

END OF QUESTION PAPER

Practice Exam Paper 1

BLANK PAGE

Collins

GCSE (9–1)
Combined Science (Biology)
Paper 2 (Higher Tier)

H

You may use:	Time allowed: 1 hour 10 minutes
• a ruler • a scientific calculator.	

Instructions

- Use black ink. You may use a HB pencil for graphs and diagrams.
- Answer **all** the questions.
- Write your answer to each question in the space provided.
- Additional paper may be used if required.

Information

- The total mark for this paper is **60**.
- The marks for each question are shown in brackets [ ].
- Quality of extended response will be assessed in questions marked with an asterisk (*).

Practice Exam Paper 2

SECTION A

Answer **all** the questions.

You should spend a maximum of 20 minutes on this section.

1 An egg cell from a human female usually contains

A one X chromosome

B two X chromosomes

C one X and one Y chromosome

D one Y chromosome

Your answer ☐ **[1]**

2 Carla uses a quadrat to count the number of dandelion plants growing in different areas of a field.

Her results are shown below.

Quadrat number	Number of dandelions
1	7
2	9
3	8
4	8
5	6

Each quadrat measures one square metre.

The total area of the field is 60 square metres.

What is the estimated number of dandelions in the field?

A 98

B 228

C 380

D 456

Your answer ☐ **[1]**

3 The diagram shows some stages in genetic engineering.

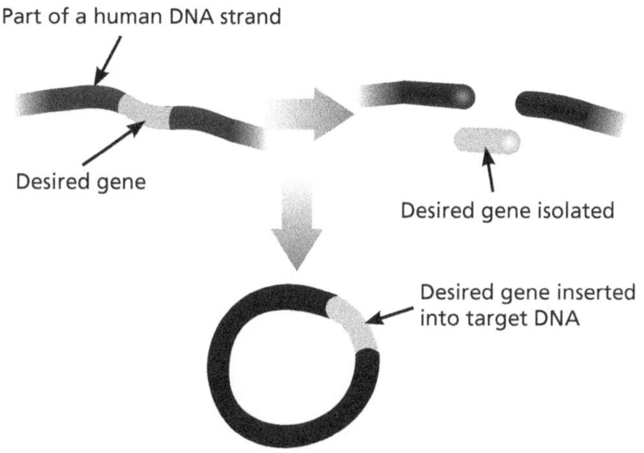

Part of a human DNA strand

Desired gene

Desired gene isolated

Desired gene inserted into target DNA

How is the desired gene inserted into the target DNA?

A using restriction enzymes

B using protease enzymes

C using ligase enzymes

D using lipase enzymes

Your answer ☐ [1]

4 Thyroxine is a hormone that helps to regulate metabolism.

Which statement about thyroxine is correct?

A When thyroxine levels become too low, production of thyroid stimulating hormone stops.

B Thyroxine is produced by the pituitary gland.

C Production of thyroxine causes the metabolic rate to lower.

D Thyroxine is produced when there are increased levels of thyroid stimulating hormone.

Your answer ☐ [1]

Turn over

5 An association between two individuals in which the association benefits both is called:

A mutualism

B competition

C commensalism

D parasitism   Your answer ☐ [1]

6 An example of a discrete variable is:

A height

B blood group

C weight

D width of hand span   Your answer ☐ [1]

7 The food web below is from a marine environment.

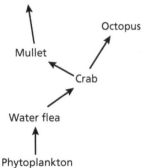

What is most likely to happen if there is overfishing of mullet?

A Dolphin numbers and crab numbers will decrease.

B Dolphin numbers will increase and crab numbers will decrease.

C Dolphin numbers decrease and crab numbers will increase.

D There will be no change in the numbers.   Your answer ☐ [1]

8 A person who has been infected with human papilloma virus (HPV) has an increased risk of developing:

A tuberculosis

B influenza

C autism

D cervical cancer   Your answer [ ] **[1]**

9 The stages below are part of the process of selective breeding. They are not in the correct order.

1. Offspring that produce a lot of meat are selected.

2. A cow and a bull that provide good quantities of meat are chosen as the parents.

3. The offspring are bred together.

4. The process is repeated over many generations.

5. The parents are bred with each other.

What is the correct order?

A 1, 3, 5, 2, 4

B 2, 5, 1, 3, 4

C 5, 1, 4, 2, 3

D 2, 5, 3, 1, 4   Your answer [ ] **[1]**

10 Which of the following statements about stem cells is **not** true?

A Stem cells are undifferentiated cells.

B Stem cells are not found in plants.

C It is possible to grow new nervous tissue using stem cells.

D Stem cells can be obtained from human bone marrow.   Your answer [ ] **[1]**

Turn over

Practice Exam Paper 2

SECTION B

Answer **all** the questions.

11   Look at the article from a nature magazine about ospreys.

Although the osprey population has increased over recent years in the UK, they remain the fourth rarest bird of prey in the UK. They are at risk because they feed on fish and, therefore, depend on high water quality of rivers and lakes. Their eggs are often stolen by collectors and migrating ospreys are at risk from being shot by hunters.

The Lake District Osprey Project has been running for several years to encourage these birds to breed in the UK. In 2001, a pair of ospreys nested beside Bassenthwaite Lake. They were the first wild osprey to breed in the Lake District for over 150 years. The Forestry Commission built a nest to encourage the birds to stay, which they did. Once the eggs were laid, wardens kept a round the clock watch to prevent disturbance and deter egg thieves. Bassenthwaite Lake is a National Nature Reserve and provides valuable habitats for a variety of wildlife.

(a) Ospreys were once endangered in the UK.

What does endangered mean?

....

.... **[1]**

(b) Give **one** way in which humans have caused the ospreys to become endangered.

....

....

.... **[1]**

(c) How has the osprey been protected and encouraged to breed in the Lake District?

....

.... **[1]**

(d) Suggest a difficulty with running projects such as the one described.

....

.... **[1]**

12 The graph below shows the smoking habits and incidence of lung cancer for people living in a European country.

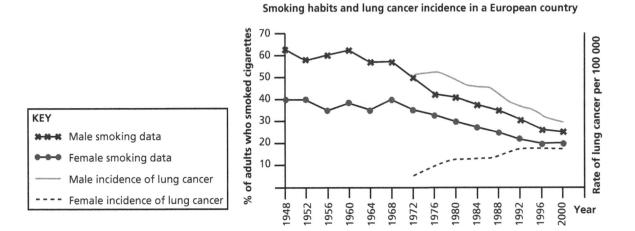

Smoking habits and lung cancer incidence in a European country

KEY
✻✻✻ Male smoking data
●●● Female smoking data
——— Male incidence of lung cancer
- - - - Female incidence of lung cancer

(a) Describe the patterns in male and female smoking habits from 1948 to 2000.

....

.... [2]

(b) Calculate the percentage difference in female smokers between 1968 and 1988.
Show your working.

percentage difference = ... [3]

(c) Does the data support the hypothesis that smoking causes lung cancer?
Give a reason for your answer.

....

....

.... [2]

(d) A healthy heart is essential for good health.

How does each of the following affect the cardiovascular system?

(i) A diet high in fat

....

.... **[1]**

(ii) A diet high in salt

....

.... **[1]**

Turn over

13 Below is one of the stages in the carbon cycle.

carbon in dead animals and plants ⟶ carbon dioxide in the atmosphere

(a) What is the name of this process?

.... [1]

(b) Explain the role of detritivores in this process.

....

....

.... [2]

(c) Nitrogen is another element that is recycled.

What substance must nitrogen be converted into so that plants can make use of it?

.... [1]

14 The table contains information about three mammals.

Animal	Binomial name	Diet
Horse	*Equus ferus*	Eats grasses and plant materials. Has only one simple stomach.
Deer	*Dama dama*	Eats shoots, fresh grass, twigs, fruit. Has a complex stomach and can bring food back up into its mouth to chew again.
Zebra	*Equus zebra*	Eats mainly grasses. Has one simple stomach.

(a) Which two animals are most closely related?
    Give **two** reasons for your answer.

    ....

    ....

    ....
                                                                                              [2]

The table gives some information about donkeys and horses.

Donkeys	Horses
Stubborn but hard working	Hard workers
Sure footed	Delicate legs
Difficult to train	Easy to train
Slower runners	Fast runners
Average life span 30–50 years	Average life span 25–30 years
Ferments food in a hind gut so less prone to colic	Cannot vomit so prone to colic
Not very intelligent	Intelligent

A mule is a cross between a donkey and a horse.
Mules are often used to carry heavy loads on their backs in mountainous regions.

(b) List the qualities required from the donkey and the horse to make the mule suitable for this work.

    (i) Qualities from the donkey

    ....

    ....

    ....
                                                                                              [2]

(ii) Qualities from the horse

....

....

.... **[2]**

(c) A donkey has 62 chromosomes and a horse has 64 chromosomes.

(i) Explain why mules cannot breed.

....

....

.... **[2]**

(ii) It is wrong to call a mule a 'species'.

Explain why.

....

.... **[1]**

(d) Scientists believe the modern day horse evolved from a horse with four toes.
The four toes were useful for spreading the load on a soft forest floor.
As the ground hardened, a large central toe, similar to the hoof of today's horses, was better for running over open country.

Suggest how the modern day horse with a hoof could have evolved through natural selection.

....

....

....

....

....

.... **[4]**

15 The diagram shows sections of a pair of chromosomes from a cat, which contain genes for eye colour, ear shape and fur pattern.

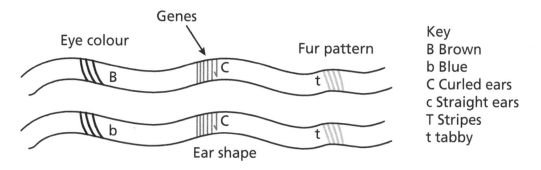

Key
B Brown
b Blue
C Curled ears
c Straight ears
T Stripes
t tabby

(a) Describe the phenotype of the cat for these three characteristics.

....

....

.... **[3]**

(b) The cat above mates with a cat that is heterozygous for fur pattern.

(i) Complete the genetic cross diagram.

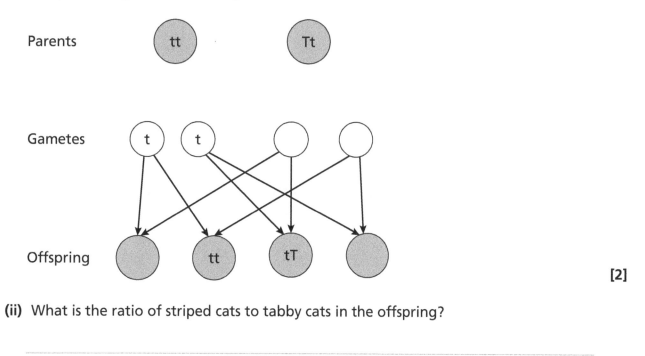

[2]

(ii) What is the ratio of striped cats to tabby cats in the offspring?

....

.... **[1]**

(c) Polycystic kidney disease (PKD) is an inherited disease that affects cats.
Fluid-filled cysts are present in the kidneys from birth, but are so small initially that they go undetected.
As the cat gets older, the cysts become larger and eventually cause kidney failure.
In older cats, PKD can be detected by a scan of the kidneys.
PKD is controlled by a single pair of genes.
The allele for PKD is dominant.

A cat breeder is concerned that some of his kittens may have inherited PKD.

 (i) How could he test to see if they had the condition?

....

.... **[1]**

 (ii) Both parents of the kittens have developed PKU.

   Draw a diagram to show the results of a cross if both parents are heterozygous for PKU.

[3]

 (iii) In a litter of eight kittens, how many kittens would you expect to
    have PKU if both parents were heterozygous for the condition?

.... **[1]**

16 The diagram shows a bacterium called *P. aeruginosa*, which often causes infections of post-surgical wounds.

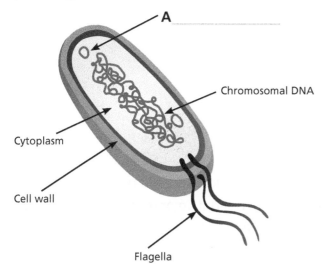

A

Chromosomal DNA

Cytoplasm

Cell wall

Flagella

(a) The structure labelled **A** is DNA that is separate from the chromosomal DNA.

On the diagram, write the name of structure **A**.   [1]

A patient has developed a post-surgical wound infection.
Some pus from the wound is examined and found to contain *P. aeruginosa*.
Doctors need to know which antibiotic will be best to treat the patient.
They spread the bacteria onto some agar in a Petri dish and add two filter paper discs that have been soaked in different antibiotics.
After incubation at 37 °C they examine the Petri dish.
The diagram shows the actual size of the dish after incubation.

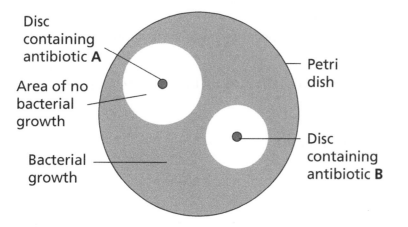

Disc containing antibiotic **A**

Area of no bacterial growth

Bacterial growth

Petri dish

Disc containing antibiotic **B**

If the area of no bacterial growth around the disc (including the disc) is greater than 200 mm², the antibiotic can be used successfully to treat the infection.

(b) Work out if either or both of the antibiotics could be used to treat the infection.
Use area of a circle = $3.14 \times r^2$. You must clearly state your answer.

_____

_____

_____

_____ **[3]**

(c) Suggest **one** other factor that might determine which antibiotics should be used
to treat an infection, other than whether the organism is sensitive or resistant to the antibiotic.

_____

_____ **[1]**

(d) Explain the action of phagocytic white blood cells when _P. aeruginosa_ first enters
a wound.

_____

_____

_____

_____ **[3]**

(e) _P. aeruginosa_ can cause problems in hospitals because it easily develops resistance to a
range of antibiotics.
Scientists would like to develop a vaccination that could be given to people at risk of
post-surgical infections.

What would scientists use in the vaccination?

_____

_____ **[1]**

END OF QUESTION PAPER

Collins

GCSE (9–1)
Combined Science (Chemistry)

Paper 3 Higher Tier

H

You must have:

- the Data Sheet (page 455)

You may use:

- a scientific calculator
- a ruler

Time allowed: 1 hour 10 minutes

Instructions

- Use black ink. You may use an HB pencil for graphs and diagrams.
- Answer **all** the questions.
- Write your answer to each question in the space provided.
- Additional paper may be used if required.

Information

- The total mark for this paper is **60**.
- The marks for each question are shown in brackets **[ ]**.
- Quality of extended responses will be assessed in questions marked with an asterisk (*).

Practice Exam Paper 3

SECTION A

Answer **all** the questions.

You should spend a maximum of 20 minutes on this section.

1 Athina wants to separate a mixture of amino acids.
  She carries out thin layer chromatography.
  The amino acid samples are placed onto silica gel and a solvent is added.

  After two hours the silica gel is sprayed with ninhydrin, which makes the amino acids become visible.

  What is the name given to the **silica gel** in this experiment?

  A gas phase

  B mobile phase

  C solid phase

  D stationary phase

  Your answer ☐ [1]

2 The diagram below shows the chromatogram before and after spraying with ninhydrin.

  What is the R_f value for sample **3**?
  Use a ruler to help you.

  A 0.16

  B 0.27

  C 0.81

  D 0.95

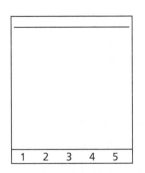

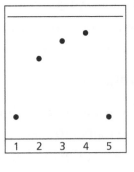

Before

After

  Your answer ☐ [1]

3   The diagrams show the electronic structure of four different atoms.

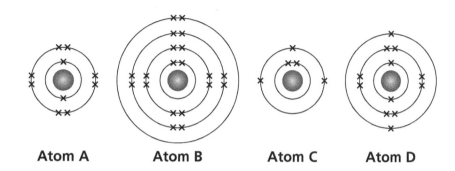

| Atom A | Atom B | Atom C | Atom D |

Which two atoms are in the same **group**?

A   atom A and atom B

B   atom B and atom D

C   atom C and atom D

D   atom A and atom C

Your answer ☐ [1]

4   An element, **X**, has an atomic mass of 88 and the atomic number 38.

Which of the following is correct for element **X**?

A   No. of protons = 38   No. of neutrons = 50   No. of electrons = 50

B   No. of protons = 50   No. of neutrons = 38   No. of electrons = 38

C   No. of protons = 38   No. of neutrons = 38   No. of electrons = 50

D   No. of protons = 38   No. of neutrons = 50   No. of electrons = 38

Your answer ☐ [1]

Turn over

5 During the electrolysis of aqueous copper sulfate, what is made at the cathode?

   A sulfate

   B hydrogen

   C copper

   D copper hydroxide

Your answer [ ] **[1]**

6 Which of these shows the balanced symbol equation for the reaction between sodium and oxygen to make sodium oxide?

   A $Na + O_2 \rightarrow NaO_2$

   B $4Na + O_2 \rightarrow 2Na_2O$

   C $S + O_2 \rightarrow SO_2$

   D $4S + O_2 \rightarrow 2S_2O$

Your answer [ ] **[1]**

7 What is the best description of the particles in a solid?

	Distance between particles	Movement of particles
A	Close together	In continuous random motion
B	Close together	Vibrating about a fixed point
C	Far apart	In continuous random motion
D	Far apart	Vibrating about a fixed point

Your answer [ ] **[1]**

8 Which of these is the **best** explanation of what is meant by a weak acid?

   A There is a large amount of acid and a small amount of water.

   B There is a small amount of acid and a large amount of water.

   C The acid is completely ionised in solution in water.

   D The acid is partially ionised in solution in water.

   Your answer [ ] **[1]**

9 Limestone is the common name for calcium carbonate.
   Limestone reacts with hydrochloric acid.

   Which of the following shows the correct balanced symbol equation for the reaction between limestone and hydrochloric acid?

   A $CaCO_3 + HCl \rightarrow CaCl + H_2O$

   B $CaCO_3 + HCl \rightarrow CaCl + H_2O + CO_2$

   C $CaCO_3 + 2HCl \rightarrow CaCl_2 + H_2O + CO_2$

   D $CaCO_3 + 2HCl \rightarrow CaCl_2 + H_2O + CO_3$

   Your answer [ ] **[1]**

10 Which of the following shows the reaction profile for an exothermic reaction?

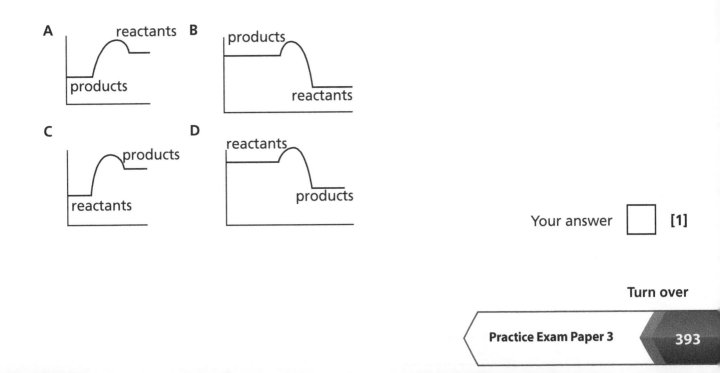

   Your answer [ ] **[1]**

Turn over

Practice Exam Paper 3

SECTION B

Answer **all** the questions.

11 Look at the graph of boiling points of elements against their atomic numbers for Period 2 of the periodic table.

(a) What does the graph tell you about the boiling point as you go across a **period**?

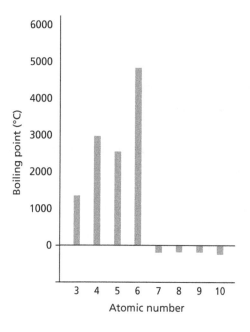

_____

_____ [2]

(b) Explain why the properties of elements change across a period.

_____

_____

_____

_____ [2]

(c) Particle **X** is a particle with the atomic number of 18 and the electronic structure 2.8.8.
Particle **Y** is a particle with the atomic number of 20 and the electronic structure 2.8.8.

Which particle is an atom and which is an ion?
Explain your answer.

....

....

.... **[2]**

(d) A diagram of a hydrogen atom is shown below.

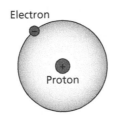

Electron

Proton

(i) Explain what the **problems** are in representing the atom in this way.

....

....

....

.... **[3]**

(ii) Niels Bohr was the scientist who devised this particular model of the atom.

Explain how his model **improved** upon earlier models.

....

.... **[2]**

Turn over

12   Look at the equation. It shows the combustion of hydrogen.

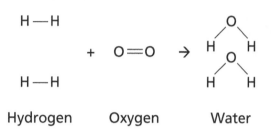

Hydrogen   Oxygen   Water

The table shows the bond energies of the bonds involved.

Bond	Bond energy (kJ/mol)
H–H	432
O=O	494
O–H	464

(a) Write down the:

number of bonds broken = ...

number of bonds made = ...   [1]

(b) Describe the energy changes when bonds are broken and made.

....

....   [2]

(c) Calculate the energy change for this reaction.
    Show your working.

energy change = ... kJ/mol   [3]

13   Look at the diagram. It shows the electrolysis of copper sulfate solution.

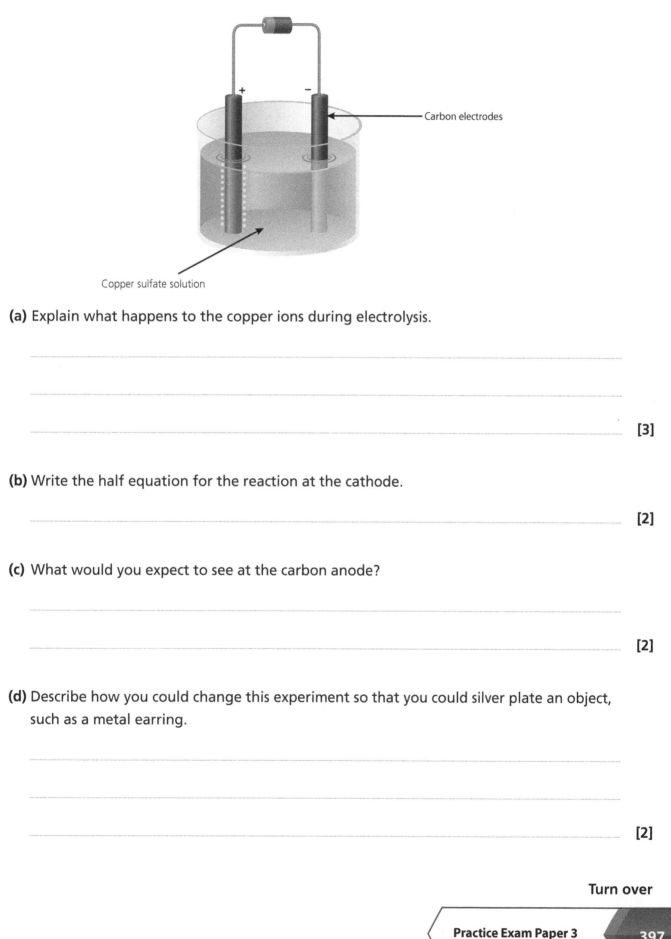

(a) Explain what happens to the copper ions during electrolysis.

....

....

.... [3]

(b) Write the half equation for the reaction at the cathode.

.... [2]

(c) What would you expect to see at the carbon anode?

....

.... [2]

(d) Describe how you could change this experiment so that you could silver plate an object, such as a metal earring.

....

....

.... [2]

Turn over

14   A student is investigating the reaction of three acids, **X**, **Y** and **Z**.

Acid	pH	Dissociation constant
X	1	7.6×10^{-4}
Y	2	1.8×10^{-5}
Z	5	8.7×10^{-10}

(a)*Suggest what the dissociation constant means in relation to acids **X**, **Y** and **Z** and explain in detail the differences between the different types of acid.

Explain how you could test the strengths of the acids experimentally.

_____

_____

_____

_____

_____

_____

_____

_____

_____

_____

_____

_____   **[6]**

(b) How much stronger is acid **X** compared to acid **Y**?

_____   **[1]**

15   Black and white film photography uses film with silver chloride crystals.
      The silver chloride crystals are sensitive to light.

   (a) The silver in silver chloride solution can be removed in a displacement reaction.
       Copper is added to the silver chloride solution, AgCl.
       Silver and copper chloride, $CuCl_2$, are produced.

       Write a balanced **symbol** equation for this reaction.

       .... [2]

   (b) (i) In the reaction, copper atoms become copper ions, Cu^{2+}, and silver ions, Ag^+, become
            silver atoms.

            Write a balanced ionic equation for this reaction.

            .... [2]

      (ii) Write a balanced half equation to show what happens to copper ions in this reaction.
           Use e^- to represent an electron.

           .... [2]

      (iii) Write a balanced half equation to show what happens to silver ions in this reaction.
            Use e^- to represent an electron.

            .... [2]

(c) Explain why the reaction between copper and silver chloride is not only a displacement reaction but also a reduction / oxidation reaction.

Use ideas about electrons in your answer.

[3]

16   A teacher carries out a thermite reaction.
     The thermite reaction is shown below.

     aluminium + iron(III) oxide → iron + aluminium oxide

   (a) Write a **balanced symbol** equation to show this reaction.

   .... [2]

   (b) Suggest **two** safety precautions a person carrying out this reaction should take.

   ....

   ....

   .... [2]

   (c) The thermite reaction is an example of an oxidation / reduction reaction.

   Write the formula of the species that is:

   oxidised ....

   reduced .... [2]

END OF QUESTION PAPER

BLANK PAGE

Collins

GCSE (9–1)
Combined Science (Chemistry)

H

Paper 4 Higher Tier

You must have:

- the Data Sheet (page 455)

You may use:

- a scientific calculator
- a ruler

Time allowed: 1 hour 10 minutes

Instructions

- Use black ink. You may use an HB pencil for graphs and diagrams.
- Answer **all** the questions.
- Write your answer to each question in the space provided.
- Additional paper may be used if required.

Information

- The total mark for this paper is **60**.
- The marks for each question are shown in brackets **[ ]**.
- Quality of extended responses will be assessed in questions marked with an asterisk (*****).

Practice Exam Paper 4

SECTION A

Answer **all** the questions.

You should spend a maximum of 20 minutes on this section.

1   Which of the following is a property of all Group 0 elements?

   A They have eight electrons in the outermost shell.

   B They are chemically unreactive.

   C They form positive ions.

   D They are less dense than air.

Your answer ☐ **[1]**

2   Look at the fractionating column.

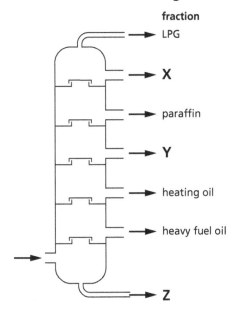

Which fraction is **X**?

   A petrol

   B aviation fuel

   C bitumen

   D alcohol

Your answer ☐ **[1]**

3 Look at the diagrams showing the electron structures of four elements.

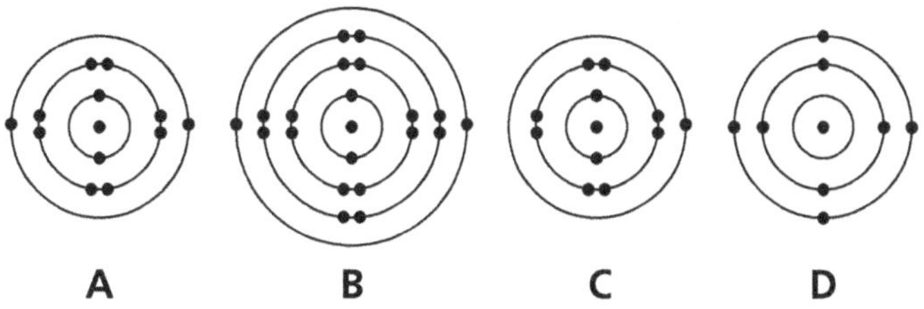

A   B   C   D

Which element is in **Period 4** of the periodic table?   Your answer [ ]   **[1]**

4 Potassium metal is reacted with water.

Which of the following statements is **true**?

A Potassium chloride is formed. The potassium is a negatively charged ion.

B Potassium chloride is formed. The potassium is a positively charged ion.

C Potassium hydroxide is formed. The potassium is a positively charged ion.

D Potassium hydroxide is formed. The potassium is a negatively charged ion.

Your answer [ ]   **[1]**

5   Wayne is investigating the reaction between hydrogen peroxide and a catalyst.

He collects the oxygen gas produced and plots the graph shown below.

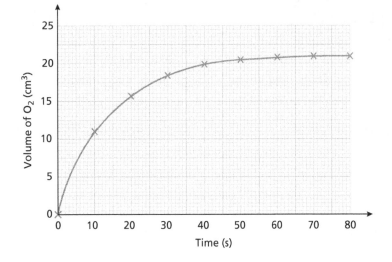

What is the rate of oxygen production between 0 and 5 s?

A   23 cm³/s

B   7 cm³/s

C   1.4 cm³/s

D   0.7 cm³/s

Your answer ☐ [1]

6 Miranda is investigating the breakdown of hydrogen peroxide into water and oxygen.

$$2H_2O_2 \rightarrow 2H_2O + O_2$$

The reaction is extremely slow.

Miranda takes four black powders which she adds to the hydrogen peroxide.
She records the amount of gas produced.

Look at the graph of her results.

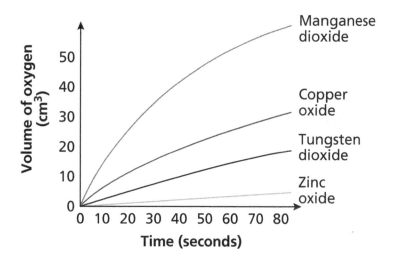

Which of the catalysts tested would be the best for catalysing this reaction?

A copper oxide

B manganese dioxide

C tungsten dioxide

D zinc oxide

Your answer [1]

7 Which statement is true for a **Group 0** element?

   A It dissolves in water to form bleach.

   B It is a metal.

   C It is an inert gas.

   D It reacts with an acid to form a salt and hydrogen gas.

                                           Your answer ☐ [1]

8 Which of the following is the name given to the process by which fractions of crude oil are converted into smaller, more useful molecules?

   A fractional distillation

   B cracking

   C chromatography

   D filtration

                                           Your answer ☐ [1]

9   Look at the following reaction:

$$2H_2(g) + O_2(g) \rightleftharpoons 2HO(l)$$

If the concentration of H_2 was **lowered**, which of the following would be true?

A   The reaction is exothermic in the direction products to reactants.

B   The yield of H_2O would increase.

C   The amount of O_2 would decrease.

D   The equilibrium would shift to the left.

Your answer [ ]   [1]

10   The factors below are considered when making a tablet computer.

Which factor would **not** be suitable for inclusion in a life cycle assessment?

A   the environmental impact of disposal of the tablet

B   the version of the operating system used

C   the energy required to disposed of the tablet

D   the energy required to make the tablet

Your answer [ ]   [1]

Turn over

SECTION B

Answer **all** the questions.

11 **(a)** Give the electronic structure of potassium.

.... **[1]**

(b) Draw a diagram to show the metallic bonding in potassium.

[2]

(c) A small piece of potassium is added to water.

Describe what observations you would make when a small piece of potassium is added to a trough of water.

....

....

.... **[2]**

(d) Write the word equation to show the reaction between water and potassium.

.... [2]

(e) Describe how the reactivity of the two metals above potassium in the reactivity series will differ to potassium.

....

....

.... [2]

12   The table below shows the composition of the Earth's atmosphere at 3 billion and 2 billion years ago (bya).

Gas	3 bya (%)	2 bya (%)	Present (%)
Nitrogen	91	82	
Oxygen	0	10	
Carbon dioxide	9	8	

(a) Complete the table to show the present day proportion of gases.   [1]

(b) Describe how the percentages of the gases in the atmosphere have changed between 3 bya and 2 bya.

Give reasons for the changes.

....

....

....   [3]

(c) The activity of humans is believed to have caused the recent increased production of greenhouse gases.

Sketch a graph to show the correlation between carbon dioxide emissions and air temperature.

[4]

(d)* Methane and carbon dioxide are greenhouse gases.

Methane is a far more powerful greenhouse gas than carbon dioxide.

Explain the greenhouse effect and discuss the consequences that increased levels of carbon dioxide and methane have on the greenhouse effect.

Suggest ways to reduce greenhouse gas emissions.

[6]

Turn over

13 Look at the flow chart. It shows the steps for the manufacture of sulfuric acid using the contact process.

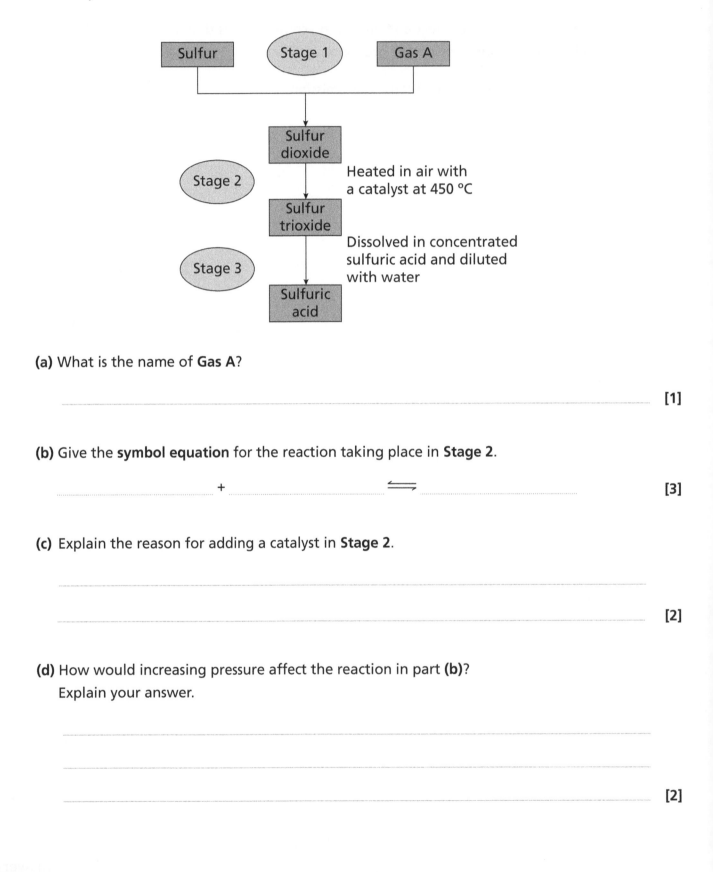

(a) What is the name of **Gas A**?

.... [1]

(b) Give the **symbol equation** for the reaction taking place in **Stage 2**.

.... + ... ⇌ .... [3]

(c) Explain the reason for adding a catalyst in **Stage 2**.

....

.... [2]

(d) How would increasing pressure affect the reaction in part **(b)**?
Explain your answer.

....

....

.... [2]

14 The available land for growing crops worldwide is decreasing.
There is land available, but often it is highly polluted with chemicals.

(a) Phytoextraction is one method of extracting the metals from contaminated soil.

Give **two advantages** and **two disadvantages** of phytoextraction.

advantages

....

.... **[2]**

disadvantages

....

.... **[2]**

(b) Approximately 800 million people in the world do not have access to potable water.

What is potable water and why is it important to increase the amount available?

....

....

.... **[2]**

15 Ethan is reacting phosphorus trichloride with chlorine.

The symbol equation is shown below:

$$PCl_3(g) + Cl_2(g) \rightleftharpoons PCl_5(g)$$

It is a dynamic equilibrium reaction.
It is exothermic in the direction reactants to products.

(a) Explain what is meant by the term 'dynamic equilibrium'.

.... **[1]**

(b) Ethan wants to maximise the quantity of phosphorus pentachloride produced.
He decides to increase the pressure.

Is Ethan correct?
Explain your answer.

....

.... **[2]**

(c) Ethan increases the temperature of the system.

Complete the sentences by filling in the missing words.

As the temperature increases, the amount of PCl_3 and Cl_2 .... whilst

the amount of PCl_5 .... **[1]**

(d) Sketch a graph to represent the dynamic equilibrium
reached from the forward and reverse reactions
using the axes provided.

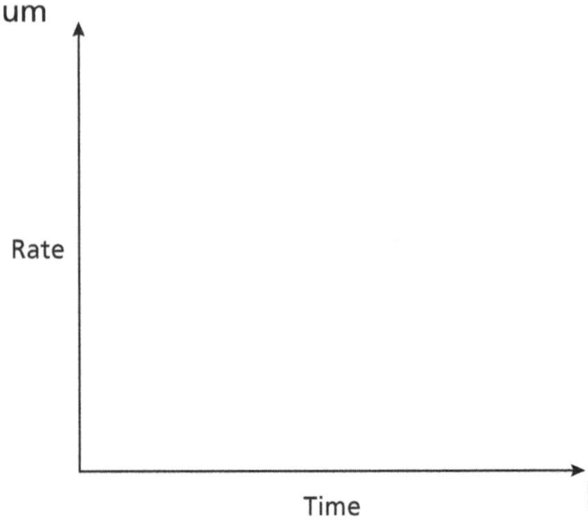

Rate

Time **[2]**

16 Magnesium reacts very slowly with cold water to produce an alkaline solution and hydrogen gas.

(a) Write a balanced symbol equation to show the reaction between magnesium and cold water.

.... [1]

(b) Describe **two** changes that could lead to the production of more hydrogen gas.

....

....

.... [2]

(c) Magnesium is in Group 2 of the periodic table.

Describe the **difference** in the reaction that would be seen with the element immediately before magnesium in the periodic table.

.... [1]

(d) Magnesium cannot be extracted from its ore with carbon.

Explain why this is the case and suggest how magnesium can be extracted from its ore.

....

....

....

....

.... [3]

END OF QUESTION PAPER

Practice Exam Paper 4

BLANK PAGE

Collins

GCSE (9–1)
Combined Science (Physics)

Paper 5 (Higher Tier)

H

Time allowed: 1 hour 10 minutes

You must have:

- the Data Sheet (page 456)

You may use:

- a scientific calculator
- a ruler

Instructions

- Use black ink. You may use a HB pencil for graphs and diagrams.
- Answer **all** the questions.
- Write your answer to each question in the space provided.
- Additional paper may be used if required.

Information

- The total mark for this paper is **60**.
- The marks for each question are shown in brackets [ ].
- Quality of extended response will be assessed in questions marked with an asterisk (*).

SECTION A

Answer **all** the questions.

You should spend a maximum of 20 minutes on this section.

1   Identify which of these liquids has the lowest density.

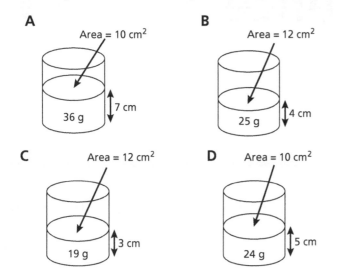

A Area = 10 cm²   7 cm   36 g

B Area = 12 cm²   4 cm   25 g

C Area = 12 cm²   3 cm   19 g

D Area = 10 cm²   5 cm   24 g

Your answer ☐ [1]

2   A balloon containing helium is released from a fairground.
As it rises, it expands, the mass remaining constant.

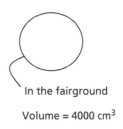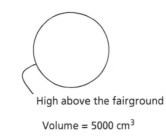

In the fairground
Volume = 4000 cm³

High above the fairground
Volume = 5000 cm³

The density of helium, measured in the balloon, high above the fairground was found to be 0.14 kg/m³.

Calculate the density of the helium in the balloon in the fairground.

A 0.130 kg/m³

B 0.175 kg/m³

C 0.182 kg/m³

D 0.112 kg/m³

Your answer ☐ [1]

3 Which of the following measurements is a good approximation for the size of an atom?

A 1×10^{-11} m

B 1×10^{-10} m

C 1×10^{-9} m

D 1×10^{-6} m   Your answer ☐ **[1]**

4 Compasses are used to plot the magnetic field lines between two permanent magnets, with identical but opposing north poles (see below).

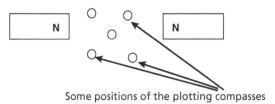

Some positions of the plotting compasses

Identify which of the following is the correct plot for the magnetic field in this case.

A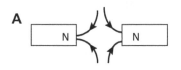

B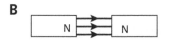

C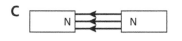

D 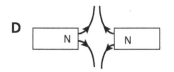   Your answer **[1]**

5 A wire is connected to a battery and an open switch.
Part of the wire **XY** is placed between opposite poles of a magnet.
That part of the wire **XY** is free to move.

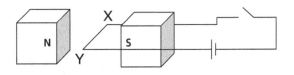

Identify the direction in which the wire will move when the switch is closed.

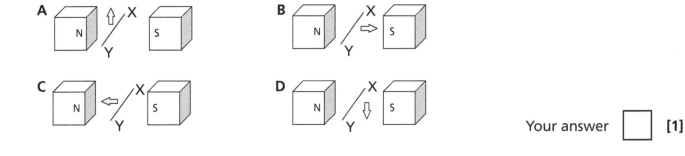

Your answer ☐ **[1]**

6 A conductor of length 0.5 m, carrying a current of 2.0 A is placed in a magnetic
field of strength 2.0×10^{-2} T.

Calculate the force that acts on the conductor.

Use the following equation:
force on a conductor carrying a current = magnetic field strength × current × length

A 2.0×10^{-2} N

B 0.5×10^{-2} N

C 4.0×10^{-2} N

D 20×10^{-2} N

Your answer ☐ **[1]**

7 A toy truck of mass 6 kg is travelling at 0.4 m/s along a track.

It collides with another truck of mass 4 kg that is travelling in the same direction at 0.2 m/s, as shown in the diagram.

They stick together after the collision. *v* is the velocity of both trucks stuck together.

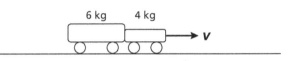

Identify which of the following is the correct value of *v*.

A 0.30 m/s

B 0.60 m/s

C 0.24 m/s

D 0.32 m/s
Your answer [ ] **[1]**

8 A train travels 240 km in 3 hours.

Calculate how long it would take to travel 80 km at half the speed.

A 2 hours

B 6 hours

C 1 hour

D 4 hours
Your answer [ ] **[1]**

Turn over

9   A suspended spring, of spring constant 12 N/m, is stretched 40 cm when a load is added.

Calculate how much work is done.

Use the formula:
energy transferred in stretching = 0.5 × spring constant × (extension)^2

A   480 J

B   2.4 J

C   1.92 J

D   0.96 J   Your answer ☐ **[1]**

10   Identify the correct value for the resistance of resistor **R** in the circuit below.

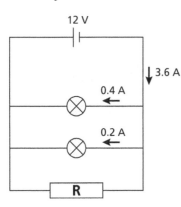

A   3.33 Ω

B   30.00 Ω

C   4.00 Ω

D   20.00 Ω   Your answer ☐ **[1]**

SECTION B

Answer **all** the questions.

11   A student sets up the apparatus below to measure the specific latent heat of ice.

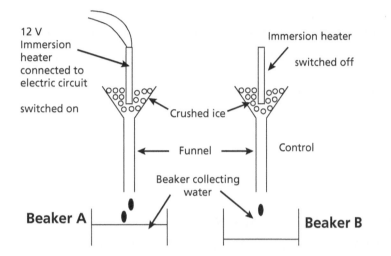

(a) Sketch a circuit diagram, including an ammeter and a voltmeter, to show how the immersion heater should be connected to the 12 V supply so that the power of the immersion heater can be calculated.

[2]

(b) Suggest extra apparatus that might be needed and the measurements that should be taken to successfully calculate the specific latent heat of ice.

[4]

Turn over

(c) Explain why the student set up a control experiment.

....

.... **[1]**

(d) The student repeats the experiment five times to have more confidence in the conclusion. One of the calculations gives too high a value for the specific latent heat of ice.

Suggest **one** error in measurement that might have caused this.

....

.... **[1]**

(e) In the experiment, 0.1 kg of ice melted.
The specific latent heat of fusion of ice = 3.34×10^5 J/kg.

Calculate how much energy is required from the heater to produce this change in state.

Use the following equation:
energy = mass × specific latent heat of fusion of ice

energy = ... J   **[2]**

12 In an experiment, trolley **X** with mass 0.5 kg is travelling along a horizontal track at a constant speed of 12 m/s.

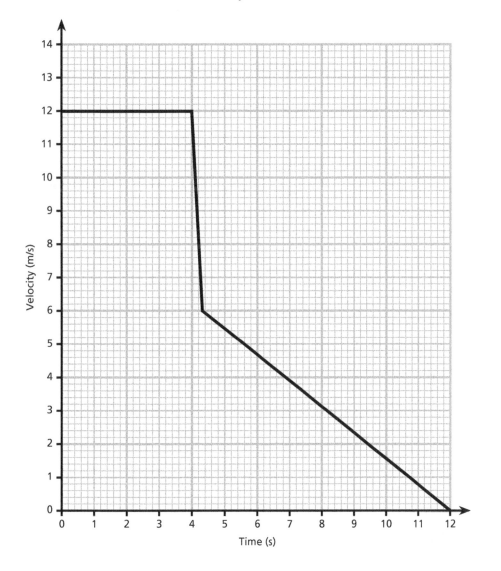

It collides with a stationary trolley **Y** of identical mass.
The two trolleys stick together after the collision and travel horizontally along the track.

The graph below shows the motion of trolley **X**.

(a) Calculate the distance travelled by trolley **X** in the first 3 seconds.

distance = _____ m **[1]**

(b) (i) Calculate the deceleration that occurred when the two trolleys stuck together.

deceleration = _____ m/s² **[2]**

(ii) Calculate the force acting on the two trolleys that caused the deceleration.

force = _____ N **[2]**

(c) Calculate the total momentum of the two trolleys after 7.5 seconds.

momentum = _____ kg m/s **[2]**

(d) On the grid below, sketch the velocity–time graph for trolley **Y** during the experiment.

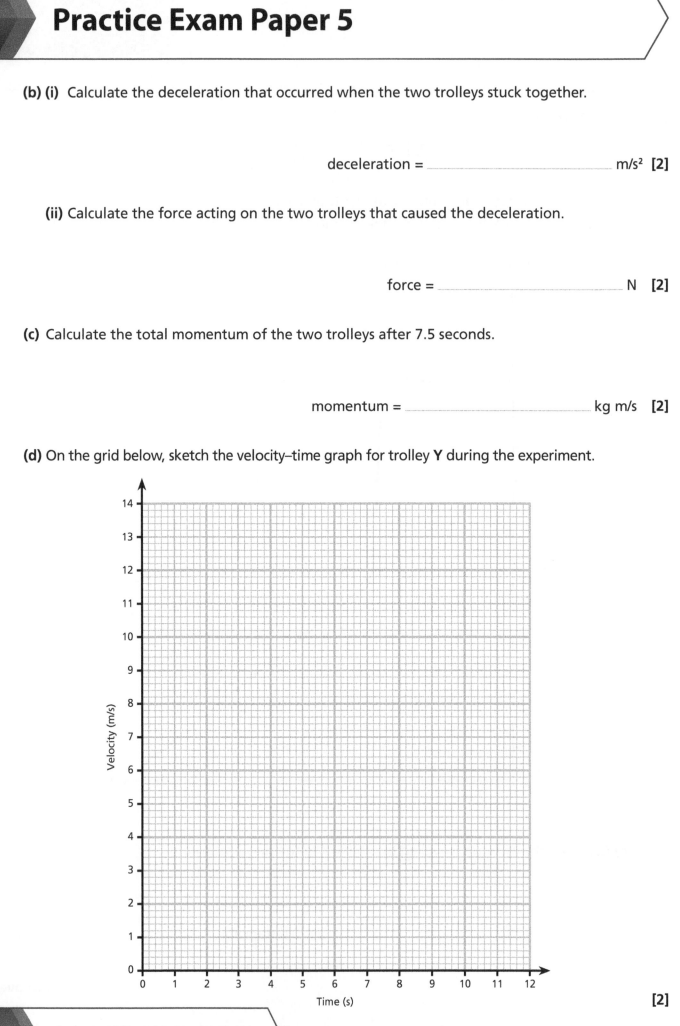

[2]

13 A student is investigating the behaviour of a thermistor.
    They set up the circuit shown here.

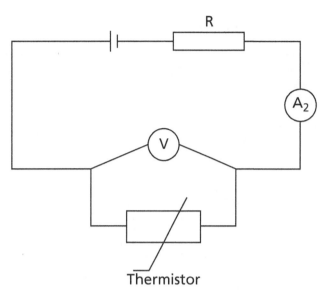

(a)* The student was given a kettle, two thermometers, two beakers and some crushed ice.

Explain how the student could use this apparatus to demonstrate the behaviour of a thermistor at different temperatures.

....

....

....

....

....

.... [4]

(b) The student replaced the voltmeter with a multimeter to directly record the resistance of the thermistor at different temperatures.

The results were recorded in the table below.

Temperature (°C)	Resistance (kΩ)			
	Reading 1	Reading 2	Reading 3	Mean
0	28.1	28.1	28.2	28.1
10	20.4	20.3	20.2	20.3
20	14.1	14.3	14.1	14.2
30	11.1	11.1	11.2	11.1
40	8.9	9.1	9.2	9.1
50	7.9	9.1	7.9	7.9
60	7.1	7.2	7.1	7.1
70	6.6	6.5	6.4	6.5
80	6.1	5.9	5.9	5.9
90	5.5	5.4	5.5	5.5

(i) There is an anomaly in the table of results.

Identify the anomaly and explain how the student should deal with it.

....

....

.... **[2]**

(ii) Write down a conclusion you can you draw from the table of results.

....

.... **[1]**

(c) The student then connected up the following circuit.
The same thermistor was used.

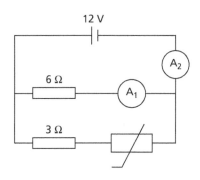

(i) Calculate the reading on ammeter A_2 if the thermistor has a resistance of 9 Ω.

Answer: .... A   **[2]**

(ii) The thermistor is placed in a beaker of ice.

Describe what difference, if any, the student would notice on the ammeter readings.

....

....   **[1]**

14 **(a)** A student rubs a plastic comb with a woollen cloth. The comb becomes positively charged.

(i) Explain how the comb becomes charged.

....

.... **[2]**

(ii) Suggest a way to show that the comb is charged.

....

.... **[2]**

(iii) For your suggestion in part **(ii)**, what would you expect to see and why?

....

.... **[2]**

(b) A charged metal sphere is resting on an insulated base.
A wire is connected to an ammeter and a metal water pipe.
The student touches the sphere with the other end of the wire.
A charge flows as shown below.

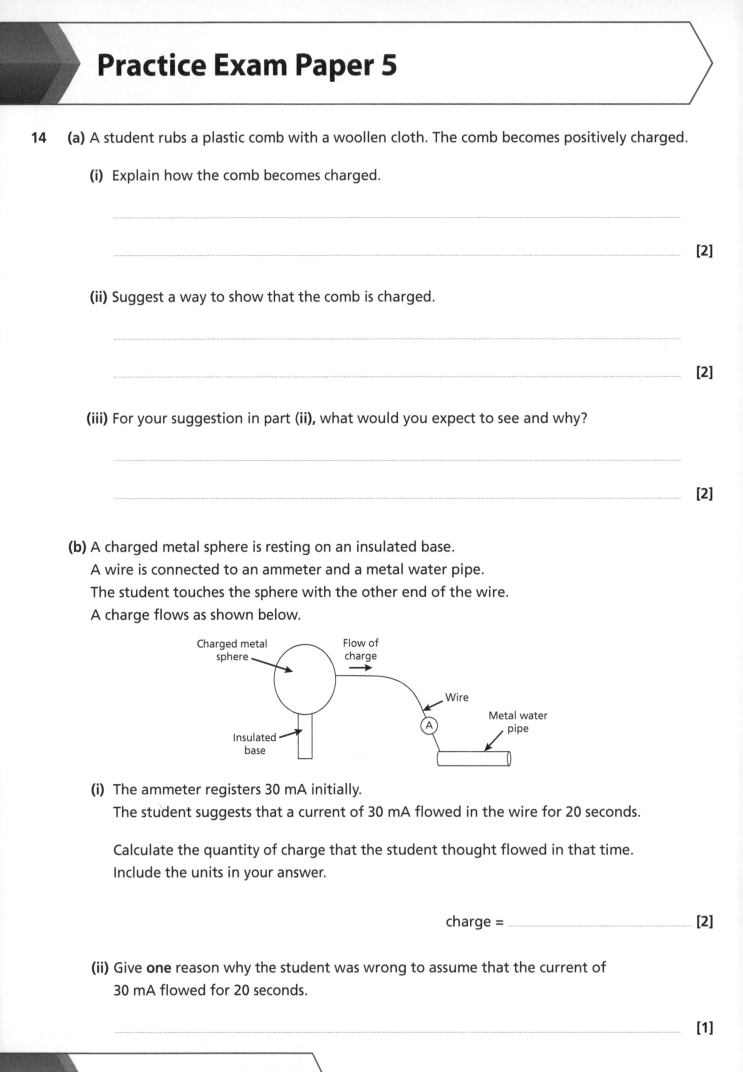

(i) The ammeter registers 30 mA initially.
The student suggests that a current of 30 mA flowed in the wire for 20 seconds.

Calculate the quantity of charge that the student thought flowed in that time.
Include the units in your answer.

charge = ... **[2]**

(ii) Give **one** reason why the student was wrong to assume that the current of
30 mA flowed for 20 seconds.

.... **[1]**

15   In an experiment, a spring was suspended vertically from a horizontal beam.
     Marked weights, 1 N each, were attached to the bottom of the spring, one at a time.
     A metre rule was used to measure the extension when each weight was added.

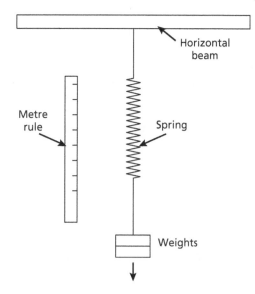

The results were recorded in the table below.

Force (N)	Extension (cm)
1	1.6
2	3.2
3	4.8
4	6.4
5	8.1
6	9.6
7	11.2
8	12.8
9	14.4
10	15.9
11	19.7
12	28.1

(a) Use the data to describe the relationship between force and extension for the spring
    when the weight was increased from 1 N to 10 N.

....

.... **[1]**

Turn over

(b) Calculate the spring constant for the spring using the extension when 4 N was the weight.

spring constant = ... N/m   [2]

(c) Calculate how much work was done when the spring was stretched by the 4 N weight.
Use the formula:

energy transferred in stretching = 0.5 × spring constant × (extension)²

work = ... J   [2]

(d) Suggest why the extension does not follow the same relationship once the weight exceeds 11 N.

....

....   [1]

16 A student is investigating how the resistance of a wire varies with length.
They set up the following circuit. A metre rule was placed alongside the wire.

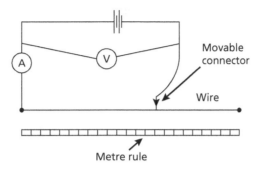

(a) Explain why the reading on the voltmeter remained constant throughout the experiment.

....

.... **[1]**

(b) The student used an 18 V battery and obtained the following results:

Resistance (Ω)	Length of wire (cm)
25	100
22	90
18	75
15	60
12	50
10	40
9	30
8.5	20

Calculate the power of the wire when it was 30 cm long.

power = ... W **[2]**

Turn over

17   A student is investigating the behaviour of a solenoid.

A slinky is used as the solenoid, with a soft iron core in the middle.

The circuit is set up as shown in the diagram below.

A magnetic field probe, connected to a data logger, will be used to measure the strength of the magnetic field.

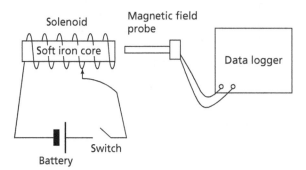

When the switch is closed, the data logger records a magnetic field of 0.2 T.

Suggest **three** changes that could be made to the apparatus in order to increase the magnetic field.

....

....

....

.... **[3]**

END OF QUESTION PAPER

Collins

GCSE (9–1)
Combined Science (Physics)
Paper 6 (Higher Tier)

H

You must have:

- the Data Sheet (page 456)

You may use:

- a scientific calculator
- a ruler

Time allowed: 1 hour 10 minutes

Instructions

- Use black ink. You may use a HB pencil for graphs and diagrams.
- Answer **all** the questions.
- Write your answer to each question in the space provided.
- Additional paper may be used if required.

Information

- The total mark for this paper is **60**.
- The marks for each question are shown in brackets [ ].
- Quality of extended response will be assessed in questions marked with an asterisk (*).

Practice Exam Paper 6

SECTION A

Answer **all** the questions.

You should spend a maximum of 20 minutes on this section.

1 Radium-88 is a radioactive element that decays to radon by emitting an alpha particle. The equation is shown below.

$$^{226}_{88}Ra \longrightarrow ^{X}_{Y}Rn + ^{4}_{2}He$$

Identify which are the correct values of **X** and **Y**.

	X	Y
A	226	89
B	88	227
C	222	86
D	86	222

Your answer ☐ **[1]**

2 A driver is travelling at 30 m/s in a car on a test track.
The driver has to carry out an emergency stop.
The thinking distance and braking distances are measured.

Which statement is **true** if there had been three more people in the car?

A The braking distance would be larger.

B The thinking distance would be larger.

C The overall stopping distance would remain the same.

D The braking distance would remain the same.

Your answer ☐ **[1]**

3   Sunlight shining on a solar cell has a power of 1000 W/m².
The solar cell is 20% efficient and has an area of 0.5 m².

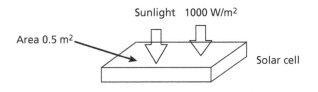

Sunlight 1000 W/m²

Area 0.5 m²

Solar cell

How much electrical energy would the solar cell produce per second?

A   80 J

B   800 J

C   400 J

D   100 J   Your answer [ ]   **[1]**

4   A car is involved in an accident.
The driver is wearing a seatbelt.
An airbag inflates on the steering wheel during the accident.

Which of the following statements is **not** correct about how to reduce the force acting on the driver during the accident?

A   A longer accident time causes a smaller force on the body of the driver.

B   The seatbelt stretches by a small amount, increasing the accident time.

C   The seatbelt keeps the driver's body firmly in the seat, reducing the force.

D   The airbag compresses as the driver's body moves forward.   Your answer [ ]   **[1]**

Turn over

5 In an experiment a signal generator is connected to a loudspeaker.
A microphone connected to an oscilloscope is placed in front of the loudspeaker,
as shown below.

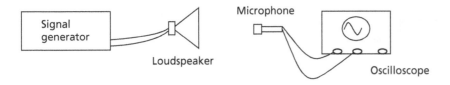

Traces on the oscilloscope, obtained for two different settings on the signal generator,
are shown below.

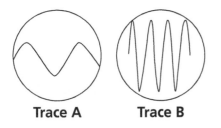

Trace A   **Trace B**

Identify which of the following have changed from trace **A** to trace **B**.

A Only the amplitude has increased.

B The amplitude and frequency have increased.

C Only the frequency has increased.

D The amplitude has increased but the frequency has decreased.   Your answer [ ] **[1]**

6 Protactinium-233 has a half-life of 27 days.
A 64 g sample of the element is left in a cupboard.

How much protactinium would remain after 108 days?

A 16 g

B 4 g

C 8 g

D 32 g   Your answer [ ] **[1]**

7 A radioactive source emits two types of radiation.

The paths of the particles passing through an electric field were tracked and shown below.

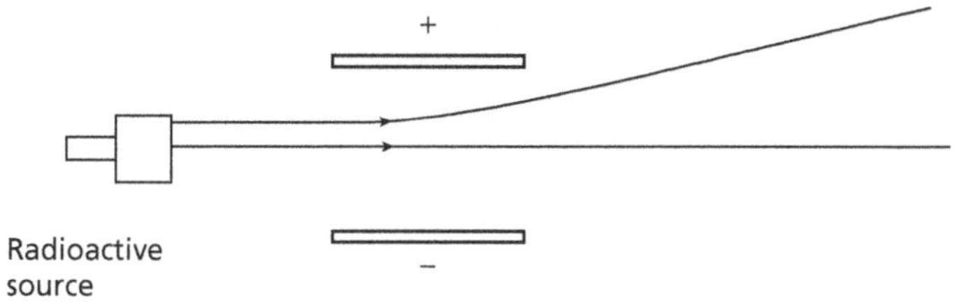

Radioactive
source

Identify the types of radiation shown in the diagram.

A beta particles and gamma rays

B alpha particles and beta particles

C alpha particles and gamma rays

D beta particles and protons   Your answer [ ]   **[1]**

8 A 250 V mains hair drier draws a current of 10 A from the supply.

How much energy would it use if it was switched on for 2 minutes?

A 2500 J

B 5000 J

C 3×10^5 J

D 3×10^4 J   Your answer [ ]   **[1]**

Turn over

9   The table shows some properties of waves in the electromagnetic spectrum.

Which row is **incorrect**?

	Electromagnetic radiation	Wavelength	Uses
A	ultraviolet	short	controlling algae in fish tanks
B	gamma	very short	smoke detectors
C	infrared	long	TV remote controls
D	radio waves	very long	astronomy

Your answer [1]

10   The following statements are about the plug of a kettle when connected to the mains and switched on.

Which of the statements is **false**?

A   The live wire in a plug is brown.

B   There is a potential difference between the live wire and neutral wire.

C   The fuse is in the live wire.

D   The earth wire is part of the circuit.   Your answer [ ]   [1]

SECTION B

Answer **all** the questions

11 Maxine, who is a teacher, uses a Geiger counter to measure the activity (rate of decay) of a radioactive source.

The results are recorded and a graph plotted using them, as shown below.

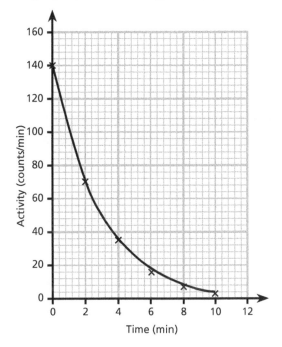

(a) (i) Use the graph to determine the half-life of the source.

half-life = _____ min   **[2]**

(ii) Suggest why, although there is a clear decay curve, not all of the points are on the curved line of best fit.

_____

_____

_____

_____

[2]

Turn over

(b) Maxine now uses another radioactive source.

She places different absorbers between the radioactive source and the Geiger counter, as shown below.

The counts per minute are recorded on the Geiger counter for each absorber.

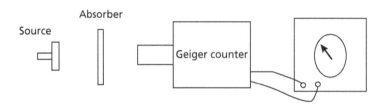

Maxine records her results in the table.

Absorber	Count rate (counts/min)
None	160
Paper	158
Aluminium	48
Lead	15

(i) Identify the types of radiation emitted from the source.

.... **[2]**

(ii) Explain why there was still a reading with lead as the absorber.

.... **[1]**

(iii) Radiation is dangerous.

Describe how Maxine should carry out this experiment safely.

....

....

.... **[2]**

12 A teacher was demonstrating waves by shaking a slinky.
The pattern is shown below.

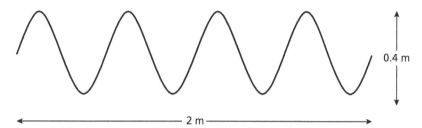

The frequency of the wave was 2 Hz.

(a) What type of wave is the teacher demonstrating?

Answer: .... **[1]**

(b) (i) State the amplitude of the wave.

amplitude = ... m **[1]**

(ii) State how the amplitude relates to energy.

....

.... **[2]**

(iii) Using the diagram above, calculate the speed of this wave.

wave speed = ... m/s **[3]**

Turn over

(c) A white light source is used to shine light through a prism.
A spectrum is seen on the white screen.

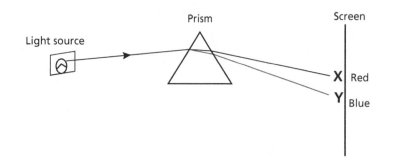

Explain why a coloured spectrum is seen on the white screen.

....

....

....

....

[3]

13 **(a)** Toni is investigating suitable insulators for her house.

She puts water at 100 °C into a copper can insulated with tissue.

She measures and records the temperature every minute for 10 minutes.

There are several different insulators available.

Toni's results are shown in the table below.

Time (min)	Temperature (°C)
0	100
1	90
2	78
3	68
4	60
5	55
6	49
7	45
8	42
9	38
10	35

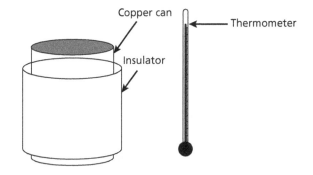

(i) Plot the readings on the graph below.

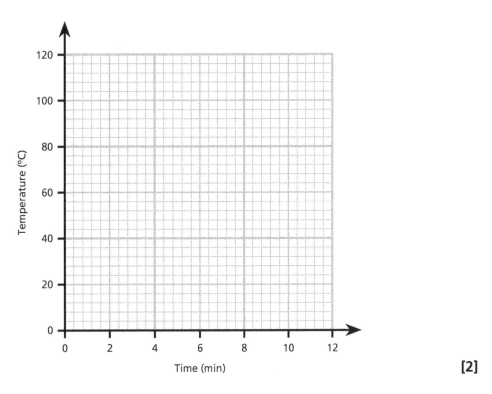

Time (min)

[2]

Turn over

(ii)* Describe, by giving experimental details, how Toni should complete her experiment to conclude which is the best insulator.

_____

_____

_____

_____

_____

_____

[5]

(b) There are heat losses in a power station.

Below is diagram showing these heat losses.

They are calculated for every 100 joules stored in the coal.

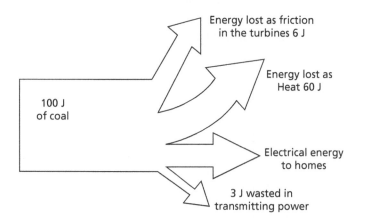

Calculate how much electrical energy reaches the homes for every 100 joules of coal in this power station.

Answer: _____ J per 100J of coal   **[2]**

14 A circuit was set up, as shown below, to transmit energy over a long distance.

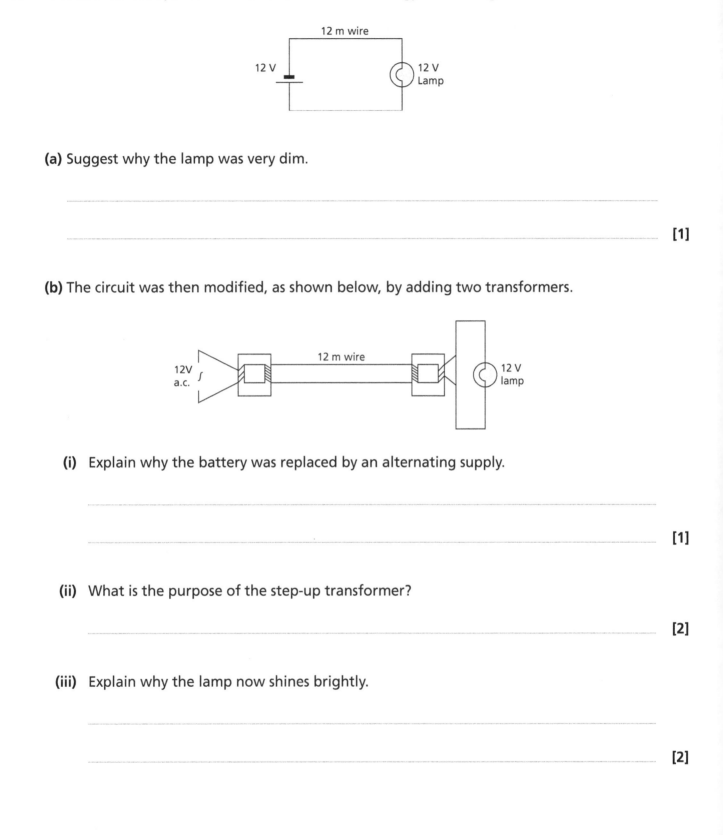

(a) Suggest why the lamp was very dim.

_____

_____ **[1]**

(b) The circuit was then modified, as shown below, by adding two transformers.

    (i) Explain why the battery was replaced by an alternating supply.

_____

_____ **[1]**

    (ii) What is the purpose of the step-up transformer?

_____ **[2]**

    (iii) Explain why the lamp now shines brightly.

_____

_____ **[2]**

Turn over

c) The National Grid works in a similar way to the model power line.

Instead of a 12 V lamp, what would be a typical value for the potential difference and frequency of a mains lamp?

.... [1]

15 A model wind turbine is being tested.

A fan is used to generate the wind.

An anemometer is used to measure the wind speed.

As the blades rotate, they drive a dynamo, producing a potential difference.

A resistor is connected with the dynamo, as shown below.

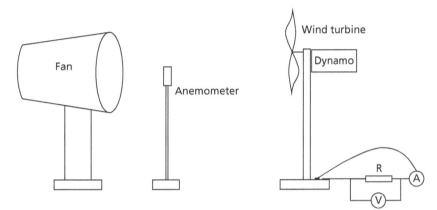

The power of the turbine, generated from the dynamo, was determined from the current through the resistor and the potential difference across it.

(a) From the results a graph was plotted of power output against wind speed.

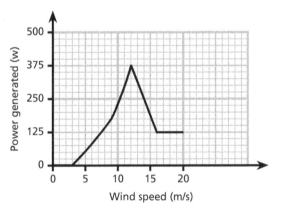

(i) At what wind speed was the output power at its maximum value?

Answer: _____ **[1]**

(ii) If the wind speed is too high, friction in the dynamo or vibrations in the blades of the turbine can cause energy losses.

Between which wind speeds was the output power falling?

_____ **[2]**

Turn over

(iii) Give **one** disadvantage of using this type of wind turbine to power a 375 W floodlight on a remote farm.

_____ [1]

(b) Solar panels are used by some farmers rather than wind turbines.

(i) Suggest **two** advantages of using solar panels to generate electricity rather than fossil fuels.

_____

_____

_____ [2]

(ii) State **two** reasons why a farmer might decide not to install solar panels on their farm.

_____

_____

_____ [2]

16 A 60 kg driver is travelling in a car at 12 m/s.

A child runs out from the kerb and the driver has to do an emergency stop.

The graph below shows the changes in the car's velocity.

It covers the time from when the driver first saw the child ($t = 0$).

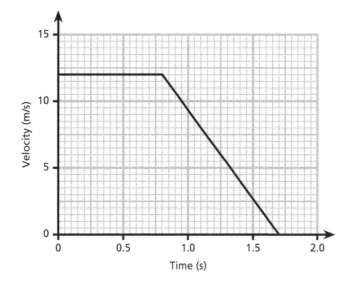

(a) What is the distance (thinking distance) that the driver covered before applying the brakes?

thinking distance = ... m   **[2]**

(b) Calculate the deceleration of the driver once the brakes were applied.

deceleration = ... m/s^2 **[2]**

(c) Suggest **one** factor that might have increased the braking distance.

....

.... **[1]**

Turn over

(d) The graph on the previous page showed the thinking and braking distances for an alert driver.

Sketch a graph below to show how it would have appeared had the driver been under the influence of alcohol or very tired.

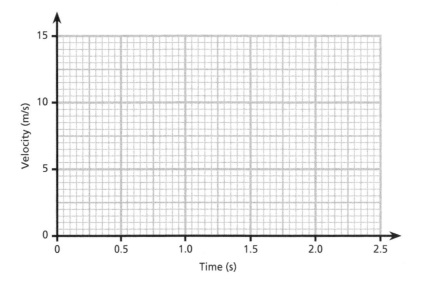

[2]

END OF QUESTIONS

Periodic Table

(1)	(2)											(3)	(4)	(5)	(6)	(7)	(0)
	2																18
1																	2
H																	**He**
hydrogen																	helium
1.0																	4.0
3	4											5	6	7	8	9	10
Li	**Be**											**B**	**C**	**N**	**O**	**F**	**Ne**
lithium	beryllium											boron	carbon	nitrogen	oxygen	fluorine	neon
6.9	9.0											10.8	12.0	14.0	16.0	19.0	20.2
11	12											13	14	15	16	17	18
Na	**Mg**											**Al**	**Si**	**P**	**S**	**Cl**	**Ar**
sodium	magnesium	3	4	5	6	7	8	9	10	11	12	aluminium	silicon	phosphorus	sulfur	chlorine	argon
23.0	24.3											27.0	28.1	31.0	32.1	35.5	39.9
19	20	21	22	23	24	25	26	27	28	29	30	31	32	33	34	35	36
K	**Ca**	**Sc**	**Ti**	**V**	**Cr**	**Mn**	**Fe**	**Co**	**Ni**	**Cu**	**Zn**	**Ga**	**Ge**	**As**	**Se**	**Br**	**Kr**
potassium	calcium	scandium	titanium	vanadium	chromium	manganese	iron	cobalt	nickel	copper	zinc	gallium	germanium	arsenic	selenium	bromine	krypton
39.1	40.1	45.0	47.9	50.9	52.0	54.9	55.8	58.9	58.7	63.5	65.4	69.7	72.6	74.9	79.0	79.9	83.8
37	38	39	40	41	42	43	44	45	46	47	48	49	50	51	52	53	54
Rb	**Sr**	**Y**	**Zr**	**Nb**	**Mo**	**Tc**	**Ru**	**Rh**	**Pd**	**Ag**	**Cd**	**In**	**Sn**	**Sb**	**Te**	**I**	**Xe**
rubidium	strontium	yttrium	zirconium	niobium	molybdenum	technetium	ruthenium	rhodium	palladium	silver	cadmium	indium	tin	antimony	tellurium	iodine	xenon
85.5	87.6	88.9	91.2	92.9	95.9		101.1	102.9	106.4	107.9	112.4	114.8	118.7	121.8	127.6	126.9	131.3
55	56	57-71	72	73	74	75	76	77	78	79	80	81	82	83	84	85	86
Cs	**Ba**	lanthanides	**Hf**	**Ta**	**W**	**Re**	**Os**	**Ir**	**Pt**	**Au**	**Hg**	**Tl**	**Pb**	**Bi**	**Po**	**At**	**Rn**
caesium	barium		hafnium	tantalum	tungsten	rhenium	osmium	iridium	platinum	gold	mercury	thallium	lead	bismuth	polonium	astatine	radon
132.9	137.3		178.5	180.9	183.8	186.2	190.2	192.2	195.1	197.0	200.5	204.4	207.2	209.0			
87	88	89-103	104	105	106	107	108	109	110	111	112		114		116		
Fr	**Ra**	actinides	**Rf**	**Db**	**Sg**	**Bh**	**Hs**	**Mt**	**Ds**	**Rg**	**Cn**		**Fl**		**Lv**		
francium	radium		rutherfordium	dubnium	seaborgium	bohrium	hassium	meitnerium	darmstadtium	roentgenium	copernicium		flerovium		livermorium		

Key

atomic number
symbol
name
relative atomic mass

Physics Data Sheet

change in thermal energy = mass $\times$ specific heat capacity $\times$ change in temperature

thermal energy for a change in state = mass $\times$ specific latent heat

$(\text{final velocity})^2 - (\text{initial velocity})^2 = 2 \times \text{acceleration} \times \text{distance}$

energy transferred in stretching = $0.5 \times$ spring constant $\times (\text{extension})^2$

potential difference across primary coil $\times$ current in primary coil =
potential difference across secondary coil $\times$ current in secondary coil

HT force on a conductor (at right-angles to a magnetic field) carrying a current =
magnetic field strength $\times$ current $\times$ length

Answers

Workbook Answers

You are encouraged to show all your working out, as you may be awarded marks for method even if your final answer is wrong.

Page 294 – Cell Structure

1. (a) Four correctly drawn lines **[3]**
   (2 marks for two correct lines;
   1 mark for one)
   Nucleus – Contains genetic material
   Cell membrane – Contains receptor molecules
   Chloroplast – Contains chlorophyll
   Mitochondrion – Contains enzymes for respiration
   (b) Chloroplasts **[1]**
2. Higher resolution than light microscope **[1]**; Uses magnets to focus an electron beam **[1]**
3. (a) Stain them **[1]**
   (b) i) White blood cell / phagocyte **[1]**
       ii) Red blood cell **[1]**
   (c) 4 × 100 = 400, ×400 magnification **[1]**

Page 295 – What Happens in Cells

1. (a) Carbohydrates **[1]**
   (b) fats / lipids **[1]**; fatty acids **[1]**; glycerol **[1]** (fatty acids and glycerol can be given in any order)
   (c) A change in pH **[1]**
2. (a) DNA **[1]**
   (b) In the nucleus **[1]**
   (c) Double helix **[1]**
3. Acid in stomach lowers the pH **[1]**; and denatures the enzyme **[1]**

Page 296 – Respiration

1. (a) Bar chart drawn with each axis labelled **[1]**; appropriate scale **[1]**; two data sets clearly shown **[1]**; accurately drawn bars **[1]**

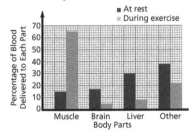

   (b) Muscles will require more glucose and oxygen **[1]**; to release more energy during exercise **[1]**; these are delivered to cells by the blood **[1]**

> Energy is **released**. Do not state that it is made or produced. This will lose you the mark.

2. (a) Fungi **[1]**
   (b) cell wall **[1]**; vacuole **[1]**
   (c) glucose → ethanol + carbon dioxide **[3]** (1 mark for each correct chemical)
3. (a) glucose + oxygen **[1]**; → carbon dioxide + water **[1]**

> When writing equations, make sure the reactants are to the left of the arrow and the products to the right.

   (b) In the blood **[1]**

Page 297 – Photosynthesis

1. (a) Photosynthesis **[1]**
   (b) Glucose **[1]**
   (c) Respiration **[1]**; makes other molecules **[1]** (Accept named molecules, e.g. cellulose)
   (d) At night, plants respire and produce carbon dioxide **[1]**; in the day, the plants respire but they will be also be photosynthesising, using up some carbon dioxide **[1]**

> Remember, plants **respire** as well as **photosynthesise**.

2. (a) water **[1]**; oxygen **[1]**
   (b) $6H_2O$ **[1]**; $6O_2$ **[1]**
   (c) Chloroplasts **[1]**
   (d) Requires / takes in energy **[1]**
   (e) Temperature **[1]**; concentration of carbon dioxide **[1]**

Page 298 – Supplying the Cell

1. (a) B, E, A, C, D **[1]**
   (b) Mitosis **[1]**
   (c) Repair of tissues **[1]**
2. (a) Diffusion: ← **[1]**
   (b) Active transport: → **[1]**
   (c) Osmosis: → **[1]**
3. (a) bone marrow **[1]**; embryos **[1]**
   (b) X = ciliated cell **[1]**; has cilia **[1]**; which move dirt and microorganisms up and out of respiratory system / move ova along the fallopian tubes **[1]**; Y= red blood cell **[1]**; concave shape increases the surface area **[1]**; for absorption of oxygen **[1]**; OR no nucleus **[1]**; so more room for carrying oxygen **[1]**

Page 299 – The Challenges of Size

1. (a) Root hair cell **[1]**
   (b) Leaf **[1]**
2. (a) Six correctly drawn lines **[6]**
   (1 mark for each correct line)

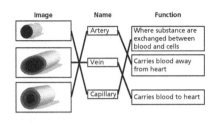

   (b) Veins **[1]**
   (c) Veins **[1]**
   (d) Glucose **[1]**; Oxygen **[1]**

> **Arteries** carry blood **away** from the heart. Remembering that both 'arteries' and 'away' start with the letter **A** should help you recall the right answer.

Page 300 – The Heart and Blood Cells

1. (a) Ventricles **[1]**
   (b) Arteries **[1]**; carry blood away from heart **[1]** OR X = pulmonary artery **[1]**; and Y = aorta **[1]**
   (c) Lungs **[1]**
   (d) Thick muscular walls **[1]**; for strong contractions **[1]**; to send blood under high pressure to body **[1]**
2. **Any three from:** carbon dioxide **[1]**; urea **[1]**; amino acids **[1]**; glucose **[1]**; antibodies **[1]**; hormones **[1]**

Page 301 – Plants, Water and Minerals

1. (a) Plant X **[1]**
   (b) Plants take in carbon dioxide for photosynthesis through stomata **[1]**; but stomata are closed to prevent water loss **[1]**
   (c) Turgid **[1]**
   (d) Transpiration **[1]**
   (e) i) Increased wind speed increases water loss **[1]**
       ii) Increased temperature increases water loss **[1]**
   (f) Guard cells open and close stomata **[1]**
   (g) Underneath of leaf shaded / not in direct sunlight **[1]**; therefore less water will be lost through evaporation **[1]**

Page 302 – Coordination and Control

1. Four correctly drawn lines **[3]** (2 marks for two correct lines; 1 mark for one)
   Sensory neurone – Carries impulses from receptor to relay neurone
   Receptor – Detects a stimulus
   Brain – Responsible for coordinating response
   Motor neurone – Carries impulse from relay neurone to effector
2. Relay **[1]**

3. stimulus [1]; electrical [1]; sensory [1]; synapse [1]; neurotransmitter [1]; receptors [1]
4. A very quick response [1]

Page 303 – The Endocrine System

1. (a) Pituitary [1]
   (b) Insulin [1]; glucagon [1]
   (c) Thyroid-stimulating hormone [1]
   (d) Kidneys [1]
   (e) Testes [1]
   (f) In the blood [1]
2. They are chemicals [1]; Their response is long lasting [1]
3. (a) Pancreas [1]
   (b) Liver [1]
   (c) Negative feedback [1]

Page 304 – Hormones and Their Uses

1. (a) Progesterone [1]
   (b) Oestrogen OR FSH [1]
2. (a) Follicle-stimulating hormone (FSH) [1]; luteinising hormone (LH) [1]
   (b) Any two from: Against some religious views [1]; can have side effects [1]; doesn't protect against STDs [1]
   (c) Any six from: Advantages: allows infertile couples to have own children [1]; can screen embryos for genetic diseases [1]; embryos not used can be used in research [1] Disadvantages: expensive [1]; less than 50% success rate [1]; can result in multiple births [1]; ethical issues with what to do with unwanted embryos [1]

Page 305 – Maintaining Internal Environments

1. (a) Denatured / destroyed [1]
   (b) Rate of reaction decreases / stops [1]
2. Type 1 has an earlier onset, whereas type 2 tends to develop later in life [1]; in type 1 the pancreas does not produce insulin, whereas in type 2 the insulin is produced but not effective [1]; type 1 requires insulin injections, whereas type 2 can usually be controlled by diet [1] (Accept any other sensible comparisons)
3. (a) Blood plasma [1]
   (b) Liver [1]
   (c) Pancreas [1]
   (d) Liver [1]
4. (a) Liver correctly marked on diagram with a X [1]
   (b) Pancreas correctly marked on diagram with a Z [1]

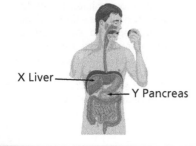
X Liver
Y Pancreas

Page 306 – Recycling

1. (a) animals [1]; plants [1]; microorganisms / decomposers [1]
   (b) Glucose [1]
2. (a) Lightning [1]
   (b) Proteins [1]
3. (a) Feeding [1]
   (b) Decomposition [1]

Page 307 – Interdependence

1. (a) Bird and mouse / hawk and fox [1]
   (b) Any one from: rabbit [1]; bird [1]; mouse [1]
   (c) Grass / barley [1]
2. (a) Disease [1]; lack of food / prey [1]; introduction of new predator / hunted [1]
   (b) Any three from: mates [1]; space / territory [1]; food [1]; water [1]
   (c) Any three from: light [1]; water [1]; space [1]; nutrients [1]
3. (a) Mutualism [1]
   (b) The oxpecker bird is not harming the buffalo [1]

Page 308 – Genes

1. genes [1]; DNA [1]; double helix [1]; haploid [1]
2. (a) Blonde hair [1]; blue eyes [1]
   (b) A good swimmer / thin / height [1]
3. Three correctly drawn lines [2] (1 mark for one correct line)
   tt – Homozygous recessive
   Tt – Heterozygous
   TT – Homozygous dominant
4. $\frac{1178}{8}$ [1]; = 147.25 [1]; = 147.3cm to 1 d.p. [1]

Page 309 – Genetics and Reproduction

1. (a) Tall [1]
   (b) Correct genotypes for both parents [1]; correct outcomes of cross [1] (Accept either a Punnett square or cross diagram)

		Parent 1	
		T	t
Parent 2	T	T T	T t
	t	T t	t t

2. (a) 70 [1]
   (b) 35 [1]
3. Correct genotypes for both parents [1]; correct outcomes of cross [1]; identifying XX as female and XY as male offspring [1] (Accept a Punnett square or a cross diagram)

		Mother	
		X	X
Father	X	X X FEMALE	X X FEMALE
	Y	X Y MALE	X Y MALE

Page 310 – Natural Selection and Evolution

1. (a) The snow leopard and tiger have evolved from a common ancestor [1]; who lived less time ago compared with the common ancestor of snow leopard and clouded leopard [1]
   (b) DNA analysis [1]
2. variation [1]; mutation [1]; environment [1]; survive [1]; offspring [1]
3. In a population of bacteria, a mutation occurs [1]; which gives rise to a bacterium that is resistant to penicillin [1]; when the patient is treated with penicillin, bacteria that are resistant to penicillin survive [1]; and multiply [1]; those bacteria that are not resistant will be killed by the antibiotic [1]

Page 311 – Monitoring and Maintaining the Environment

1. Loss of habitat [1]; loss of biodiversity [1]; loss of potential medicines and drugs [1]; rainforests important in cycling carbon / removing excess carbon dioxide from the atmosphere through photosynthesis [1]
2. (a) Provides money for developing schools / health services [1]
   (b) Income from tourism goes towards park / trust [1]; and is used to manage park / maintain habitats / enforce poaching laws [1]

Page 312 – Investigations

1. (a) Capture–mark–recapture method [1]
   (b) $\frac{12 \times 10}{6}$ [1]; $= \frac{120}{6} = 20$ [1]
2. (a) Any two from: size of pot [1]; amount / mass / type of compost [1]; amount / volume of acid rain used to water [1]; size of beans [1]; light [1]; temperature [1]; length of roots and shoots [1]
   (b) B [1]
   (c) Plant more beans at each pH [1]; and calculate mean result [1]

Page 313 – Feeding the Human Race

1. (a) Restriction [1]
   (b) Ligase [1]
   (c) Plasmids easily inserted into bacteria [1]; bacteria multiply rapidly [1]
   (d) (i) Any insects trying to eat leaves will be killed so the crop isn't damaged [1]
       (ii) Could be used in developing countries where people are malnourished [1]
2. (a) It kills insects other than aphids [1]
   (b) The β-farnesene will deter aphids from eating the wheat / the aphids will think the wheat is dangerous [1]
   (c) Take DNA from an aphid [1]; cut the β-farnesene gene from the aphid DNA [1]; using restriction enzymes [1]; insert the gene into the wheat DNA [1]; using ligase enzymes [1]

Page 314 – Monitoring and Maintaining Health

1. AIDS [1]; Measles [1]; Cholera [1]
2. All microorganisms correctly matched [4]; (2 marks if two are correctly matched; 1 mark if one is correctly matched)

Virus	10μm	Thrush
Bacteria	1μm	Malaria
Protozoa	100μm	Tuberculosis
Yeast	0.1μm	Chickenpox

3. (a) HIV / human immunodeficiency virus [1]
   (b) **Any three from:** bodily fluids during unprotected sex [1]; mother to baby during pregnancy or birth [1]; mother to baby in breast milk [1]; blood transfusion [1]; sharing needles [1]
   (c) People with AIDS have weakened immune system [1]; so are prone to bacterial infections [1]

Page 315 – Prevention and Treatment of Disease

1. (a) Help blood to clot [1]; forms a scab, which prevents entry of microorganisms [1]
   (b) Microorganisms and small particles of dirt are trapped by mucus [1]
2. (a) **Any two from:** computer simulations [1]; tests on cells grown in laboratory [1]; tests on animals [1]; tests on healthy people [1]
   (b) The sugar pill (placebo) is a control [1]; to compare the effect of using the drug and using no drug (with no psychological influences) [1]
3. (a) Waxy cuticle [1]; cell wall [1]
   (b) **Any two from:** wind [1]; water [1]; animal (insect) vectors [1]

Page 316 – Non-Communicable Diseases

1. Advantage: potential for people to change lifestyle / get early treatment / regular screening [1]; Disadvantage: could lead to depression / people unable to get jobs / insurance (accept any other sensible answer) [1]
2. (a) $BMI = \frac{74}{1.62 \times 1.62}$ [1] $= 28.2$ [1]
   (b) **Any two from:** heart disease [1]; stroke [1]; diabetes [1]; cancer [1]; joint / muscular problems / arthritis [1]
3. B [1]

Page 317 – Particle Model and Atomic Structure

1. C [1]
2. B [1]
3. (a) A = carbon-13 [1]; B = 6 [1]; C = 8 [1]; D = 6 [1]
   (b) proton [1]
   (c) 60 [1]
4. **Any three from:** there is no indication of the differences in attraction between particles [1]; no difference in size of particles [1]; between the particles is empty space [1]; diagram is only in 2D, not 3D [1]

Page 318 – Purity and Separating Mixtures

1. Each price is for a different purity [1]; the most expensive is the purest grade of glucose / no contaminants [1]
2. D [1]
3. A_r: Mg = 24, O = 16, H = 1, 24 + (2 × 1) + (2 × 16) [1]; = 58 [1]
4. A_r: C = 12, H = 1, O = 16, $\frac{60}{12} = 5$, $\frac{12}{1} = 12$, $\frac{16}{16} = 1$ [1]; $C_5H_{12}O$ [1]
5. Sandy should carry out evaporation / crystallisation [1]; pour the ink into a conical flask / suitable glass container and then gently heat [1]; the liquid component of the ink will evaporate leaving the dried ink pigments [1]

Page 319 – Bonding

1. D [1]
2. A [1]
3. Covalent bonds occur where one electron in the outer shell of each atom [1]; is shared by each of the atoms (two shared electrons = one covalent bond) [1]; ionic bonds occur where the metal element donates an electron or electrons and the non-metal receives the electron(s) [1]; both atoms end up with a full outer shell [1]
4. C [1]
5. A correctly drawn diagram (2.8.7) [1]

Page 320 – Models of Bonding

1. (a) A correctly drawn dot and cross diagram [1]; and displayed formula [1]

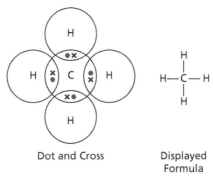

Dot and Cross   Displayed Formula

   (b) Ball and stick models can give an indication of bond angles / the 3D shape of the molecule. [1]
2. A correctly drawn diagram [1]

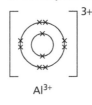

Al^{3+}

Page 321 – Properties of Materials

1. C [1]
2. Graphite is made up of repeating carbon atoms that are held together with covalent bonds [1]; it takes a lot of energy to separate carbon atoms covalently joined [1]; water is a covalent molecule made up of only two hydrogens and an oxygen [1]; the water molecules are all joined together by weak forces, making them easy to separate from each other [1]
3. A [1]
4. The carbon atoms are each covalently joined to four other carbon atoms [1]; this makes the carbon atoms extremely difficult to separate from each other [1]
5. (a) Drawing showing each carbon atom bonded to four other carbon atoms [1]; in the correct structure [1]

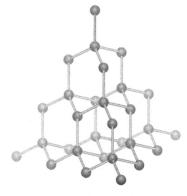

   (b) Diamond cannot conduct electrons because there are no free electrons [1]; as each atom is covalently bonded to four other carbon atoms [1]

Page 322 – Introducing Chemical Reactions

1. (a) Be = 1; Al = 2; O = 4 [1]
   (b) Al = 2; O = 5; Si = 1 [1]
   (c) Ba = 1; P = 2; O = 6 [1]
   (d) C = 18; H = 36; O = 2 [1]
2. B [1]
3. $MgCO_3 + 2HNO_3 \rightarrow Mg(NO_3)_2 + CO_2 + H_2O$ [2] (1 mark for the correct chemical symbols; 1 mark for correct balancing numbers)

Page 323 – Chemical Equations

1. C [1]
2. The measurement [1]; of the relative amounts of reactants and products in chemical reactions [1]
3. (a) $2Br^- \rightarrow Br_2 + 2e^-$ [1]

Page 321 – Properties of Materials

3. (a) A correctly drawn diagram [1]

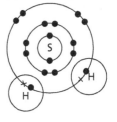

   (b) 2 [1]

(b) $H_2 \rightarrow 2H^+ + 2e^-$ **[1]**
(c) $Fe^{3+} + 3e^- \rightarrow Fe$ **[1]**
4. **(a)** Spectator ions are ions that are present on both sides of a reaction / the ions do not change / they remain the same. **[1]**
(b) Na^+ and SO_4^{2-} **[1]**
(c) $Cu^{2+}(aq) + CO_3^{2-}(aq) \rightarrow CO_3(s)$ **[2]** (1 mark for correct balancing; 1 mark for correct charges)

Page 324 – Moles and Mass

1. B **[1]**
2. C **[1]**
3. A_r: Be = 9, Al = 27, O = 16 **[1]**; formula mass = $(1 \times 9) + (2 \times 27) + (4 \times 16)$ **[1]**; = 127 **[1]**
4. Reaction equation: $2H_2(g) + O_2(g) \rightarrow 2H_2O(l)$, 5mol of H_2 will lead to 5mol of H_2O **[1]**; M_r: $H_2O = (2 \times 1) + 16 = 18g/mol$ **[1]**; $5mol \times 18g/mol = 90g$ **[1]**

> Look at the equation and work out the ratio of reactant to product. Here it is 2mol of H_2 to 2mol of H_2O, which is a 1:1 ratio. This can then be multiplied to find any amount.

5. $\dfrac{\text{atomic mass of Rb}}{\text{Avogadro's constant}} = \dfrac{85g}{6.022 \times 10^{23}}$ **[1]** = $1.4 \times 10^{-22}g$ **[1]**

Page 325 – Energetics

1. Exothermic reactions release energy to the surroundings **[1]**; and cause a temperature rise **[1]**; in this case, the energy given out is used to make light **[1]**
2. A **[1]**
3. An endothermic reaction takes in energy from the environment **[1]**; this means that there will be a temperature drop **[1]**
4. **(a)** exothermic **[1]**
(b) endothermic **[1]**
(c) exothermic **[1]**
(d) endothermic **[1]**

Page 326 – Types of Chemical Reactions

1. **(a)** chlorine **[1]**
(b) oxygen **[1]**
(c) bromine **[1]**
(d) oxygen **[1]**
2. **(a)** iron **[1]**
(b) carbon **[1]**
(c) $Fe_2O_3(s) + 3CO(g) \rightarrow 2Fe(s) + 3CO_2(g)$ **[2]** (1 mark for correct reactants; 1 mark for correct products)
3. **(a)** HCl(aq) **[1]**
(b) $HNO_3(aq)$ **[1]**
(c) $CH_3CH_2COOH(aq)$ **[1]**
(d) HBr(aq) **[1]**
4. B **[1]**

Page 327 – pH, Acids and Neutralisation

1. Using universal indicator is a qualitative measurement **[1]**; it is possible for two people to arrive at two different pH levels **[1]**; a data logger gives a precise, quantitative measurement **[1]**
2. B **[1]**; D **[1]**; E **[1]**
3. **(a)** C **[1]**
(b) D **[1]**
(c) A **[1]**
(d) B **[1]**

Page 328 – Electrolysis

1. **(a)** C **[1]**
(b) Above zinc, but below carbon **[1]**
2. Metals below carbon can be removed from their ores by displacement by carbon **[1]**; more reactive metals will not be displaced **[1]**; electrolysis separates the molten ions **[1]**

Page 329 – Predicting Chemical Reactions

1. B **[1]**
2. C **[1]**
3. Hold a glowing splint at the end of a test tube containing the gas **[1]**; if the gas is oxygen, the splint will reignite **[1]**
4. Both are in Period 2, which means their outer electrons are close to the nucleus / they have a strong attraction for electrons **[1]**; oxygen requires 2 more electrons to fill the outer shell, whilst fluorine requires just 1 extra electron **[1]**; it is easier to attract 1 electron than 2, so fluorine is more reactive **[1]**

Page 330 – Controlling Chemical Reactions

1. A **[1]**
2. C **[1]**
3. **Any two from:** use a catalyst **[1]**; increase temperature **[1]**; increase pressure **[1]**; increase concentration **[1]**
4. **(a)** B **[1]**
(b) The larger the surface area **[1]**; the greater the opportunity for the reactants to come together and have a successful reaction **[1]**

Page 331 – Catalysts and Activation Energy

1. **(a)** A steeper line equals a faster rate of reaction **[1]**
(b) 45°C **[1]**
(c) $74cm^3$ **[1]**
(d) 8min (accept 7.5 min) **[1]**
(e) Accept any time under 1.5min **[1]**
2. A catalyst speeds up the rate of reaction **[1]**; it does this by lowering the activation energy **[1]**; so the reaction needs less energy to start **[1]**; it does this without being involved directly / it is not a reactant **[1]**

Page 332 – Equilibria

1. A reaction that only goes in one direction / cannot go back / cannot re-form reactants **[1]**

2. C and D **[1]**
3. C **[1]**
4. It will have no effect (the number of moles of gas is the same on each side) **[1]**

> Look at the amount of moles of gas on either side. If they are the same, pressure will not have an effect.

Page 333 – Improving Processes and Products

1. ore **[1]**
2. C **[1]**
3. **(a)** Wear eye protection / safety goggles **[1]**; tilt the tube away from your body **[1]**
(b) copper(II) oxide + carbon → carbon dioxide + copper **[2]** (1 mark for each correct product)
(c) $2Cu_2O(s) + C(s) \rightarrow CO_2(g) + 4Cu(s)$ **[2]** (1 mark for correct reactants; 1 mark for correct products)

Page 334 – Life Cycle Assessments and Recycling

1. Recycling reduces the amount of waste going to landfill **[1]**; the materials that would be lost are still in circulation **[1]**; energy costs of dumping the product are reduced **[1]**
2. **(a)** Obtaining raw materials **[1]**; manufacture **[1]**; use **[1]**; disposal **[1]** (Accept in any order)
(b) Carbon fibre **[1]**; because oil takes millions of years to form / oil cannot be replaced **[1]**

Page 335 – Crude Oil

1. A = refinery gases **[1]**
B = naphtha **[1]**
C = diesel **[1]**
D = bitumen **[1]**
2. Short chain hydrocarbons only have a few intermolecular forces keeping them together **[1]**; long chain hydrocarbons have a lot of intermolecular forces keeping the chains together, making it much more difficult to separate them **[1]**
3. Cracking is taking long chain hydrocarbons and breaking them into smaller, more useful short-chained molecules **[1]**
4. B **[1]**

Page 336 – Interpreting and Interacting with Earth's Systems

1. **(a)** The numbers in the table will have been rounded (up and down) **[1]**
(b) D, E, A, B, C **[2]** (1 mark for A after D; 1 mark for C after B)
2. C **[1]**

Page 337 – Air Pollution and Potable Water

1. Four correctly drawn lines **[3]** (2 marks for two correct lines; 1 mark for one)
Carbon monoxide – air poisoning,
Sulfur dioxide – kills plants and aquatic

life, Nitrogen oxides – photochemical smogs, Particulates – coat surfaces with soot
2. (a) A = large solids [1];
       B = small particulates [1]
   (b) C = bacteria [1]; D = chlorine [1]
3. Correctly drawn apparatus and process [1]; all correctly labelled [1]

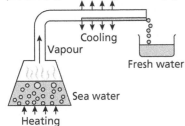

Cooling
Vapour
Fresh water
Sea water
Heating

Page 338 – Matter, Models and Density

1. The forces are weak   [1]
2. The nucleus repelled [1]; positive alpha particles [1]
3. The aluminium atoms are more closely packed in the solid form (than when in the liquid state) [1]; so the density is higher [1]
4. mass = density × volume, mass = $(8.96 \times 10^3) \times 2$ [1]; $= 1.792 \times 10^4$kg [1] (Accept 17 920kg)
5. The electron   [1]
6. Bohr's model had the electrons in orbits [1]; around the nucleus [1]
7. A correct sketch of molecules in steam [1]; and water [1]

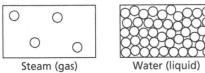

Steam (gas)   Water (liquid)

Page 339 – Temperature and State

1. **Any one from:** temperature [1]; mass [1]; the number of aluminium atoms [1]; the structure of aluminium atoms [1]
2. A reversible [1]; physical change [1]
3. The forces between the solid / liquid's atoms or molecules [1]; the motion of the atoms or molecules [1]
4. (a) Energy = 0.1 × (2.26 × 10⁶J) [1]; $= 2.26 \times 10^5$J [1] (Accept 226 000J or 226kJ)
   (b) Energy = 0.1 × 4200 × 65 [1]; $= 2.73 \times 10^4$J [1] (Accept 27 300J or 27.3kJ)

> The change in temperature is 100 – 35 = 65°C. You need to use this value in your calculation.

   (c) There is greater energy from steam changing state to water than boiling water cooling [1]; when steam comes in contact with skin, both energies are added together [1]; as the change of state and cooling both take place [1]; and greater energy release to the skin causes greater damage [1]

Page 340 – Journeys

1. Speed is a scalar quantity as it only has magnitude [1]; velocity is a vector as it has both magnitude and direction [1]
2. Distance = average speed × time [1]; distance = 30 × 40 = 1200 m [1]
3. Acceleration = $\dfrac{\text{change in velocity}}{\text{time}}$
   $= \dfrac{(70 - 30)}{8}$ [1]; = 5m/s² [1]
4. Velocity and acceleration   [1]

> An object on a circular path is constantly changing direction, so acceleration and velocity also change because they are vector quantities involving direction.

5. kinetic energy = 0.5 × mass × (speed)² [1]; = 0.5 × 50 × 8² [1]; = 1600J [1]

Page 341 – Forces

1. A gravitational force can only be attractive; an electrical force can be repulsive or attractive   [1]
2. A body stays still or keeps moving at constant velocity [1]; unless an external force acts on it [1]
3. There is an attractive force from the Sun pulling on the Earth [1]; there is also an attractive force from the Earth pulling on the Sun [1]; these forces are equal in magnitude but opposite in direction [1]
4. At terminal velocity, the force of gravity downwards (weight) is balanced by air friction (resistive force) [1]; producing an equal and opposite upward force [1]; so no net force is acting / the resultant force is zero [1]
5. Inertia is a measure of how difficult it is [1]; to change the velocity of an object [1]
6. The satellite is constantly changing direction in orbit, so the velocity (a vector quantity) is changing [1]; speed (a scalar quantity) doesn't depend on direction [1]

Page 342 – Force, Energy and Power

1. Momentum = 310kg × 20m/s [1]; $= 6.2 \times 10^3$kg m/s [1]
2. Momentum is conserved [1]; the total momentum of the rocket and water is zero when the rocket is at rest [1]; the water leaving the rocket has momentum in a downwards direction [1]; so the rocket must have an equal momentum in the opposite direction [1]
3. Potential energy = 60 × 10 × 12 [1]; = 7200J [1]
4. Power = $\dfrac{150 \times 10 \times 12}{90}$ [1]; = 200W [1]
5. The work done by the person has to overcome friction [1]; as well as to move the piano [1]

Page 343 – Changes of Shape

1. A material that is deformed when a force is applied and does not return

to its original shape when the force is removed   [1]
2. Jupiter has greater mass than Mars [1]
3. spring constant = $\dfrac{\text{force}}{\text{extension}} = \dfrac{1}{2 \times 10^{-2}}$ [1]; = 50N/m [1]

> You must be able to recall the following equation and rearrange it to make the spring constant the subject:
> force = spring constant × extension

4. The straight line part of the graph shows the spring obeying Hooke's law and behaving elastically [1]; the point at which the graph just stops showing proportionality (begins to curve) is the elastic limit [1]; after that point, the graph curves, showing plastic behaviour [1]
5. Energy = 0.5 × spring constant × (extension)² = 0.5 × 80 × $(30 \times 10^{-2})^2$ [1]; = 3.6J [1]

Page 344 – Electric Charge

1. An electric field   [1]
2. A magnetic field [1]; and heat [1] (Accept produces light)
3. The electrons are not free to move [1]; so charge cannot flow [1]
4. charge = current × time = 5 × 10 [1]; = 50C [1]
5. (a) Electrons are transferred from the rod to the cloth [1]; having lost electrons, the rod is now positively charged [1]; having gained electrons, the cloth is now negatively charged [1]
   (b) Electrons are attracted up to the plate at the top [1]; leaving both the stem and gold leaf positively charged [1]; the leaf is repelled by the stem and rises [1]

Page 345 – Circuits

1. Resistance   [1]
2. (a) A (filament) light bulb / lamp   [1]
   (b) The graph is linear for low values of current and obeys Ohm's law [1]; then as the current increases, the graph becomes non-linear as the wire in the bulb (lamp) heats up [1]; this increases the resistance [1]; due to more collisions between the moving electrons and (increasingly) vibrating metal atoms [1]
3. (a) A diode   [1]
   (b) Current can only flow one way through it   [1]

Page 346 – Resistors and Energy Transfers

1. *Either* ratio of resistors is 12 : 4 = 3 : 1 [1]; as potential difference has the same ratio, potential difference = $\dfrac{36}{3}$ = 12V [1]
   OR using
   current = $\dfrac{\text{potential difference}}{\text{resistance}} = \dfrac{36}{12}$ = 3A, so potential difference across 4Ω resistor = 3 × 4 [1]; = 12V [1]

2. $\frac{1}{R} = \frac{1}{24} + \frac{1}{8}$ [1];

$R = 6\Omega$ [1]

3. The current has two paths to flow through the combination [1]; and the potential difference is the same for each path [1]; so the total current flowing into the combination is greater than for each individual branch [1]; consequently the combined resistance for the combination is lower than for each separate branch [1]

> $V = IR$. V is constant so if I increases, as it does with two parallel branches, then R must decrease.

4. Total resistance = 2 + 3 + 5 [1];
$V = IR$: 20 = current × 10, current = 2A [1]

Page 347 – Magnetic Fields and Motors

1. The compass needle is no longer affected by the magnetic field of the coil, as no current is flowing through the wires of the coil [1]; therefore, the needle moves to point towards the Earth's magnetic north pole [1]
2. (a) There will be a greater density of iron filings (in the strongest region of the magnetic field) [1]
   (b) The field lines would appear closer together [1]; as a greater current is flowing in the coil [1]; strengthening the magnetic field [1]
3. force = 2.4 × 10^{-2} × 3 × 0.5 [1];
   = 3.6 × 10^{-2}N [1]

Page 348 – Wave Behaviour

1. A longitudinal wave [1]
2. (a) velocity = $\frac{0.9}{3}$ [1]; = 0.3m/s [1]
   (b) frequency = $\frac{velocity}{wavelength} = \frac{0.3}{0.01}$ [1];
   = 30Hz [1]
3. (a) Measure the distance to the wall and double it [1]; then divide that number by the time between the claps to get the speed of sound [1] (Accept a correct answer written as an equation, i.e. speed of sound = $\frac{distance\ to\ wall \times 2}{time\ between\ claps}$)
   (b) Dividing the time for 10 claps by 10 gives a more precise value for the time the sound took to reflect [1]; because the error is divided by 10 [1] (1 mark only for just saying it is more precise; a reason must be given for the second mark)

> The error in starting and stopping a stopwatch is the same whether timing 1 clap or 10 claps.

4. Velocity (speed) [1]; and wavelength [1]

Page 349 – Electromagnetic Radiation

1. Frequency [1]
2. The dentist takes many X-rays in a

day [1]; the risk to the health of the dentist is high [1]
3. X-rays pass through the body tissue [1]; but are absorbed by bones [1]; any crack or gap in the bones would lead to some X-rays in that region reaching the detector, giving an image of the fracture [1]
4. speed = frequency × wavelength [1];
   frequency = $\frac{3 \times 10^8}{1500}$ [1]; = 2 × 10^{5}Hz [1] (Accept 200 000Hz)
5. Infrared [1]
6. If absorbed by cells of the body [1]; they ionise atoms and can cause chemical changes that can be harmful [1] (Accept they can cause cells to become cancerous)

Page 350 – Nuclei of Atoms

1. Protons [1]; and neutrons [1]
2. Isotopes [1]
3. Ionising radiation can cause atoms to lose electrons [1]; and become ions [1]
4. Beta particles [1]; and gamma rays [1]
5. X = 90 [1]; Y = 231 [1]
6. (a) Both have six protons [1]; but the carbon-12 nucleus has six neutrons and the carbon-14 nucleus has eight neutrons [1]
   (b) The charge on both nuclei is the same, +6 [1]

Page 351 – Half-Life

1. Half-life is the time taken for the number of radioactive nuclei in a sample to halve [1]
2. (a) 52 hours is 4 half-lives [1]; 64g halved 4 times (64 → 32 → 16 → 8 → 4) [1]; = 4g [1]
   (b) $\frac{4}{64} = \frac{1}{16}$ [1]
3. Irradiation is the exposure of an object to ionising radiation [1]
4. (a) $x = 93$ [1]; $y = 4$ [1]
   (b) It doesn't need to be handled or replaced [1]; so avoids exposure to alpha particles [1]

Page 352 – Systems and Transfers

1. (a) Energy in the form of heat [1]; is transferred from the saucepan to the water [1]
   (b) The water warms up as it gains energy [1]
2. The system gains energy from outside (by heating or work done on it) [1]; or loses energy to the outside (by transferring heat or doing work) [1]
3. Heat = 1.2 × 4200 × 80J [1];
   = 4.032 × 10^{5}J [1] (Accept 403 200J)
4. (a) Gravitational potential energy [1]; changes to kinetic energy [1]; which changes to electrical energy [1]
   (b) Heat / thermal [1]

Page 353 – Energy, Power and Efficiency

1. It is transferred in the form of thermal energy in the transformer [1]

2. A kilowatt-hour is the energy that a 1 kilowatt appliance transfers [1]; in 1 hour [1]
3. **Any two from**: reduce air flow between the inside and outside of the house [1]; insulate the loft [1]; use double glazing [1]; use carpets [1] (Accept any other sensible suggestion)
4. (a) power = 250 × 2 [1]; = 500W [1]
   (b) energy = 500 × 10 × 60 [1];
   = 3 × 10^{5}J [1] (Accept 300 000J)
5. (a) power = 250 × 8 [1]; = 2000W [1]
   (b) energy = 2000 × 400 [1];
   = 8.0 × 10^{5}J [1] (Accept 800 000J)

Page 354 – Physics on the Road

1. Thinking distance [1]; and braking distance [1]
2. (a) acceleration (deceleration) = $\frac{change\ in\ velocity}{time} = \frac{25}{10}$ [1];
   = 2.5m/s^2 [1]

   (b) force = 1500 × 2.5 [1]; = 3750N [1]
3. (a) One student holds the ruler vertically [1]; the other puts their open hand level with the 0cm mark, so that the ruler is between their fingers and thumb [1]; the student drops the ruler [1]; the other student catches it [1]; the length at which the ruler was caught is recorded [1]; the reaction time is calculated [1]
   (b) Any value between 0.15 and 0.3 seconds [1]

Page 355 – Energy for the World

1. A non-renewable energy source is one that once it is used cannot be replaced [1]; in a reasonable time scale [1] (Accept: A natural resource that cannot be grown or produced [1]; at the same time as it is consumed [1])
2. (a) Stored chemical energy (in coal) [1]; changes to heat in the boiler [1]; which produces steam [1]; which gives the engine kinetic energy [1]
   (b) **Any two from**: burning alcohol still emits carbon dioxide [1]; alcohol is highly flammable so it is a hazard to people near the engine [1]; if farms produce alcohol, it would reduce the number of fields for food production [1]; there would be difficulty in transporting it, either by road or on the train [1]; extra emissions would be a factor with transportation [1] (Accept any other sensible response)
3. (a) The amount of uranium in rocks on Earth is decreasing because of its decay [1]
   (b) **Any one advantage from**: it doesn't emit carbon dioxide or particulates into the atmosphere [1]; a small amount of fuel produces an enormous quantity of energy [1] (Accept any similar sensible response)

Any one disadvantage from:
waste products are radioactive and dangerous **[1]**; waste products need to be transported for reprocessing **[1]**; radioactive waste is problematic to dispose of **[1]**; radioactive waste has a long half-life so will remain dangerous for hundreds of years **[1]** (Accept any similar sensible response)

Page 356 – Energy at Home

1. An a.c. supply produces a current that keeps rapidly changing direction **[1]**; the current from a d.c. supply only goes one way **[1]**
2. 230V **[1]** (Accept 240V or 250V); 50Hz frequency **[1]**
3. (a) A step-down transformer **[1]**

   (b) $\frac{12}{V_s} = \frac{60}{2400}$ **[1]**; $V_s = \frac{12 \times 2400}{60} = 480V$ **[1]**

   (c) In transformer A, power in = power out **[1]**; $12 \times 2 = 480 \times$ current in transmission lines **[1]**; current = 0.05A **[1]**

> Transformers are almost 100% efficient.

Page 357–372 – Practice Exam Paper 1

Section A
1. B **[1]**
2. A **[1]**
3. B **[1]**
4. A **[1]**
5. D **[1]**
6. D **[1]**
7. C **[1]**
8. C **[1]**
9. B **[1]**
10. B **[1]**

Section B
11. (a) Bubbles could be different sizes **[1]**; individual bubbles difficult to count if produced quickly **[1]**

    (b) Can measure volume of gas produced **[1]**

    (c) A correct curve drawn **[1]**

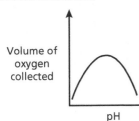

    (d) Idea of enzyme shape **[1]**; active site where substrate fits **[1]**; enzyme specificity (other substrates will not fit) **[1]**

12. (a) Surface area = 4 × 4 × 6 = 96cm² **[1]**; Volume = 4 × 4 × 4 = 64cm³ **[1]**; SA: V ratio = $\frac{96}{64}$ = 1.5 : 1 **[1]**

    (b) The 2cm / smallest cube **[1]**; because it has the largest surface area to volume ratio **[1]**

    (c) Large multicellular animals have a small surface area to volume

ratio which decreases the rate of diffusion **[1]**; so they need specialised systems which take substances nearer to where they are needed, reducing the diffusion distance **[1]**

13. (a) 0.66 + 0.62 + 0.61 = 1.89 **[1]**; mean = $\frac{1.89}{3}$ **[1]**; = 0.63 **[1]**

    (b) A = sensory neurone **[1]**; B = motor neurone **[1]**; C = effector **[1]**

    (c) (i) Axes labelled and correct **[1]**; all points plotted correctly **[1]**; line of best fit (not a line from one point to next) **[1]**

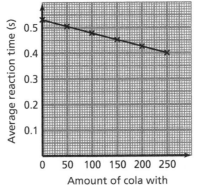

Amount of cola with caffeine (cm³)

        (ii) No, there could be something in the cola other than caffeine that affected reaction times, therefore, control needs to be cola not water **[1]**

14. (a) Some deoxygenated blood will pass through hole from right side to left side **[1]**; and be sent round body without going to the lungs **[1] OR** some oxygenated blood will pass from left side to right side **[1]**; so less oxygenated blood goes to the body **[1]**

    (b) The heart will have to work harder to force blood out, causing chest pain / feeling weak or dizzy when exercising **[1]**

    (c) (i) Inside red blood cells **[1]**; joined to haemoglobin / as oxyhaemoglobin **[1]**

        (ii) **Any two from:** hormones **[1]**; antibodies **[1]**; glucose **[1]**; amino acids **[1]**; minerals **[1]**

15. (a) 6, 6, 6 **[1]** (All three need to be correct for mark)

    (b) (i) Carbon dioxide gas was produced **[1]**

        (ii) Sugar / glucose is a substrate for respiration **[1]**

    (c) Aerobic respiration happens in the mitochondria but anaerobic respiration happens in the cytoplasm **[1]**; aerobic is the complete breakdown of glucose but anaerobic is incomplete **[1]**; aerobic requires oxygen, anaerobic does not **[1]**; aerobic higher yield of ATP / more energy than anaerobic **[1]**

16. (a) Water moves into root hair cells **[1]**; moving from area of high water concentration to area of low

concentration **[1]**; by osmosis **[1]**

    (b) Net water movement would be out of plant **[1]**; because concentration of water would be higher in plant cells than outside **[1]**

    (c) temperature / carbon dioxide levels **[1]**

    (d) Glucose moves by diffusion **[1]**; from areas of high concentration to areas of low concentration **[1]**; via the phloem tissue **[1]**; minerals enter plant by active transport **[1]** and move through plant due to the transpiration stream **[1]**; from roots to leaves through the xylem tissue **[1]**

> In the exam, if you see an asterisk (*) next to a question, it means that the quality of your response will be assessed. This means you must give all your ideas in a clear and logical way to avoid losing marks.

Page 373–388 – Practice Exam Paper 2

Section A
1. A **[1]**
2. D **[1]**

> $\frac{(7 + 9 + 8 + 8 + 6)}{5} = 7.6$
> $7.6 \times 60 = 456$

3. C **[1]**
4. D **[1]**
5. A **[1]**
6. B **[1]**
7. C **[1]**
8. D **[1]**
9. B **[1]**
10. B **[1]**

Section B
11. (a) At risk of extinction **[1]**

    (b) **Any one from:** hunting / taking eggs **[1]**; polluting lakes / rivers (which reduced food source) **[1]**

    (c) **Any one from:** nest built for breeding pairs **[1]**; wardens protect nests from disturbance / egg thieves **[1]**

    (d) **Any one from:** intense on labour **[1]**; high labour costs **[1]**; difficult to 'police' area **[1]**; has to be long-term project **[1]**

12. (a) The number of male and female smokers has decreased steadily **[1]**; since 1968 **[1]**

    (b) 40 – 25 = 15 **[1]**; $\frac{15}{40} \times 100$ **[1]**; = 37.5% **[1]**

    (c) Yes – because as % of males who smoked decreased **[1]**; so did incidence of cancer in males **[1] OR** No – because as % females who smoked decreased **[1]**; the incidence of cancer increased **[1]**

    (d) (i) Fat deposits build up and block arteries / increase risk of heart attack and strokes **[1]**

        (ii) Increases blood pressure **[1]**

13. (a) (Microbial) respiration [1] (Accept decomposition)
  (b) Detrivores break matter into smaller pieces [1]; to increase surface area [1]
  (c) Nitrates [1]

> Sometimes students get confused and think that the decaying matter gives off carbon dioxide. It doesn't. **Respiration** of the **microorganisms / decomposers** which feed on the decaying matter is where the carbon dioxide comes from.

14. (a) Horse and zebra: share same genus / both *Equus* [1]; both have simple stomach, whereas deer has complex stomach [1]
  (b) (i) Donkey: sure footed, less prone to colic, long life span, hardworking [2] (1 mark for more than two qualities; 2 marks for all qualities)
    (ii) Horse: intelligent [1]; easy to train [1]
  (c) (i) Donkey gametes will have 31 chromosomes and horse gametes 32, which means mule cells will have 63 chromosomes [1]; 63 chromosomes cannot be halved in meiosis to produce gametes so mules are sterile [1]
    (ii) Because species can breed with each other to produce fertile offspring [1]
  (d) A mutation produced a horse with a larger central toe [1]; this horse was better at running so more likely to survive [1]; and breed [1]; and pass on genes to offspring [1]
15. (a) A tabby pattern [1]; with brown eyes [1]; and curled ears [1]
  (b) (i) Gametes: T and t [1]; offspring (in order): Tt and tt [1]
    (ii) 1 : 1 [1]
  (c) (i) Genetic screening [1]
    (ii) Correct genotypes of parents [1]; correct genotypes of offspring [1]; correct phenotypes of offspring [1] (Accept any correct genetic diagram)

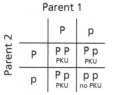

Parent 1

		P	p
Parent 2	P	P P PKU	P p PKU
	p	P p PKU	p p no PKU

    (iii) 6 [1]

> ¾ have PKU, so in a litter of 8 that would be 6 kittens with PKU.

16. (a) Plasmid [1]
  (b) Antibiotic A can be used, B cannot [1];

Disc A: radius in range of 8.5–9.5mm, so $8.5 \times 8.5 = 72.25 \times 3.14 = 226.9mm^2$ [1];
Disc B: radius in range 6.5–7.5mm, so $6.5 \times 6.5 = 42.25 \times 3.14 = 132.7mm^2$ [1]
  (c) Cost / if patient allergic to antibiotic [1]
  (d) White blood cell engulfs the bacterium [1]; destroyed within white blood cell [1]; mention of phagocytosis or lysozyme or egestion of debris [1]
  (e) Dead / weakened form of bacterium [1]

Page 389–402 – Practice Exam Paper 3

Section A

1. D [1]
2. C [1]
3. B [1]
4. D [1]
5. C [1]
6. B [1]
7. B [1]
8. D [1]
9. C [1]
10. D [1]

Section B

11. (a) The boiling point tends to increase as you go across the period from Groups 1 to 4 [1]; with a drop in Groups 5, 6, 7 and 0 [1]
  (b) **Any two from:** the period indicates the number of electron shells around the nucleus [1]; as you move across a period, the elements on the left-hand side are more likely to be metals, as they donate electrons [1]; those on the right are non-metals, as they receive electrons [1]; there is a change from metallic bonding to covalent bonding [1]; as you move right across the period in the metals section, the metals become less reactive relative to each other [1]; as you move left in the non-metals section, the non-metals become less reactive relative to each other [1]
  (c) Particle X is an atom, Y is an ion [1]; the atomic number indicates the number of protons. If the number of electrons, which can be found from the electronic structure, matches the atomic number it must be neutral, i.e. an atom [1] (Or reverse argument)
  (d) (i) The atom is not to scale [1]; the proton is drawn too large [1]; and the electron is too close to the nucleus [1]
    (ii) One early model had the charges distributed evenly throughout the atom [1]; Bohr had electrons in discrete shells / energy levels [1]
12. (a) broken = 3, made = 4 [1]
  (b) When bonds are broken: endothermic / energy is taken

---

in [1]; when bonds are made: exothermic / energy is released [1]
  (c) energy needed to break bonds = +1358kJ/mol [1]; energy released when new bonds form = –1856kJ/mol [1]; energy change = –498kJ/mol / 498kJ/mol [1]
13. (a) The copper ions have a 2+ charge [1]; they will move to the negative electrode / cathode where they will pick up 2 electrons [1]; this means that the outer shell will have 2 electrons in it / it is the same as the elemental form of copper [1]
  (b) $Cu^{2+}(aq) + 2e^-$ [1]; $\rightarrow Cu(s)$ [1]
  (c) Bubbles [1]; of oxygen gas [1]
  (d) Replace the solution with silver nitrate [1]; replace the cathode with the earring [1]
14. (a)* **Answer to include discussion of the dissociation constant, for example:** the dissociation constant is likely to be related to how easily an acid dissociates into ions [1]; there are weak and strong acids and those that are strong dissociate easily (e.g. H_2SO_4) [1]; so strong acids have a larger dissociation constant [1]; (accept alternative comments about weak acids)
    Answer to also include discussion of the experiment, for example: the strengths of the acids could be compared by titration with an alkali / by adding a volume of alkali to a known volume of acid gradually until the acid is neutralised [1]; the stronger the acid, the greater the volume of alkali needed to neutralise it [1]; this is repeated until concordant results are achieved [1]

> Some exam questions have an asterisk (*) alongside the number. This means that marks are going to be awarded based on the level of response that you give. The examiner is looking for a factual explanation using scientific words. To gain full marks, your answer must be written in a logical and coherent way, with all the key points clearly explained, supported and developed.

  (b) 10 times stronger [1]
15. (a) $Cu + 2AgCl \rightarrow 2Ag + CuCl_2$ [2] (1 mark for correct formulae; 1 mark for balancing molecules)
  (b) (i) $Cu + 2Ag+ \rightarrow Cu^{2+} + 2Ag$ [2] (1 mark for correct reactants; 1 mark for correct products)
    (ii) $Cu \rightarrow Cu^{2+} + 2e^-$ (accept $Cu - e^- \rightarrow Cu^{2+}$) [2] (1 mark for correct reactants; 1 mark for correct products)
    (iii) $2Ag^+ + 2e^- \rightarrow 2Ag$ (accept $2Ag^+ \rightarrow 2Ag - 2e^-$) [2] (1 mark for correct reactants; 1 mark for correct products)

(c) Copper atoms lose electrons / silver ions gain electrons [1]; the copper is oxidised [1]; and the silver ions are reduced [1]

16. (a) $2Al(s) + Fe_2O_3(s) \rightarrow 2Fe(s) + Al_2O_3(s)$ [1]

(b) **Any two from:** eye protection [1]; carry out behind a safety screen [1]; use a fuse to start the reaction [1]; do not do the reaction inside [1]

(c) oxidised = Al [1]; reduced = Fe [1]

Page 403–418 – Practice Exam Paper 4

Section A

1. B [1]
2. A [1]
3. B [1]
4. C [1]
5. C [1]
6. B [1]
7. C [1]
8. B [1]
9. D [1]
10. B [1]

Section B

11. (a) 2.8.8.1 [1]

(b) Diagram that shows positive ions arranged in solid-like formation [1]; indication of sea of delocalised electrons [1]

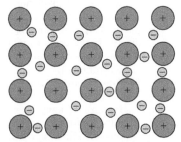

(c) The potassium moves around rapidly on the surface of the water [1]; and a gas ignites with a lilac flame [1]

(d) potassium + water → potassium hydroxide + hydrogen [2] (1 mark for each correct product)

(e) The higher up the group [1]; the less reactive they will be [1]

12. (a) $N_2 = 78$, $O_2 = 21$, $CO_2 = 0.04$ (accept 0.03) [1]

(b) The amount of nitrogen has dropped by 9%, because of the increase in oxygen levels [1]; there was no oxygen 3bya but 2bya 10% of the atmosphere was oxygen, due to the appearance of photosynthesising plants [1]; the amount of carbon dioxide remained relatively constant, this was due to volcanic activity [1]

(c) A graph with x-axis labelled Time (Years) and y-axes labelled Amount of CO_2 and Air Temperature (may be on either side) [1]; sketched lines for air temp and CO_2 with labels or key [1]; shape resembles a hockey stick [1]; timing indicates CO_2 levels increased [1]

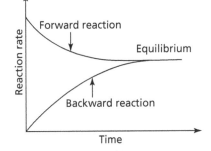

(d)*Answer must explain the effect of CO_2 and CH_4 on the greenhouse effect (trapping increased heat energy which is unable to escape the Earth) and make at least two suggestions on how to reduce emissions, e.g.: The greenhouse effect is the term used for the ability of the atmosphere to trap some of the infrared radiation [1]; this causes the temperature of the Earth to increase to a higher level [1]; methane and carbon dioxide are greenhouse gases and increased levels of these gases in the atmosphere increases the amount of infrared radiation trapped, so the temperature of the Earth increases [1]; to reduce the greenhouse effect we need to find alternative sources of energy that do not produce carbon dioxide or methane [1]; for example, if we were to use hydrogen fuel cells instead of fossil fuels such as petrol, then there would be zero emissions [1]; methane is a more powerful gas than carbon dioxide, so minimising organic waste in landfill or reducing the amount of livestock, such as cows, would help to reduce the greenhouse effect [1]

13. (a) Oxygen [1]

(b) $2SO_2$ [1] + O_2 [1] $\rightleftharpoons$ $2SO_3$ [1]

(c) The reaction is slow [1]; the catalyst lowers the activation energy [1]

(d) Increasing the pressure would increase the amount of SO_3 [1]; this is because there are fewer molecules of gas on the product side of the equilibrium / the system will adjust to balance the increased pressure on the reactant side [1]

14. (a) Advantages **(any two from)**: low cost [1]; no maintenance [1]; wildlife can flourish at the site [1]; aesthetically pleasing [1]
Disadvantages **(any two from)**: takes a long time [1]; limited to contamination that can be accessed by the plant roots [1]; burning / disposing of the plants could release contamination [1]; there is a maximum concentration of contaminant the plants can handle [1]

(b) Potable water is water that is fit for human consumption [1]; lives will be saved if more people have access to clean, drinkable water [1]

15. (a) The rate of forward reaction equals the rate of the reverse reaction [1]

(b) Yes - increasing the pressure shifts the reaction to the side with the fewer molecules [1]; in this case there are two molecules on the reactant side and only one on the product side [1]

(c) increases, decreases [1]

(d) Two curves drawn leading to equilibrium [1]; labelled correctly [1]

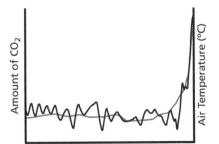

16. (a) $Mg(s) + 2H_2O(l) \rightarrow Mg(OH)_2(aq) + H_2(g)$ [1]

(b) Increase the temperature of the water [1]; increase the surface area to volume ratio / add the magnesium in small pieces / as a powder [1]

(c) The reaction would be faster / more vigorous [1]

(d) Magnesium is higher than carbon in the reactivity series [1]; electrolysis [1]; of the molten ore is required [1]

Pages 419–436 – Exam Practice Paper 5

Section A

1. D [1]
2. B [1]
3. B [1]
4. D [1]
5. D [1]
6. A [1]
7. D [1]
8. A [1]
9. D [1]
10. C [1]

Section B

11. (a) A circuit drawn with a battery, ammeter in series [1]; and a voltmeter connected in parallel to the heater/ battery, as shown [1] (or similar arrangement)

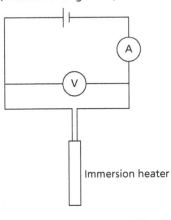

(b) **Apparatus:** balance to measure the mass of the water collected (accept a measuring cylinder to calculate mass from the volume if known density of water is referred to, but not on its own) **[1]**; stopwatch to measure the time the heater was on **[1]**
Measurements: mass of water collected due to the immersion heater (by subtracting the mass in the control beaker B from the mass collected in beaker A) **[1]**; time the heater was on **[1]**

(c) To measure the mass of ice that melted due to the room temperature and not due to the heater **[1]**

(d) Too low a measurement of the mass of water collected or too high a reading of the time the heater was on **[1]**

(e) energy = $0.1 \times (3.34 \times 10^5)$ **[1]**; 3.34×10^4J **[1]** (Accept 33 400J)

12. (a) From the graph: distance travelled = 12m/s × 3s = 36m **[1]**

(b) (i) deceleration = $\frac{6-0}{12-4.3}$ **[1]**; = 0.78m/s² **[1]**
(Accept acceleration = –0.78m/s²)

(ii) force = 1 × 0.78 **[1]**; = 0.78N **[1]** (Accept force = –0.78 N)

(c) Taken from graph: velocity after 7.5s = 3.5m/s **[1]**; total momentum = 3.5kgm/s **[1]**

(d) Line drawn to show $v = 0$ from 0 to 4 seconds, with a steep gradient rising to $v = 6$ between 4 and 4.3 seconds **[1]**; line falling from $v = 6$ to $v = 0$ between 4.3 seconds and 12 seconds **[1]**

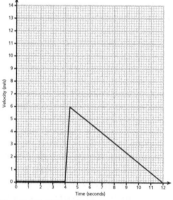

13. (a)* **For the full 4 marks, your answer must provide a detailed explanation of how to carry out the experiment, including which measurements are to be taken and recorded with reference to calculating the resistance, e.g.** Pour hot water in one beaker and put crushed ice in another, place a thermometer in each beaker **[1]**; lower the thermistor into each beaker in turn, measure and record the temperature and the corresponding readings on the voltmeter and ammeter **[1]**; keep the thermometer and the thermistor in the beaker of hot water and repeat the measurements as the water

cools **[1]**; the resistance at each temperature can now be calculated and a conclusion drawn **[1]**

(b) (i) The 'reading 2' at 50°C is too high at 9.1Ω **[1]**; ignore the anomaly and just use the other two readings to calculate the average resistance at 50°C OR repeat the faulty measurement of resistance at 50°C **[1]**

(ii) An inverse relationship: as the temperature of the thermistor rises, its resistance decreases **[1]**

(c) (i) 2A + 1A **[1]**; 3A **[1]**

> Using Ohm's law, 2A passes through the 6Ω resistor. The thermistor and 3Ω resistor have a combined resistance of 3 + 9 = 12Ω. Using Ohm's law, 1A flows through the 3Ω resistor.

(ii) Reading on A_1 is unchanged, reading on A_2 decreases **[1]**

14. (a) (i) Electrons are transferred from the comb to the woollen cloth when it is rubbed **[1]**; removal of electrons results in the comb becoming positively charged **[1]**

(ii) Comb held near a running tap **[1]**; will attract the water **[1]** OR comb brought near an uncharged electroscope **[1]**; will cause the leaf to rise **[1]** OR comb will pick up small pieces of paper or attract hair **[1]**; if held over it **[1]** (Accept any other similar suggestion)

(iii) Mention of opposite charges attracting, like charges repelling **[1]**; the explanation must include a mention of electrostatic induction **[1]**

(b) (i) charge = current × time **[1]**; = $\frac{30}{1000} \times 20 = 0.6$ coulombs **[1]**

(ii) **Any one from:** the potential difference decreased as charge flowed off the sphere **[1]**; the current flowing would decrease with time **[1]**; the current would not be constant **[1]** (or similar answer based on current falling with time) **[1]**

15. (a) Force is proportional to extension / the relationship is linear **[1]**

(b) force = spring constant × extension,
4 = spring constant × 0.064 **[1]**;
spring constant = 62.5N/m **[1]**

(c) energy transferred = $0.5 \times 62.5 \times (0.064)^2$ **[1]**; = 0.128J (Accept 0.13J) **[1]**

(d) The elastic limit of the spring has been exceeded **[1]**

16. (a) The voltmeter measured the potential difference of the battery **[1]**

(b) Current at 30cm = $\frac{18}{9}$ = 2A **[1]**;
power = VI = 18 × 2 = 36W **[1]**

17. Increase the number of turns on the solenoid **[1]**; decrease the length of the solenoid (keeping the number of turns constant) **[1]**; increase the current flowing through the solenoid **[1]** (Accept: increase the number of batteries or potential difference)

Pages 437–454 Exam Practice Paper 6

Section A
1. C   **[1]**
2. A   **[1]**
3. D   **[1]**
4. C   **[1]**
5. B   **[1]**
6. B   **[1]**
7. A   **[1]**
8. C   **[1]**
9. B   **[1]**
10. D   **[1]**

Section B
11. (a) (i) From the graph in 2.1 minutes the count rate drops from 140 to 70, then 70 to 35 in 2.1 minutes **[1]**; 2.1min **[1]** **(Accept 2min)**

(ii) Decay is a random process for each nucleus (giving slight variations in counts/min) **[1]**; there are so many nuclei that overall there is a pattern **[1]**

(b) (i) Beta **[1]**; and gamma **[1]**

(ii) The background count (due to natural radiation in the environment) **[1]**

(iii) **Any two from:** without handling the source (using forceps) **[1]**; keeping as much distance from the source as possible **[1]**; using the source for a minimum time **[1]**

12. (a) Transverse waves **[1]**

(b) (i) $\frac{0.4}{2}$ = 0.2m **[1]**

(ii) The greater the amplitude, the greater the amount of energy **[1]**; the wave is carrying **[1]**

(iii) wavelength = $\frac{2m}{4 \text{ waves}}$ = 0.5m **[1]**; speed = frequency × wavelength **[1]**; = 2 × 0.5 = 1m/s **[1]**

(c) White light is made up of all the colours **[1]**; each colour has a different wavelength (accept frequency) **[1]**; each wavelength (frequency) is refracted through a different angle by the prism, producing the spectrum **[1]**

13. (a) (i) Correctly plotted points **[1]**; with line of best fit (a curve) **[1]**

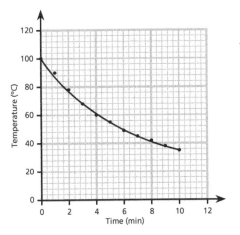

(ii)* **For the full 5 marks, your answer must provide a detailed explanation of how the experiment should be carried out, including apparatus to be used, measurements to be taken and a clear explanation of how the conclusion can be made, e.g.** Replace the tissue paper with a different insulator, keeping the thermometer in the copper can **[1]**; place the same mass of water as before at 100°C in the beaker **[1]**; start the stopwatch and record the temperature every minute up to 10 minutes **[1]**; repeat this process for each of the insulators keeping the starting mass of water and its temperature of 100°C the same **[1]**; Toni can conclude that the insulator with the smallest change in temperature is the best **[1]**

(b) total energy lost = 6 + 60 + 3 = 69J **[1]**; electrical energy to the homes = 100 − 69J = 31J per 100J of coal **[1]**

14. (a) Energy will be lost from the wires due to the heating effect of the current **[1]**

(b) (i) Transformers only work with an alternating voltage **[1]**

(ii) The step-up transformer increases the potential difference in the 20m wire **[1]**; and decreases the current in the same ratio **[1]**

(iii) The current in the 12m wire is very much smaller **[1]**; there is minimal energy loss due to heating **[1]**

> Remember heat lost = I^2R, so keeping the current small is vital to minimise heat losses.

(c) 230V, 50Hz **[1]**

15. (a) (i) 12m/s **[1]**

(ii) Between 12m/s **[1]**; and 16m/s **[1]**

> The gradient of the graph is negative where the output power is falling, as the wind speed increases.

(iii) The output power falls at speeds greater / lower than 12m/s (Accept: or below 12m/s) dimming the lamp OR the lamp would only be at maximum brightness at a wind speed of 12m/s **[1]**

(b) (i) **Any two from:** solar panels don't emit greenhouse gases **[1]**; solar panels don't produce particulates **[1]**; solar energy is renewable **[1]**; there are no waste products to be disposed of from solar energy

[1]; they have a long lifespan, require little maintenance, and can be used in remote areas where there is no mains electricity (or other sensible response) **[1]**

(ii) **Any two from:** solar cells are expensive to buy **[1]**; large areas of land are needed for high-output solar cells **[1]**; the intensity of sunlight varies during the day and season, and so will the output of the solar panels **[1]**; there is no output from the solar panels at night **[1]**; they are not very attractive to look at (visual pollution) **[1]**; output can be affected by the weather **[1]**

16. (a) From the graph: velocity = 12m/s, time = 0.8 seconds **[1]**; distance travelled = 12 × 0.8 = 9.6m **[1]**

(b) Change in velocity = 12m/s − 0, time for the change = 0.9 seconds **[1]**; deceleration = $\frac{12}{0.9}$ = 13.3m/s² **[1]**

(c) **Any one from:** a wet road **[1]**; worn brakes **[1]**; worn tyres **[1]**; poor road surface **[1]**; increased load in the car due to passengers **[1]** (or any other suitable answer) **[1]**

(d) A graph similar to below, having a longer thinking distance **[1]**; but the **same** braking distance **[1]**

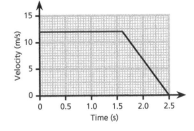

Notes

GCSE Combined Science Workbook

Notes

GCSE Combined Science Workbook

Acknowledgements

The authors and publisher are grateful to the copyright holders for permission to use quoted materials and images.

p16 toeytoey/Shutterstock.com; p16 b extender_01/Shutterstock.com; p38 Crisan Rosu/Shutterstock.com; p43 Evgeniia Abisheva/Shutterstock.com; 47 Patricia Chumillas/Shutterstock.com; p63 Matej Hudovernik/Shutterstock.com; p141 Dariush M/Shutterstock.com; p144 daulon/Shutterstock.com; p294 toeytoey/Shutterstock.com; p370 bmnarak/Shutterstock.com; p378 CLS Design/Shutterstock.com

Published by Collins
An imprint of HarperCollins*Publishers* Ltd
1 London Bridge Street
London SE1 9GF

HarperCollins*Publishers*
1st Floor, Watermarque Building,
Ringsend Road, Dublin 4, Ireland

ISBN 9780008160814

First published 2016
This edition published 2021

10 9 8 7 6 5 4

British Library Cataloguing in Publication Data.

A CIP record of this book is available from the British Library.

Commissioning Editor: Emily Linnett and Fiona Burns
Biology author: Fran Walsh
Chemistry author: Eliot Attridge
Physics authors: David Brodie and Trevor Baker
Project Manager: Rebecca Skinner
Project Editor: Hannah Dove
Cover Design: Sarah Duxbury and Kevin Robbins
Copy-editor: Rebecca Skinner
Proofreader: Aidan Gill
Typesetting and artwork: Jouve India Private Limited
Production: Lyndsey Rogers
Printed in the United Kingdom

MIX
Paper from
responsible source
FSC
www.fsc.org **FSC C007454**

This book is produced from independently certified FSC™ paper to ensure responsible forest management.

For more information visit:
www.harpercollins.co.uk/green